RELAXATION PROCESSES IN MOLECULAR EXCITED STATES

PHYSICS AND CHEMISTRY OF MATERIALS WITH LOW-DIMENSIONAL STRUCTURES

Series C: Molecular Structures

RELAXATION PROCESSES IN MOLECULAR EXCITED STATES

Edited by

J. FÜNFSCHILLING

University of Basel,
Institute of Physics

KLUWER ACADEMIC PUBLISHERS

DORDRECHT / BOSTON / LONDON

Library of Congress Cataloging-in-Publication Data

ISBN-13:978-94-010-6876-5   e-ISBN-13:978-94-009-0863-5
DOI: 10.1007/978-94-009-0863-5

Published by Kluwer Academic Publishers,
P.O. Box 17, 3300 AA Dordrecht, The Netherlands.

Kluwer Academic Publishers incorporates the publishing programmes of
D. Reidel, Martinus Nijhoff, Dr W. Junk, and MTP Press.

Sold and distributed in the U.S.A. and Canada
by Kluwer Academic Publishers,
101 Philip Drive, Norwell, MA 02061, U.S.A.

In all other countries, sold and distributed
by Kluwer Academic Publishers Group,
P.O. Box 322, 3300 AH Dordrecht, The Netherlands

printed on acid free paper

TABLE OF CONTENTS

INTRODUCTION

Relaxation phenomena of excited molecular states are abundant in all nature. They mediate such key processes as photochemical reactions or even the pathways of ordinary chemical reactions. However, for a long time the main research in electronic relaxation processes was concerned with anorganic solids, in part because of their great technological importance (photography, semiconductors ...) in part also because these compounds were the "workhorses" of the solid state physicists. In the last 30 years, there was a steadily increasing interest in organic molecular systems, first in molecular crystals and later in all forms of molecular solids (glasses, polymers, membranes, ...).

The present volume combines papers on quite different types of relaxation phenomena: the type of solid studied, the electronic states involved, the physical processes responsible for the relaxations are all different. Nevertheless, after reading this book, a more clear and complete picture of the phenomenon "relaxation" emerges that proves that this volume is more than just a collection of individual articles.

The volume starts with the paper "Spin-lattice and spin-spin relaxation in photo-excited triplet states in molecular crystals" by Jan Schmidt. Even in these seemingly simple systems of isolated guest molecules in a single crystal host, the relaxation phenomena are quite involved and a very thorough investigation is necessary to find the key relaxation processes. The end of the article provides a bridge to the following paper: it treats interactions of two molecules (dimers), where resonant interactions become important and lead to new, characteristic relaxation processes.

The same interactions are, of course, dominant in pure molecular crystals, which is the topic of the second paper: "The dynamics of one-dimensional triplet excitons in molecular crystals" by Jan Schmidt. One-dimensional crystals are certainly one of the topics the originators of the series of this book, "physics and chemistry of materials with low-dimensional structures", had in mind. Jan Schmidt presents a series of beautiful experiments on relaxation phenomena of triplet excitons and discusses in detail the intricate interplay between phonon processes, impurities (chain ends) and interchain interactions. Some of his curves could easily serve as textbook illustrations, e.g. in the chapter on the time-resolved thermalization of the k-vector, where we can follow in detail the evolution in k-space of the originally excited $k = 0$ state towards thermal equilibrium and identify thus the scattering processes responsible for this relaxation.

The third paper, "Spectral hole-burning in crystalline and amorphous organic solids. Optical relaxation processes at low temperature" by Silvia Völker treats, as does the first paper, isolated organic impurities in organic hosts, but the relaxations are monitored via fluorescence, and spin relaxation does not interfere. Hole-burning is a spectroscopic technique with extremely high resolution and is thus the ideal tool to study electronic relaxation processes. The Leiden group has brought

1

J. Fünfschilling (ed.), Relaxation Processes in Molecular Excited States, 1—2.
© 1989 *by Kluwer Academic Publishers.*

this technique to perfection, publishing many high resolution spectra that are the delight of every spectroscopist. The present paper contains much of this work and is a very thorough review of the hole-burning work both in crystalline and in amorphous hosts. The wealth of detailed information available from these data establishes hole-burning spectroscopy as one of the key spectroscopic tools to study relaxation phenomena. Especially striking are the results on the temperature dependence of the linewidth of holes in amorphous systems. The almost ubiquous $T^{1.3}$-dependence excited especially the theorists and let them devise numerous models that could explain this curious observation. Whatever the final answer will be, the hole-burning experiments have certainly proven and confirmed the existence of glass-specific structural relaxations (Two Level Systems, TLSs) that have earlier been postulated from heat capacity and ultrasound attenuation experiments in quite different systems.

In the fourth paper, "Relaxation theory applied to scattering of excitations and optical transitions in crystals and solids", Robert Silbey presents a unifying theoretical framework to describe the relaxation processes presented by the first three more experimentally oriented papers. He uses the Redfield formalism to describe spin relaxation of triplets as well as the interaction of molecules with TLSs in organic glasses. As mentioned above and discussed in detail in this paper, there exist a large number of competing theories trying to describe the temperature dependence of the linewidth of spectral holes in glasses, and the theoretical framework presented by Robert Silbey includes most of them as special cases. This paper thus not only unifies the description of the experimental results presented in the first chapters, but also a large number of different theoretical approaches used to describe the influence of structural relaxations on spectral linewidths.

Relaxation processes in molecular solids emerge from this book as well understood processes, which have an enormous potential to yield very detailed information on the dynamics and the structure of the matrix studied. Only a small fraction of this potential has been explored so far, and we expect to see in the next few years new results which will yield many more exciting new insights of such structures as glasses, membranes or polymers.

SPIN-LATTICE AND SPIN-SPIN RELAXATION IN PHOTO-EXCITED TRIPLET STATES IN MOLECULAR CRYSTALS

J. SCHMIDT

Centre for the Study of Excited States of Molecules, Huygens Laboratory, University of Leiden,
P.O. Box 9504, 2300 RA Leiden, The Netherlands.

1. Introduction

Spin-Lattice Relaxation (SLR) of paramagnetic ions in diamagnetic crystals has been extensively studied by many authors and in the past 50 years a considerable literature has been published (for a review see for instance the books of Gorter [1], Manenkov and Orbach [2], and Abragam and Bleaney [3]). The classic papers of Waller [4], Heitler and Teller [5], van Vleck [6], Kronig [7] and others have shown that the dominant spin-lattice interaction in such systems is through the thermal modulation of the crystalline electric field. This affects the electron spin via the spin-orbit coupling and induces transitions between the electron spin levels.

At very low temperatures the amplitude of the ionic vibrations are small and one considers only the terms linear in the strain. This term describes a process in which an electron spin flips from one magnetic sublevel to the other simultaneously with the creation or annihilation of a phonon with the same energy as the spin resonance frequency ω_0. The relaxation rate which results from this direct process is proportional to the temperature T, provided that $kT > \hbar\omega_0$. The direct process is usually found to dominate the SLR at temperatures of the order of 1 K or less.

At higher temperatures one must include terms quadratic in the strain in order to describe the observed relaxation rates. These terms describe the process in which a phonon with a frequency ω scatters from the spin inelastically by flipping the spin. If the ion has an even number of electrons, so the spin state is not degenerate in zero-magnetic field, the contribution to the SLR rate is proportional to T^7. If the spin state is degenerate in zero-magnetic field a SLR rate proportional to T^9 is found.

In many cases the paramagnetic ion has an excited electronic state separated from the ground state by an energy ΔE. For this case it has been shown by Orbach [8] that there is an additional contribution to the SLR rate which varies with temperature as $\exp\{-\Delta E/kT\}$ for $T \ll \Delta E$. This term arises from the thermal excition to and decay from the excited state, via the absorption and subsequent emission of a phonon, with a spin flip occurring in this process. In most cases one finds that these two two-phonon processes dominate the direct process at temperatures higher than 1 K.

Spin-spin relaxation (SSR) in dilute ionic crystals has only been studied after the

3

J. Fünfschilling (ed.), Relaxation Processes in Molecular Excited States, 3—50.
© 1989 *by Kluwer Academic Publishers.*

introduction of electron spin echo techniques. In the paper by Mims [9] it is explained that usually it is not possible to define a relaxation time T_2 in the sense of the Bloch equations. Instead, it is better to use the operational definition of a phase memory time T_M to denote the characteristic decay time of the electron spin echo envelope. In the simplest case T_M is associated with the lattice relaxation and $T_M = T_1$. In many instances however, other mechanisms associated with changes in the local magnetic field determine the value of T_M, such as spin-spin interactions among the electron spins and dipolar interactions among the nuclear spins.

SLR and SSR in photo-excited triplet states in molecular crystals have attracted relatively little attention compared with the ionic systems. The first studies on SLR, in the presence of a magnetic field, were published in the late sixties by Fischer and Denison [10], Schwoerer and Sixl [11], and Wolfe [12]. After a first report by Hall and El-Sayed [13] investigations on SLR in the absence of a magnetic field were published by Zuclich et al. [14] and Antheunis et al. [15]. These authors studied several aspects of the SLR such as the dependence on the temperature, the guest concentration, the type of host material and the influence of the presence of the magnetic field. From the results it was clear that the SLR rates strongly depend on the particular combination of guest and host molecule but it was not possible to draw a definitive conclusion about the mechanism responsible for the SLR process.

In molecular crystals it is unlikely that the SLR is caused by spin-orbit coupling of the triplet spin to the lattice as in paramagnetic systems in ionic crystals. First the crystal fields are relatively weak because the molecules are bound together by weak van der Waals forces instead of the strong Coulomb forces in ionic crystals. Second, the spin-orbit coupling in these systems is small (the g-value usually is very close to the free electron value 2.0023) and, thus, the triplet spin is only loosely coupled to the lattice.

In order to understand the coupling mechanism between triplet spins of phosphorescent molecules and the phonons in the crystal one has studied especially the temperature dependence of the SLR rates. From the results of Schwoerer et al. [16], and Wolfe [12] and also from later studies (see for instance Konzelmann et al. [17]) one concluded that one-phonon as well as two-phonon processes occur. In practice it was difficult to discriminate between a Raman- or Orbach-type process, because most experiments were performed in a limited temperature range (usually between 1.2 K and 4.2 K). It seemed most likely to attribute the observed temperature dependence of the SLR to a Raman-type process, because the interpretation in terms of an Orbach process required the assumption of an activation energy of the order of $10-20$ cm^{-1}. Since the difference in energy of the electronic states and also the vibrational energies of the molecules usually are much larger (for instance the lowest skeletal vibration of naphthalene occurs at 180 cm^{-1}) one discarded the possibility that states $10-20$ cm^{-1} above the lowest triplet state T_0 might be responsible for the observed two-phonon process.

In this chapter we show that two-phonon processes, reminiscent of Orbach processes in paramagnetic ions in ionic crystals, play an important role in the SLR process in photo-excited triplet state molecules dissolved in foreign host crystals.

In zero-magnetic field and in moderately high magnetic fields this often appears to be the dominating mechanism. The close lying states involved in this process are so-called "local or pseudo-local phonon states". They correspond with localized vibrations, mostly librations of the guest molecule in the force field of the surrounding host molecules, with energies typically in the order of $10-20$ cm^{-1}. In very high magnetic fields however there is evidence that the direct process via the emission of resonant phonons becomes the most efficient SLR process [18].

SSR in photo-excited triplet state molecules was also studied only after the introduction of electron spin echo techniques [19]. It appeared that many of the conclusions of Mims [9], regarding the causes for the phase memory time T_M in ionic systems, likewise apply in molecular crystals. For instance it was found that in most cases the dephasing was caused by flip-flop motions in the nuclear spin system [20]. Since the magnetic moment of the proton spins, which are the most abundant ones in aromatic crystals, is large, this explained the fact that T_M usually is short (in the order of 1 μs) and independent of the temperature in the liquid helium temperature range. In this chapter however, we will present an example of a system, in which the proton spins are replaced by deuterons, and where a temperature dependence of T_M is observable which can be related with the presence of a nearby excited triplet state. Thus it is concluded that Orbach-type processes may also affect the SSR in molecular crystals provided that the dephasing is not overwhelmed by the interactions with the nuclear spin system.

A beautiful model system for studying the effect of a nearby excited state on the spin relaxation processes in the lowest triplet state of a phosphorescent molecule is provided by the dimers occurring in isotopically mixed crystals of naphthalene-h$_8$ in naphthalene-d$_8$. The dimers of interest are the translationally inequivalent pairs which have their two triplet states separated by 2.5 cm^{-1} (two times the excitation transfer interaction). It appears that the SLR and SSR in the lowest triplet component is determined by an Orbach-type process between the two triplet states. Since not only the energy separation of the two states is known exactly but also the properties of their sublevels it is possible to test theoretical models for SLR and SSR resulting from this two-phonon process in great detail. The experimental results on this system and the theoretical analysis will form an important part of this chapter.

2. The Theoretical Model

In this section we shall review the theoretical model for describing the spin-spin and spin-lattice relaxation resulting from the excitation transfer between the lowest triplet state T_0 and a nearby excited state T_e. In the case of molecules dissolved in foreign host crystals T_0 is the zero-phonon triplet state and T_e corresponds with a state where in addition to the electronic excitation a quantum of a librational motion is absorbed. In the case of dimers occurring in isotopically mixed crystals T_0 and T_e correspond with the two (zero-phonon) triplet eigenstates of the pair separated by two times the exchange interaction J. In both cases the energy difference of the two states will be indicated by ΔE.

J. SCHMIDT

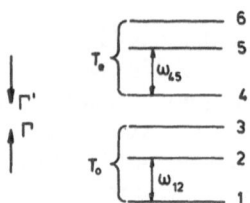

Fig. 1. The sublevels of the lowest triplet state T_0 and the excited, librational state T_e.

In Fig. 1 we have indicated the six triplet levels of interest in T_0 and in T_e. The zero-field splittings in T_e are supposed to be slightly different from those in T_0 and the three spin eigénvectors in T_e to be rotated with respect to those in T_0. Further it is assumed that there is a stochastic transfer process described by a probability Γ for excitation from T_0 to T_e and a probability Γ' for the reverse process. These two rates are related via the Boltzmann factor i.e. $\Gamma/\Gamma' = \exp\{-\Delta E/kT\}$. This excitation and decay will lead to dephasing of the magnetic resonance transitions in T_0 and a redistribution of the populations of the spin levels of T_0 i.e. to an effective SLR process.

To describe the effect of excitation and decay properly we must solve the density matrix for this 6 state system. As argued by Silbey [21] we can try to simplify the problem of solving for the 36 matrix elements by isolating the terms important for the experimental observables and treat only a subset of terms. For instance if we are interested in a certain transition in the lower state T_0 we need only to consider the related transition in the upper state with almost the same frequency. Scattering of the excitation between the two levels is assumed to occur without a change in the spin state i.e. the phonon excitation interaction is independent of the spin variables. The physical model for the scattering is then that a molecule in the lower triplet level can be excited to the higher level by absorption of a phonon. The excitation remains in the upper level for a time $\tau = (\Gamma')^{-1}$ and then returns to the lower one via the emission of a phonon. Since the resonance frequency in the upper state is different from that in the lower state a dephasing results and since the spin axes are slightly rotated spin-lattice relaxation is simultaneously induced.

Considering the transition at frequency ω_{12} in the lower state and isolating the reduced density matrix elements the following expression is found for the variables in the lower triplet state (Silbey equation 3.3)

$$\dot{\sigma}_{12} = -i\omega_{12}\sigma_{12} - \Gamma\sigma_{12} + \Gamma'\sigma_{45}.$$
$$\dot{\sigma}_{11} - \dot{\sigma}_{22} = -\Gamma(\sigma_{11} - \sigma_{22}) + \Gamma'(\sigma_{44} - \sigma_{55}). \tag{1}$$
$$\sigma_{21} = \sigma_{12}^*.$$

The time evolution of the density matrix elements for the related transition at frequency ω_{45} in the upper triplet state is described by (Silbey equation 3.4)

$$\dot{\sigma}_{45} = -i\omega_{45}\cos 2\theta\sigma_{45} + i\omega_{45}\sin 2\theta(\sigma_{44} - \sigma_{55}) - \Gamma'\sigma_{45} + \Gamma\sigma_{12}.$$
$$\dot{\sigma}_{44} - \dot{\sigma}_{55} = -i\omega_{45}\sin 2\theta(\sigma_{45} - \sigma_{54}) - \Gamma'(\sigma_{44} - \sigma_{55}) + \tag{2}$$
$$+ \Gamma(\sigma_{11} - \sigma_{22}).$$

Where θ indicates the angle of rotation of the spin states in T_e with respect to those in T_0. Using $\sigma_{12} = \sigma_{21}^*$ and introducing the new variables $r_1 = (\sigma_{12} + \sigma_{21})/2$, $R_1 = (\sigma_{45} + \sigma_{54})/2$, $r_2 = (\sigma_{12} - \sigma_{21})/2i$, $R_2 = (\sigma_{45} - \sigma_{54})/2i$, $r_3 = (\sigma_{11} - \sigma_{22})$ and $R_3 = (\sigma_{44} - \sigma_{55})$ we can convert the two expressions (1) and (2) into the well-known coupled Bloch equations introduced by Gutowsky et al. [22] and McConnell [23],

$$\dot{r}_1 = -\omega_{12}r_2 - \Gamma r_1 + \Gamma' R_1.$$
$$\dot{r}_2 = \omega_{12}r_1 - \Gamma r_2 + \Gamma' R_2. \tag{3a}$$
$$\dot{r}_3 = -\Gamma r_3 + \Gamma' R_3.$$

$$\dot{R}_1 = -\omega_{45}\cos 2\theta R_2 - \Gamma' R_1 + \Gamma r_1.$$
$$\dot{R}_2 = \omega_{45}\cos 2\theta R_1 - \omega_{45}\sin 2\theta R_3 - \Gamma' R_2 + \Gamma r_2. \tag{3b}$$
$$\dot{R}_3 = \omega_{45}\sin 2\theta R_2 - \Gamma' R_3 + \Gamma r_3.$$

The solutions to (3) are linear combinations of six exponentials, three corresponding to slowly decaying variables and three corresponding to quickly decaying variables. We can associate the three slow exponentials with the spin resonance frequency, the line width $1/T_2$ and the SLR rate $1/T_1$ of the observed lower triplet state. It is instructive to look first at the case where the rotation of the spin axes in the upper state is zero ($\theta = 0$). Then the 6×6 matrix breaks up in three 2×2 matrices and the six eigenvalues of (3) are [21],

$$\lambda\pm = \frac{i}{2}(\omega_{12} + \omega_{45}) - \left(\frac{\Gamma + \Gamma'}{2}\right) \pm \left\{\Gamma\Gamma' + \right.$$

$$\left. + \frac{1}{4}(-i\omega_{12} + i\omega_{45} - \Gamma + \Gamma')\right\}^{1/2}, \lambda_\pm^*, 0, \Gamma + \Gamma'. \tag{4}$$

The three eigenvalues related with the lowest triplet state are the observed resonance frequency ω_{obs} (Im λ_+), the spin-spin relaxation rate T_2^{-1} (Re λ_+), and the spin-lattice relaxation rate $T_1^{-1} = 0$. At low temperatures where $\Gamma/\Gamma' = \exp\{-\Delta E/kT\} \ll 1$ we find,

$$T_1^{-1} = 0$$

$$T_2^{-1} = \Gamma' \frac{(\delta/\Gamma')^2}{1 + (\delta/\Gamma')^2} \exp\{-\Delta E/kT\} \tag{5}$$

$$\varepsilon = \frac{\delta}{1 + (\delta/\Gamma')^2} \exp\{-\Delta E/kT\}$$

with $\delta = \omega_{12} - \omega_{45}$ and $\varepsilon = \omega_{obs} - \omega_{12}$.

We see that in this particular case the jumps between the two triplet states only lead to a dephasing process in the lower triplet state and to a shift of the resonance frequency. Both effects showing an activated behaviour with the temperature. In the limit that the decay rate Γ' is very large compared with the difference in resonance frequency δ, T_2^{-1} becomes very small. This makes physical sense, since under this condition each jump causes only a very small phase error. In the limit of a very small value of Γ' (long lifetime of the upper state) compared with δ each jump will cause a complete loss of phase memory and it is seen that then $T_2^{-1} = \Gamma$.

If the spin axes in T_e are not parallel to those in T_0, i.e. $\theta \neq 0$, we must diagonalize a 6 × 6 matrix. Perturbation calculations have been performed by Levinsky and Brenner [24], Verbeek and Schmidt [25], Vollmann [26], Dietz *et al.* [27], and Silbey [21] for low temperatures where again $\Gamma/\Gamma' = \exp\{-\Delta E/kT\} \ll 1$. It is found that

$$T_1^{-1} = \Gamma' \sin^2 2\theta \; \frac{(\omega_{45}/\Gamma')^2}{1 + (\omega_{45}/\Gamma')^2} \; \exp\{-\Delta E/kT\}. \tag{6}$$

The equations for T_2^{-1} and ε are considerably more complicated but if the rotation of the spin axes θ is small the expressions in good approximation are

$$T_2^{-1} = \Gamma' \left[\frac{(\delta/\Gamma')^2}{1 + (\delta/\Gamma')^2} + \frac{1}{2} \sin^2 2\theta \left\{ \frac{1}{1 + (\delta/\Gamma')^2} - \right. \right.$$

$$\left. \left. - \frac{1}{1 + (\omega_{12}/\Gamma')^2} \right\} \right] \exp\{-\Delta E/kT\}. \tag{7}$$

$$\varepsilon = \left[\frac{\delta}{1 + (\delta/\Gamma')^2} + \frac{1}{2} \frac{\omega_{12} \sin^2 2\theta}{1 + (\omega_{12}/\Gamma')^2} \right] \exp\{-\Delta E/kT\}. \tag{8}$$

When comparing the expressions (7) and (8) with those obtained for the case where $\theta = 0$ it is clear that the SLR rate T_1^{-1} is a direct consequence of the rotation of the spin axes in the upper state. It is also observed that T_2^{-1} and ε now contain two contributions. In the expression for T_2^{-1} the first is a dephasing term identical to the case $\theta = 0$ whereas the second term is a consequence of the rotation of the spin axes. A similar argument applies to the expression for ε.

In the situations that will be considered by us θ is small and we mostly limit ourselves to low temperatures. Under these conditions T_1^{-1}, T_2^{-1} and ε are found to follow an exponential behaviour with the temperature and the expressions (6), (7) and (8) will be used to analyse the experimental results. At these low temperatures the decay from T_e to T_0 only occurs via a spontaneous emission process and Γ' can be considered to be constant. However at higher temperatures T_e will also decay via stimulated processes and Γ' will show a temperature dependence. As a result T_2^{-1} first increases with the temperature and then decreases. A typical example of this behaviour is found in the case of the naphthalene dimer. For a more complete discussion of this effect the reader is referred to the chapter of Silbey [28] and the papers of Reineker and coworkers [29, 30].

3. Experimental

3.1. THE MEASUREMENT OF THE SPIN-LATTICE RELAXATION RATES

The information on SLR in photo-excited triplet state molecules has been obtained mainly by studying the recovery signal in the phosphorescence intensity, $\Delta I_{ph}(t)$, upon the sudden sweep of a microwave field through one of the resonance transitions of the triplet state. These experiments have been performed in the presence as well as in the absence of a magnetic field. In both cases we study the evolution of the population distribution of a three-level system towards thermal equilibrium. It will be clear that the rate equations describing this evolution for these two cases may be described in a similar way.

In Fig. 2 we illustrate the different dynamic processes that determine the population distribution over the triplet sublevels T_i ($i = 1, 2, 3$),

1. the populating rates P_i,
2. the total decay rates k_i consisting of a radiative part k_i^r and a non-radiative part k_i^{nr},
3. the spin-lattice relaxation rates w_{ij} from T_i to T_j.

Note that the SLR rate T_1^{-1} used in the previous section to describe the recovery of the population difference between two sublevels is related to the w_{ij} via $T_1^{-1}(ij) = w_{ij} + w_{ji}$.

The master equations that describe the evolution of the populations N_i of the three magnetic states T_i are given by

$$\dot{N}_1 = P_1 - (k_1 + w_{12} + w_{13})N_1 + w_{21}N_2 + w_{31}N_3$$
$$\dot{N}_2 = P_2 - (k_2 + w_{23} + w_{21})N_2 + w_{32}N_3 + w_{12}N_1 \qquad (9)$$
$$\dot{N}_3 = P_3 - (k_3 + w_{31} + w_{32})N_3 + w_{13}N_1 + w_{23}N_2.$$

We assume that the experiments are performed under the condition of low intensity excitation into the singlet system so that the relative populating rates P_i

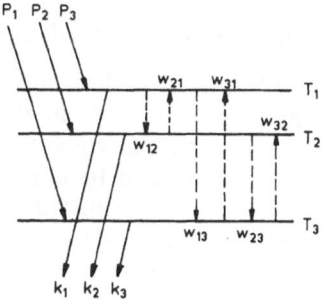

Fig. 2. The three triplet sublevels with their kinetic parameters; the populating rates P_i, the decay rates k_i and the spin-lattice relaxation rates w_{ij}.

are not affected by depletion of the ground state. Further, since the Boltzmann factor at the temperatures of interest is close to unity, we take $w_{ij} = w_{ji}$.

The rate constants determined by the optical pumping cycle, P_i and k_i, are essentially temperature independent in the low temperature range of interest whereas the SLR rates usually increase with increasing temperature. On changing the temperature one therefore generally observes a considerable variation of the ratio between the relaxation rates w_{ij} and the decay rates k_i. According to the value of this ratio one distinguishes three temperature ranges [31].

(a) *The Low Temperature Region of Isolation*

This region is determined by the condition that the relaxation rates w_{ij} are negligible relative to the decay rates k_i to the singlet ground state S_0. In this case the evolution of the phosphorescence intensity is determined by the k_i's and the P_i's and information on the relaxation rates can not be obtained.

(b) *The Intermediate Region*

Here the SLR rates w_{ij} are of the same order of magnitude as the decay rates k_i. The equations (9) then present a considerable problem since they contain all nine kinetic parameters P_i, k_i and w_{ij}. This problem may be simplified by continuous saturation of one of the zero-field transitions with a resonant microwave field [14, 15, 32]. In experimental practice however the accuracy of the measurement of the individual SLR rates critically depends on the complete saturation of all three transitions. This poses serious experimental problems because the transitions often are inhomogeneously broadened and moreover the applied microwave field is spatially inhomogeneous, especially when using a helix in zero-field experiments. Further it appears that in many systems the intermediate region corresponds to a very small temperature range. In view of these reasons we preferred to perform the experiments in the temperature range where the SLR rates w_{ij} exceed the decay rates k_i.

(c) *The High Temperature Region of Dominant Relaxation*

The advantage of working in this region is that one can neglect the effect of the populating and depopulating rates P_i and k_i in the equations (9). It is then relatively easy to obtain the mean relaxation rate $\bar{w}$ from the recovery signal in the phosphorescence intensity $\Delta I_{ph}(t)$. This can be understood as follows. Since the total population of the triplet state remains constant during the evolution only two of the three differential equations (9) are independent. As a result the recovery signal $\Delta I_{ph}(t)$ only contains two time constants λ_1 and λ_2 i.e.

$$\Delta I_{ph}(t) = C_1 \exp\{-\lambda_1 t\} + C_2 \exp\{-\lambda_2 t\}. \tag{10}$$

The constants C_1 and C_2 depend on the particular transition studied but the time constants λ_1 and λ_2 only depend on the relaxation rates w_{ij} [32],

$$\lambda_{1,\,(2)} = (w_{12} + w_{23} + w_{31}) + (-)\{w_{12}(w_{12} - w_{31}) + $$
$$+ w_{31}(w_{31} - w_{23}) + w_{23}(w_{23} - w_{12})\}^{1/2}, \tag{11}$$

and thus

$$(\lambda_1 + \lambda_2)/2 = (w_{12} + w_{23} + w_{31}) = 3\bar{w}. \tag{12}$$

An important advantage, moreover, is that the result does not depend on saturation of one of the microwave transitions.

In principle the individual relaxation rates w_{ij}, can also be obtained with the help of saturation experiments. For instance when saturating the $T_i - T_j$ transition an effective two-level system is created which will decay to the equilibrium population distribution with a rate

$$\lambda_{ij} = \tfrac{3}{2}(w_{ki} + w_{jk}). \tag{13}$$

By combining the results of three of such saturation experiments one can find the individual relaxation rates w_{ij}. In practice however large spreads are found in the values for w_{ij}, because also in this case the results depend critically on the attainment of saturation, a condition that often is difficult to obtain.

It is possible to obtain the individual SLR rates w_{ij} in the region of dominant relaxation without using saturating microwave fields. This method was developed by van Noort et al. [33] and is based on the fact that an additional piece of information is contained in the two pre-exponential factors C_1 and C_2 in (10). Here we present the analysis of this problem. Later, in Section 4.6, it will be demonstrated that in several cases reliable values for the w_{ij}'s can be obtained.

We start by considering the master equation (9) for the evolution of the populations of the triplet sublevels and rewrite it in the following form

$$\dot{\mathbf{n}} = -R\mathbf{n}, \tag{14}$$

where

$$n = \begin{pmatrix} n_1 \\ n_2 \\ n_3 \end{pmatrix} = \begin{pmatrix} N_1 - N_1^0 \\ N_2 - N_2^0 \\ N_3 - N_3^0 \end{pmatrix}, \tag{15}$$

and

$$R = \begin{pmatrix} w_{12} + w_{13} & -w_{12} & -w_{13} \\ -w_{12} & w_{12} + w_{13} & -w_{23} \\ -w_{13} & -w_{13} & w_{13} + w_{23} \end{pmatrix}. \tag{16}$$

N_i^0 is the equilibrium population of sublevel i and N_i its actual time dependent population. Note that $n_1 + n_2 + n_3 = 0$.

In the analysis we follow the method of Vega [34] and as the first step we transform equation (14) to a new coordinate system in which the vector $\mathbf{n}$ has one of its components equal to $n_1 + n_2 + n_3 = 0$. In principle this can be achieved in three ways. When studying the results of an experiment where the microwave field is swept through the transition $T_1 - T_2$ it is most convenient to use the transformation matrix

$$U_{12} = \begin{pmatrix} \dfrac{1}{\sqrt{2}} & -\dfrac{1}{\sqrt{2}} & 0 \\[2mm] \dfrac{1}{\sqrt{6}} & \dfrac{1}{\sqrt{6}} & -\dfrac{2}{\sqrt{6}} \\[2mm] \dfrac{1}{\sqrt{3}} & \dfrac{1}{\sqrt{3}} & \dfrac{1}{\sqrt{3}} \end{pmatrix} . \tag{17}$$

The corresponding matrices U_{23} and U_{13} for the other two pairs of levels can easily be obtained from (17) by interchanging the appropriate columns.

The transformation via U_{12} reduces the three coupled differential equations to two sets; the trivial identity $(d/dt)(n_1 + n_2 + n_3) = 0$ and a pair of coupled equations

$$\mathbf{v}_{12}(t) = -\sigma_{12}\mathbf{v}_{12}(t) \tag{18}$$

where

$$\mathbf{v}_{12}(t) = \begin{pmatrix} \dfrac{1}{\sqrt{2}}\,(n_1 - n_2) \\[4mm] \sqrt{\dfrac{3}{2}}\,(n_1 + n_2) \end{pmatrix} \tag{19}$$

and

$$\sigma_{12} = \begin{pmatrix} 2w_{12} + \dfrac{1}{2}\,w_{13} + \dfrac{1}{2}\,w_{23} & \dfrac{\sqrt{3}}{2}\,(w_{13} - w_{23}) \\[4mm] \dfrac{\sqrt{3}}{2}\,(w_{13} - w_{23}) & \dfrac{3}{2}\,(w_{13} + w_{23}) \end{pmatrix} . \tag{20}$$

It is seen that we are left with two differential equations; one for the sum and one for the difference of the populations n_1 and n_2.

$$\mathbf{v}_{12}(t) = \mathbf{v}_{12}(0)\exp\{-\sigma_{12}(t)\}. \tag{21}$$

To find the time dependent behaviour of the elements of $\mathbf{v}_{12}(t)$ it is most convenient to transform σ to a diagonal matrix Λ, which proves to have the λ_1 and λ_2 as elements,

$$\Lambda = \begin{pmatrix} \lambda_1 & 0 \\ 0 & \lambda_2 \end{pmatrix}. \tag{22}$$

The transformation matrix D_{12}

$$D_{12} = \begin{pmatrix} \cos \psi_{12} & \sin \psi_{12} \\ -\sin \psi_{12} & \cos \psi_{12} \end{pmatrix} \tag{23}$$

is such that

$$\sigma_{12} = \tilde{D}_{12}\Lambda D_{12} = \begin{pmatrix} \lambda_1 \cos^2 \psi_{12} + \lambda_2 \sin^2 \psi_{12} & (\lambda_1 - \lambda_2) \cos \psi_{12} \sin \psi_{12} \\ (\lambda_1 - \lambda_2) \cos \psi_{12} \sin \psi_{12} & \lambda_1 \sin^2 \psi_{12} + \lambda_2 \cos^2 \psi_{12} \end{pmatrix}. \tag{24}$$

Equation (21) then becomes

$$v_{12}(t) = \begin{pmatrix} v_{12}^1(t) \\ v_{12}^2(t) \end{pmatrix}$$

$$= \begin{pmatrix} e^{-\lambda_1 t} \cos^2 \psi_{12} + e^{-\lambda_2 t} \sin^2 \psi_{12} & (e^{-\lambda_1 t} - e^{-\lambda_2 t}) \cos \psi_{12} \sin \psi_{12} \\ (e^{-\lambda_1 t} - e^{-\lambda_2 t}) \cos \psi_{12} \sin \psi_2 & e^{-\lambda_1 t} \sin^2 \psi_{12} + e^{-\lambda_2 t} \cos^2 \psi_{12} \end{pmatrix} \times$$

$$\times \begin{pmatrix} v_{12}^1(0) \\ v_{12}^2(0) \end{pmatrix}. \tag{25}$$

The individual relaxation rates w_{ij} can now be expressed in terms of the parameters λ_1, λ_2 and ψ_{12} by comparing the elements of σ_{12} in (20) with the corresponding elements in (24). We find

$$w_{12} = \lambda_1 \left(\frac{2}{3} \cos^2 \psi_{12} - \frac{1}{6} \right) + \lambda_2 \left(-\frac{2}{3} \cos^2 \psi_{12} + \frac{1}{2} \right).$$

$$w_{13} = \lambda_1 \left(\frac{1}{3} \sin^2 \psi_{12} + \frac{1}{\sqrt{3}} \cos \psi_{12} \sin \psi_{12} \right) +$$

$$+ \lambda_2 \left(\frac{1}{3} \cos^2 \psi_{12} - \frac{1}{\sqrt{3}} \cos \psi_{12} \sin \psi_{12} \right). \tag{26}$$

$$w_{23} = \lambda_1 \left(\frac{1}{3} \sin^2 \psi_{12} - \frac{1}{\sqrt{3}} \cos \psi_{12} \sin \psi_{12} \right) +$$

$$+ \lambda_2 \left(\frac{1}{3} \cos^2 \psi_{12} + \frac{1}{\sqrt{3}} \cos \psi_{12} \sin \psi_{12} \right).$$

The constants λ_1 and λ_2 represent the two time constants in the bi-exponential recovery curve and are obtained directly from the experiments. The remaining problem is to show how the constant ψ_{12} is related to the pre-exponential factors C_1 and C_2 in (10), and the radiative decay rates of the triplet sublevels.

As we have explained before our signal is induced by the sudden sweep of a microwave field through one of the resonance transitions. The expression for the resulting change in the phosphorescence intensity is given by

$$\Delta I_{ph} = X(n_1 k_1^r + n_2 k_2^r + n_3 k_3^r). \tag{27}$$

where X is an instrument constant. When sweeping transition $T_1 - T_2$ we have at time $t = 0$, immediately following the sweep, $n_1(0) = -n_2(0)$ and $n_3(0) = 0$. Then from equation (19) it follows that $v_{12}^1(0) = n_1(0)\sqrt{2}$ and $v_{12}^2(0) = 0$. Further

$$n_1(t) = \frac{1}{\sqrt{2}} v_{12}^1(t) + \frac{1}{\sqrt{6}} v_{12}^2(t).$$

$$n_2(t) = -\frac{1}{\sqrt{2}} v_{12}^1(t) + \frac{1}{\sqrt{6}} v_{12}^2(t). \tag{28}$$

$$n_3(t) = -n_1(t) - n_2(t) = -\frac{2}{\sqrt{6}} v_{12}^2(t).$$

We can now rewrite equation (27) with the help of (28) and (25)

$$\Delta I_{ph}(t) = n_1(0)X(A \cos^2 \psi_{12} + B \cos \psi_{12} \sin \psi_{12}) e^{-\lambda_1 t} +$$
$$+ n_1(0)X(A \sin^2 \psi_{12} - B \cos \psi_{12} \sin \psi_{12}) e^{-\lambda_2 t} \tag{29}$$

with $A = k_1^r - k_2^r$ and $B = (1/\sqrt{3})(k_1^r + k_2^r - 2k_3^r)$. Comparing (29) with (10) we obtain

$$C = C_1/C_2 = (A + B \, \text{tg} \, \psi_{12})/(A \, \text{tg}^2 \, \psi_{12} - B \, \text{tg} \, \psi_{12}) \tag{30}$$

or

$$\psi_{12} = \text{arc tg } Y_{12} \quad \text{with} \quad Y_{12} = \frac{(BC + B) \pm \{(BC + B)^2 + 4A^2C\}^{1/2}}{2AC} \tag{31}$$

And thus we have expressed the third parameter ψ_{12} necessary for the determination of the w_{ij}'s in terms of the observable quantity $C = C_1/C_2$ and the relative radiative decay rates k_i^r. Unfortunately the solution of (31) yields two roots and the only way to check which one gives the right solution is to perform a second experiment on one of the other transitions. In practice however it is often possible to disregard one solution because it leads to physically unacceptable values for the w_{ij}'s.

As an illustration of the SLR measurements we present in Fig. 3 two examples of optically detected recovery curves for aniline-d_2 in p-xylene-h_{10} at $T = 3.2$ K upon a sweep through the 4.744 GHz (a) and 2.633 GHz (b) zero-field transitions. At this temperature the condition of dominant relaxation applies and the recovery

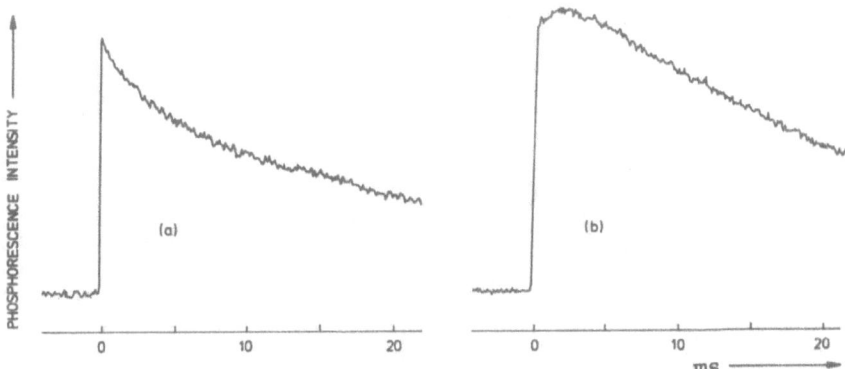

Fig. 3. (a) The change in the phosphorescence intensity of aniline-d_2 in p-xylene-h_{10} at 3.2 K upon a sweep through the 4.744 GHz zero-field transition. (b) Upon a sweep through the 2.633 GHz zero-field transition.

curve clearly shows a bi-exponential behaviour. Note that in (b) the fast component has an opposite sign compared with the curve in (a). Since for this system the radiative constants are well known it is possible to obtain reliable values for the individual relaxation rates by using the parameters λ_1, λ_2, C_1, and C_2 derived from these curves.

Finally we mention that in addition to optical detection methods measurements of SLR rates have been performed using Electron Spin Echo (ESE) techniques with microwave detection. In this case we measure a signal that is proportional to the difference of the populations of two triplet sublevels in contrast to the optical detection experiments where we measure the population evolution of the individual levels. These ESE experiments were also performed in the temperature region of dominant relaxation, by first saturating a particular transition and then inspecting the return to equilibrium. It will be clear that this recovery curve contains the same two time constants λ_1 and λ_2 (10, 11). This can be seen for instance from (25) which describes the time dependent behaviour of the population difference of the triplet levels T_1 and T_2 upon saturation of the $T_1 - T_2$ transition.

3.2. THE MEASUREMENT OF THE SPIN-SPIN RELAXATION RATES

The phase memory time T_M in the phosphorescent triplet state molecules was measured, with the help of Electron Spin Echo (ESE) techniques using microwave detection, from the decay of the two-pulse echo envelope curve. From the Bloch formalism one would expect a purely exponential decay according to $E(2\tau) = E(0)\exp\{-2\tau/T_2\}$, where τ is the interval between the two microwave pulses. In actual practice however a more complicated decay curve is often observed. For such cases we adopt the definition of Mims [9] for T_M as the time, measured from the first microwave pulse, which corresponds to an attenuation of $\exp\{-1\}$ of the echo signal.

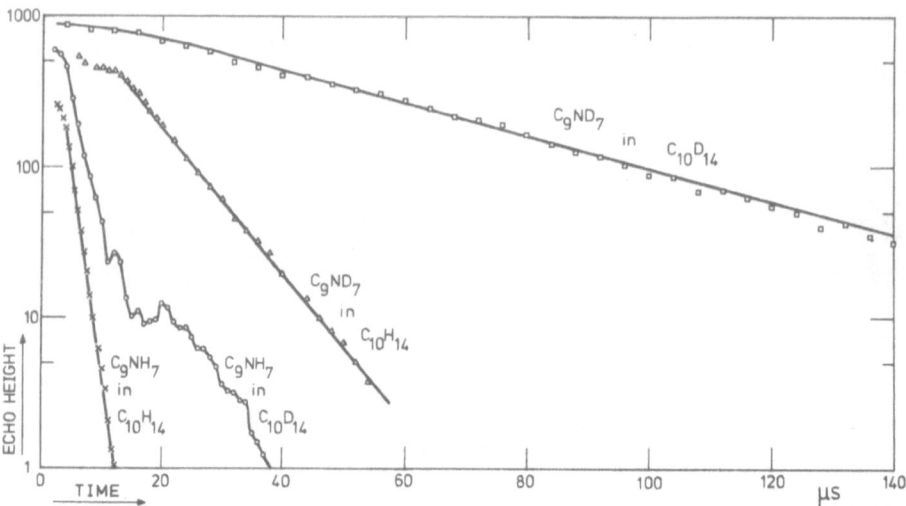

Fig. 4. The decay curves derived from two-pulse ESE experiments on the $T_z - T_x$ transition of quinoline C_9NH_7 or C_9ND_7 in a single crystal of durene $C_{10}H_{14}$ or $C_{10}D_{14}$ at $T = 1.2$ K. The echo height is plotted as a function of the time interval 2τ between the first microwave pulse and the echo signal.

The reason that the decay of the ESE signal in solids often is non-exponential is that the Bloch model only applies for spins in fast relative motion with respect to each other, e.g. paramagnetic ions in liquid solutions. In many instances in our aromatic crystals the dephasing of the triplet spins is caused by flip-flop motions in the nuclear spin system. These disturbances in the environment modulate the Larmor frequencies and randomnize the phases of the triplet spins. As an example we show in Fig. 4 the effect of deuteration of guest and host molecules on the phase memory time T_M. This figure clearly shows non-exponential decays and even that modulations are present in the decay curve. (For an explanation of this nuclear modulation effect see Mims [9]). However when the dephasing is dominated by an Orbach-type process involving a close-lying excited triplet state we find an exponential decay of the two-pulse echo envelope curve.

3.3. THE EXPERIMENTAL ARRANGEMENTS

The experiments to be discussed here have been performed in the temperature range from 0.35 K to 50 K. Between 1.1 K and 4.2 K bath cryostats with quartz windows were used. For temperatures lower than 1.1 K a special ^{3}He dewar was designed that was introduced in the ^{4}He bath cryostat (for a complete description of these two systems see for instance [35]). For temperatures above 4.2 K a Leybold Heraeus helium gas flow croystat was employed equipped with an electronic unit to stabilize the temperature to a precision of 0.1 K.

In most experiments the sample was irradiated with a 200 Watt Osram high pressure mercury arc via an appropriate set of filters. In addition the sample could be irradiated with the output of an excimer pumped dye laser or the frequency doubled output of a Nd—YAG pumped dye laser.

The SLR measurements in zero field were performed in a conventional ODMR spectrometer [36]. The SLR rates in a magnetic field and the directions of the spin-axes were determined in a home-build X-band or K-band EPR spectrometer equipped with means for optical detection [37, 38].

The zero-field and X-band ESE spectrometers have been described in detail elsewhere (see for instance [37]). In the zero-field spectrometer microwave powers up to 10 W were used. Typical $\pi/2$- and π-pulse durations are 200 and 400 ns. With a recovery time of about 800 ns of the receiving system it was possible to observe the two-pulse echo 1.5 μs after the beginning of the first microwave pulse. In the X-band ESE spectrometers microwave powers up to 300 W were used. Typical $\pi/2$- and π-pulse durations are 40 and 80 ns. In this system it is possible to observe the two-pulse echo signal 0.8 μs after the beginning of the first microwave pulse.

4. Experimental Studies on Spin-Lattice and Spin-Spin Relaxation in Triplet State Molecules in Chemically Mixed Crystals

4.1. THE SYSTEM NAPHTHALENE IN DURENE

The system naphthalene (N) in durene (D) represents the first case for which it could be proven convincingly that the presence of (pseudo) local phonon states plays an important role in the SLR process in photo-excited triplet states. In Fig. 5 we show the temperature dependence of the average SLR rate $\bar{w}$ between 2.0 K and 4.2 K in a sample of N-h_8 in D-h_{14} and in N-d_8 in D-h_{14}. These results were obtained by measuring the two time constants λ_1 and λ_2 in the decay of the phosphorescence upon a sweep through the high frequency $T_z - T_x$ zero-field transition and taking $\bar{w} = (\lambda_1 + \lambda_2)/6$ (see Section 3.1). The dependence on the temperature is well described by a Boltzmann factor $\exp\{-\Delta E/kT\}$ with an activation energy $\Delta E = (16 \pm 2)$ cm^{-1}. It is seen that within the experimental accuracy no measurable difference exists between N-h_8 and N-d_8 thus excluding the possibility that hyperfine effects play a role in the SLR process.

Several attempts to measure the individual SLR rates between the triplet sublevels did not yield reliable results. In saturation experiments (see Section 3.1) it was difficult to ensure that complete saturation of the zero-field transitions was established and in the experiments where the pre-exponential factors were also considered (see the discussion in Section 3.1c) it appeared that either one of them was very small or that the difference between the two exponential factors λ_1 and λ_2 was very small.

In addition to the temperature dependence of the SLR rates it is found that the resonance frequencies of the zero-field transitions depend on the temperature as predicted by equation 8. In Fig. 6 we present the shifts of the optically detected

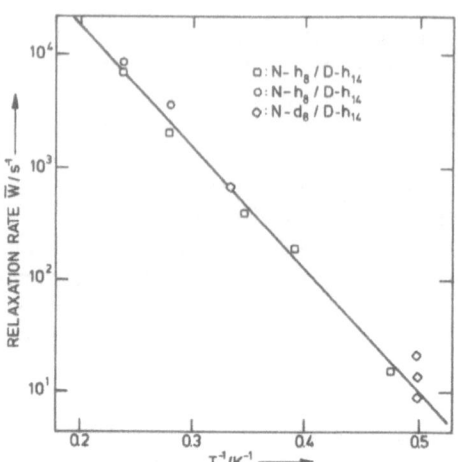

Fig. 5. The average SLR rate $\bar{w}$ as a function of the inverse temperature obtained on three different samples of naphthalene in durene in zero-magnetic field. The value of $\bar{w}$ was derived from the two time constants λ_1 and λ_2 which appear in the recovery curve of the phosphorescence intensity upon a sweep through the $T_z - T_x$ transition; $\bar{w} = (\lambda_1 + \lambda_2)/6$.

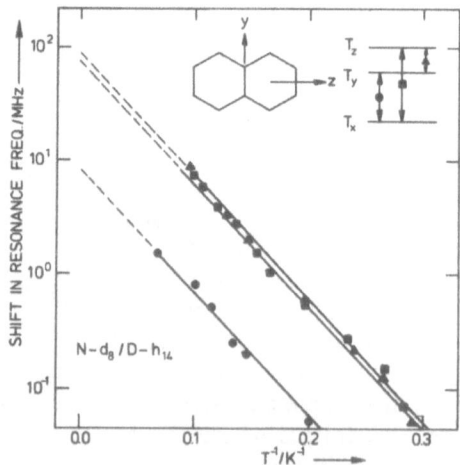

Fig. 6. The shifts of the optically detected zero-field transitions as a function of the inverse temperature in naphthalene-d_8 in durene-h_{14}. On raising the temperature the $T_x - T_y$ transition shifts to higher frequency whereas the $T_y - T_z$ and $T_z - T_x$ transitions shift to lower frequency.

transitions as a function of the inverse of the temperature. On raising the temperature the $T_y - T_x$ transition shifts to higher frequency whereas the $T_z - T_y$ and $T_z - T_x$ transitions move to lower frequency. The temperature dependence is described again with a Boltzmann factor with an activation energy $\Delta E = (16.5 \pm 2)\,\mathrm{cm}^{-1}$ [39].

The presence of a local phonon state just above T_0 in N/D is demonstrated clearly by the observation of a hot band in the phosphorescence spectrum $(12 \pm 2)\,\mathrm{cm}^{-1}$ to the high energy side of the 0—0 transition (see Fig. 7). In Fig. 8 it is seen that the relative intensity of this hot band, with respect to that of the 0—0 band, depends on the temperature with an activation energy $\Delta E = (14 \pm 3)\,\mathrm{cm}^{-1}$. When assuming that the hot band is lifetime broadened we find for the lifetime of the related local phonon state $\tau = (\Gamma')^{-1} = 1.5$ ps [40].

In the S_0 singlet ground state the energy of the local phonon is somewhat higher than in T_0 as can be seen in Fig. 9. Here phonon sidebands with a spacing of $17\,\mathrm{cm}^{-1}$ are visible in the phosphorescence spectrum to the low energy side of the 0—0 band [36].

The above observations support the idea that the SLR in the lowest triplet state T_0 is caused by thermally induced transitions between T_0 and a local phonon state T_e. Further we presume that the spin axes in the two states do not coincide. According to expression (6) the resulting SLR rates then depend on the rates of transfer Γ and Γ' between the two states and on the rotation θ of the spin axes in T_e with respect to those in T_0.

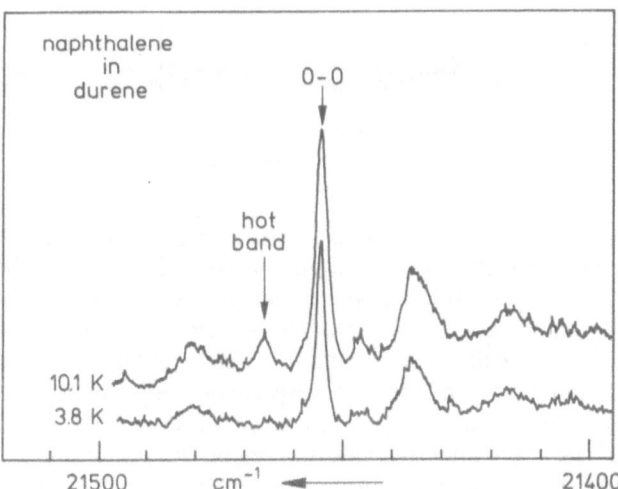

Fig. 7. The phosphorescence spectrum of naphthalene-d_8 in durene-h_{14} in the vicinity of the 0—0 band at 3.8 K and 10.1 K. In the high temperature recording a line appears on the high energy side of the 0—0 band which is attributed to the presence of a local phonon state above T_0.

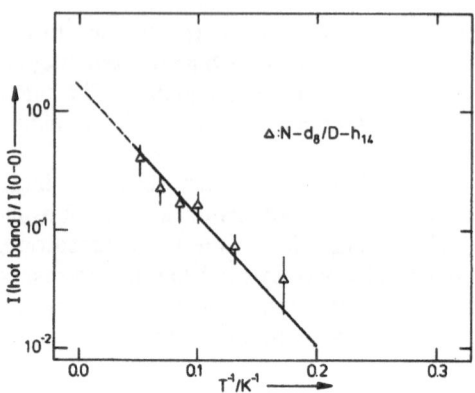

Fig. 8. The intensity of the hot band in the phosphorescence spectrum of naphthalene-d_8 in durene-h_{14}, normalized with respect to the intensity of the origin, as a function of the inverse temperature.

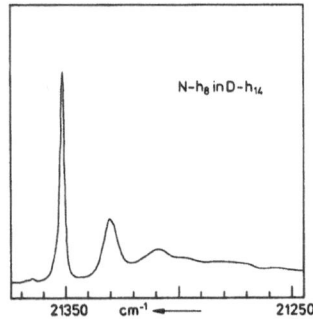

Fig. 9. The 0—0 band region in the phosphorescence spectrum of naphthalene-h_8 in durene-h_{14}. $T = 4.2$ K.

The rotation of the spin axes in the state T_e was measured with the help of optically detected EPR (ODMR) experiments as a function of the temperature between 1.2 K and 50 K [40]. These experiments are based on the idea that the directions of the spin axes, as observed in an EPR experiment, depend on the distribution of the molecules over T_0 and T_e. The results are presented in Fig. 10a and it seen that the z-axis undergoes a surprisingly large rotation of 5° in the y-z plane upon a variation of the temperature from 1.2 K to 30 K and that the x-axis rotates over 1° in the same temperature region in the x-y plane. Within the accuracy of the experiments a rotation of the z-axis in the x-z plane could not be observed. In Fig. 10b the same results are plotted on a logarithmic scale as a function of the inverse temperature. Up to 30 K the rotations can be described by

$$\theta(T) = \theta_\infty \exp\{-\Delta E/kT\}, \tag{32}$$

with $\Delta E = (14.5 \pm 2.5)\,\text{cm}^{-1}$, $\theta_\infty(y\text{-}z) = 10°$, and $\theta_\infty(x\text{-}y) = 2°$.

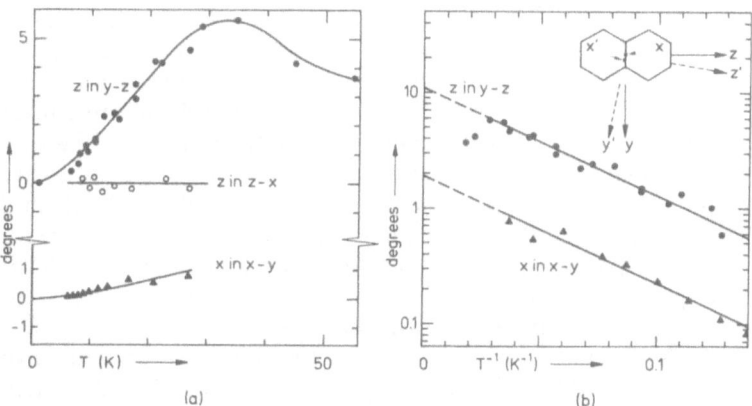

Fig. 10. (a) The change in direction of the spin axes in the triplet state of naphthalene-d_8 in durene-h_{14} within the principal planes as a function of the temperature. (b) The same results as in (a) as a function of the inverse temperature.

The results of Fig. 10 confirm beautifully that the spin axes in T_0 and T_e do not coincide. The temperature variation of the direction of the spin axes as observed in the experiment is expected to be described by the expression

$$\theta(T) = \theta_\infty \exp\{-\Delta E/kT\}/(1 + \exp\{-\Delta E/kT\}) \tag{33}$$

with θ_∞ equal to the difference in the direction of the spin axes in T_e and T_0. At low temperatures (33) reduces to expression (32) in accordance with the observations. The results show that the spin axes, upon the absorption of the vibrational quantum, rotate over a surprisingly large angle of $10°$ in the y-z plane whereas the rotation perpendicular to this plane is only $2°$.

Since for the present case the rotation of the spin axes mainly occurs in the molecular plane we may assume that the dominant SLR in T_0 will be between the T_z and T_y sublevels and thus that

$$\bar{w} = (w_{zy} + w_{yx} + w_{zx})/3 = (T_1^{-1})/6. \tag{34}$$

From the experiment we derive $T_1^{-1} = 7 \times 10^6 \exp\{-\Delta E/kT\}$. To check whether our theoretical model applies we use expression (6) with the parameters $\tau = (\Gamma')^{-1} = 1.5$ ps and $\theta_\infty(y\text{-}z) = 10°$ and find as the theoretical estimate for the SLR rate in T_0.

$$T_1^{-1} = 7 \times 10^6 \exp\{-\Delta E/kT\} \tag{35}$$

in surprisingly good agreement with the experimental finding. In the theoretical estimate we have used the observation that the frequency shift with the temperature is small and thus that ω_{45} in (6), which represents the resonance frequency in T_e, is almost equal to ω_{12} the resonance frequency of $T_z - T_y$ in T_0.

It is worth making a few comments about our interpretation. First the lifetime of T_e is so short that $\omega_{45}/\Gamma' \ll 1$. This means that the triplet spins in T_e precess over a very small angle about their new axis of quantization and that upon one excursion to T_e only a small population redistribution occurs in T_0. Second we think that the two points measured at the highest temperatures of the curves "z in y-z" in Fig. 10, which deviate from the exponential behaviour, are a manifestation of the presence of a higher lying local phonon state. Third from the EPR experiments we know that all three spin axes change their directions on thermal excitation. As a result SLR will occur between all three pairs of spin levels in T_0. These SLR rates will depend on the magnitude of the rotation of the spin axes as well as on the zero-field splittings (see equation (6)). On the basis of our information we predict that the SLR rate between T_z and T_y is the dominant one. As mentioned before we were unable to prove this experimentally.

A plausible explanation for the rotation of the spin axes is that the naphthalene molecule reorients as the result of thermal excitation of a librational motion in an asymmetric potential well. To obtain more insight in the intermolecular interactions determining the orientation of the naphthalene molecule in the durene matrix, computer calculations were performed, based on the well-known atom-atom potential method [41, 42], to find the intermolecular potential energy curve. This computation was combined with a Monte Carlo procedures so that for a given orientation of the guest molecule the lattice is allowed to relax to its lowest energy configuration [43]. At a later stage this calculation was further refined to allow the methyl hydrogens of neighbouring durene molecules to rotate about their $C—C$ bond during the librational motion of the naphthalene molecule [44]. This refinement was introduced because it soon appeared that the dominant terms in the total interaction energy were due to repulsive interactions between the hydrogens of naphthalene and the methyl hydrogens of its neighbours.

In Fig. 11 we present two potential curves of naphthalene in its metastable $^3B_{2u}$ state in the durene lattice for a rotation about its out-of-plane axis. In Fig. 11b the result is shown of the refined calculation where the methyl groups are allowed to rotate whereas in Fig. 11a the methyl groups are kept fixed. The difference of the two curves very clearly show the effect of the rotation of the methyl group on the interaction potential. It is interesting to note that the large difference between the two curves is caused by rotations of the methyl groups $\leq 12°$.

There is little doubt that the double minimum potential well is related with the large reorientation of the naphthalene molecule upon excitation to the local phonon state. Using a special mathematical procedure [45] it was found that the first librational frequency is 22.2 cm^{-1} while the mean orientation in the librational ground state and first excited state differ by 1.13°. We have indicated the positions of the ground and first excited librational state in Fig. 11b. It is observed that the first librational motion is confined to the lower part of the potential well. We think that the actual barrier height between the two minima is smaller than we have calculated and that as a result the libration corresponds with a much more asymmetric motion as suggested by the potential curve. We are inclined to believe that this is not a serious problem and that it can be solved by further refinements of the calculations.

To support the conclusion that the methyl hydrogens play an important role in

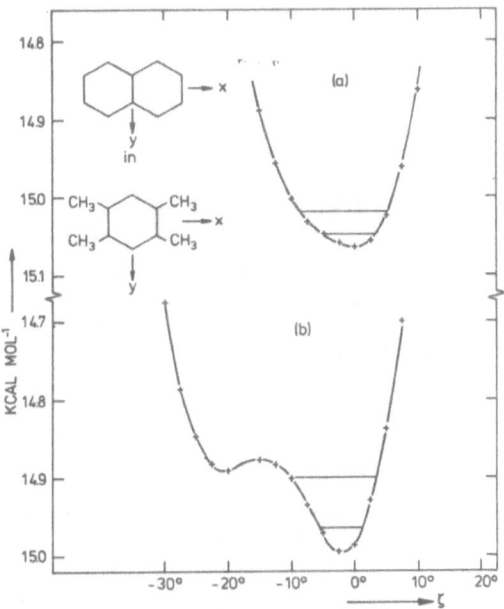

Fig. 11. (a) The potential energy curve for naphthalene in its lowest $^3B_{2u}$ state in the durene lattice for a rotation about its out-of-plane axis as calculated by the method described in [43, 44]. During this calculation the methyl groups of the neighbouring durene molecules are kept in fixed positions. (b) The results of a similar calculation where now the methyl groups of the neighbouring durene molecules are allowed to rotate about their $C-C$ axes.

the asymmetry of the librational motion, temperature dependent EPR experiments were performed on a crystal of N-d_8 in D-d_{14} to compare the rotation of the spin axes with the previous results on N-d_8 in D-h_{14}. The results are displayed in Fig. 12 where for comparison we have also indicated the results on N-d_8 in D-h_{14}. It is seen that the rotation again occurs predominantly in the plane of the naphthalene molecule but that the activation energy and the size of the rotation are considerably different. For temperatures below 30 K the rotation can be described by equation (32) with $\Delta E = (9 \pm 2)\,\text{cm}^{-1}$ and $\theta_\infty = (7 \pm 1)°$.

In agreement with the results displayed in Fig. 12 it is found that deuteration of the durene matrix has an appreciable effect on the SLR rates [44]. Unfortunately reliable values for the SLR rates in N-d_8 in D-d_{14} could only be obtained in the limited temperature range between 1.2 K and 2.5 K. At higher temperatures $\bar{w}$ is so fast that it could not be measured any more with good accuracy. The small temperature range only allowed to establish an upper limit for the activation energy, $\Delta E < 8\,\text{cm}^{-1}$ in accordance with the measurements of the direction of the spin axes. As a result of this low activation energy the average SLR rate $\bar{w} = (8 \pm 5) \times 10^2\,\text{s}^{-1}$ at 2 K, i.e. almost two orders of magnitude faster than in N-d_8 in D-h_{14}.

As a further confirmation of the effect of deuteration of the host on the

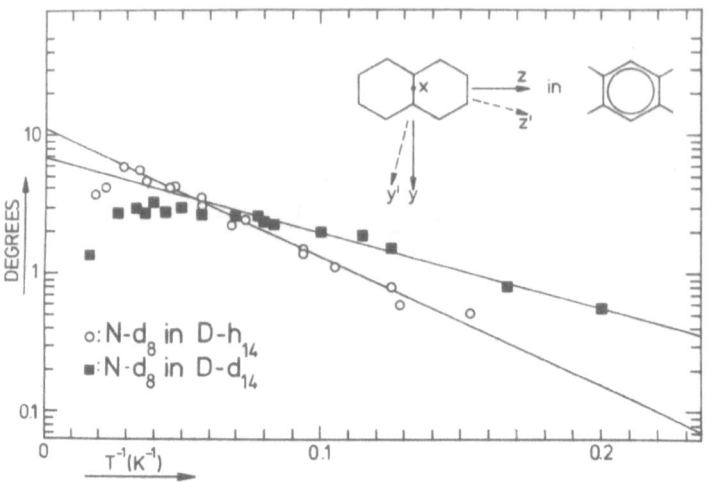

Fig. 12. The change in direction of the spin axes in the molecular plane of N-d$_8$ in D-d$_{14}$ as a function of the inverse temperature.

librational frequency of the guest the position of the hot band in the phosphore-scence spectrum of N-d$_8$ in D-d$_{14}$ was measured. This hot band was found at (7.5 ± 2) cm^{-1} to the high energy side of the 0—0 transition in very good agreement with the estimate from the temperature dependence of the directions of the spin axes and the SLR rates.

In the presence of a magnetic field one expects that the SLR rates resulting from the exchange between T_0 and T_e, likewise depend on the rotation of the spin eigenvectors in the two triplet states. To check this idea numerical calculations were performed of these rotations as a function of the direction of the magnetic field and the results were compared with experimentally determined SLR rates [46]. In Fig. 13a and Fig. 14a the results are presented of the measurements of the average SLR rate $\bar{w}$ in the triplet state of N-d$_8$ in D-h$_{14}$ as a function of the direc-tion of the magnetic field in the two principal planes x-y and y-z. The results are plotted in a polar diagram, i.e. the length of the vector from the center gives the SLR rate and its direction corresponds with the direction of the magnetic field. In the two figures the directions of the spin axes x', y' and z' in the librational state T_e, as determined from the experiments presented in Fig. 10, are also drawn.

The experiments were performed at 2.39 K by sweeping a microwave field through one of the "$\Delta m = \pm 1$" transitions and measuring the two time constants λ_1 and λ_2 in the bi-exponential recovery curve of the phosphorescence intensity. The two curves for $\bar{w}$ correspond with the two values of the magnetic field at which a "$\Delta m = \pm 1$" transition occurs for $\nu = 6$ GHz, the resonance frequency of the saturating microwave field.

The amount by which the spin eigenvectors in T_e are rotated with respect to those in T_0 depends on the strength and the direction of the applied magnetic field. It is this anisotropy that is expected to cause the anisotropy of the SLR rates.

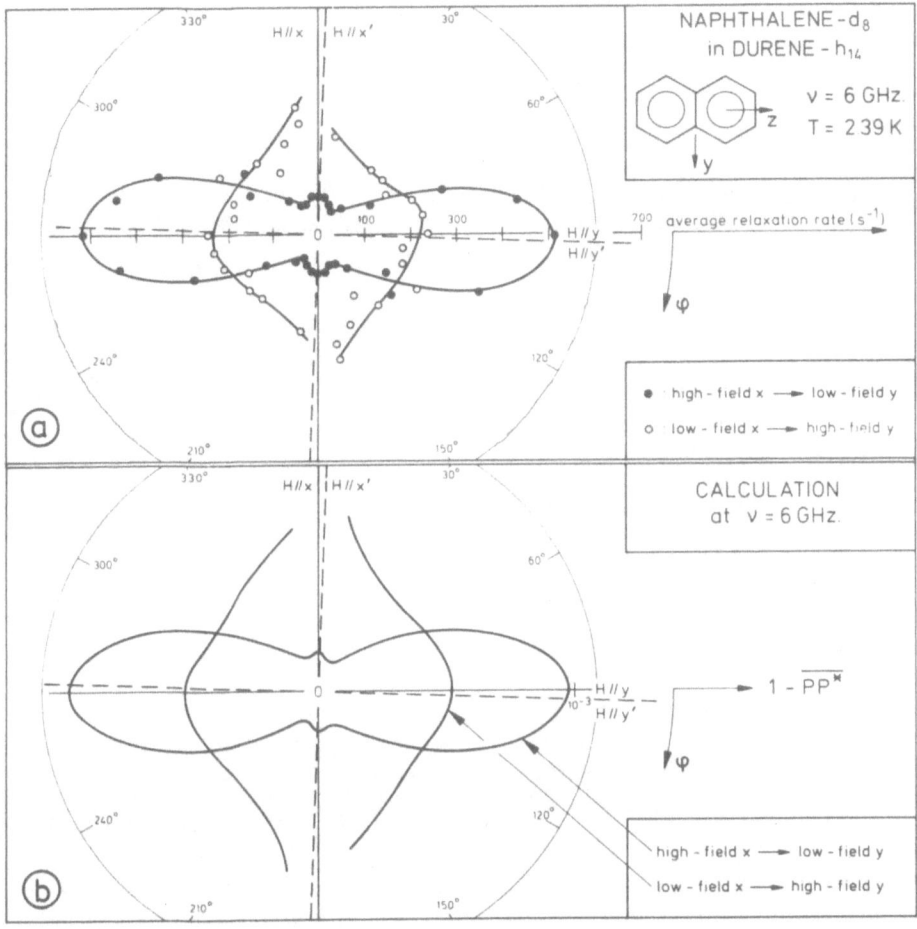

Fig. 13. (a) The average SLR rates between the triplet sublevels of the lowest triplet state of naphthalene-d_8 in durene-h_{14} as a function of the orientation of the magnetic field in the x-y plane. The rates have been obtained by saturating either the high-field or low field "$\Delta m = \pm 1$" transition at a frequency of 6 GHz and measuring the two time constants λ_1 and λ_2 in the recovery curve of the phosphorescence intensity. The value $\bar{w} = (\lambda_1 + \lambda_2)/6$ has been determined every 10° and each point is the mean of two experiments. (b) The calculated average rotation $1 - \overline{|P|^2}$ of the spin eigenstates in the local phonon state T_e with respect to those in T_0 of naphthalene in durene as a function of the orientation of the magnetic field in the x-y plane. The field strength has been taken such that the energy difference of one of the "$\Delta m = \pm 1$" transitions always is equal to 6 GHz. For the relative orientation of the spin axes in T_0 and T_e we used the values given in equation 32.

According to this idea the average rotation of the spin eigenvectors was derived by calculating the factor

$$1 - \frac{1}{3} \sum_{ii'} |P_{ii'}|^2 = 1 - \overline{|P|^2} \tag{36}$$

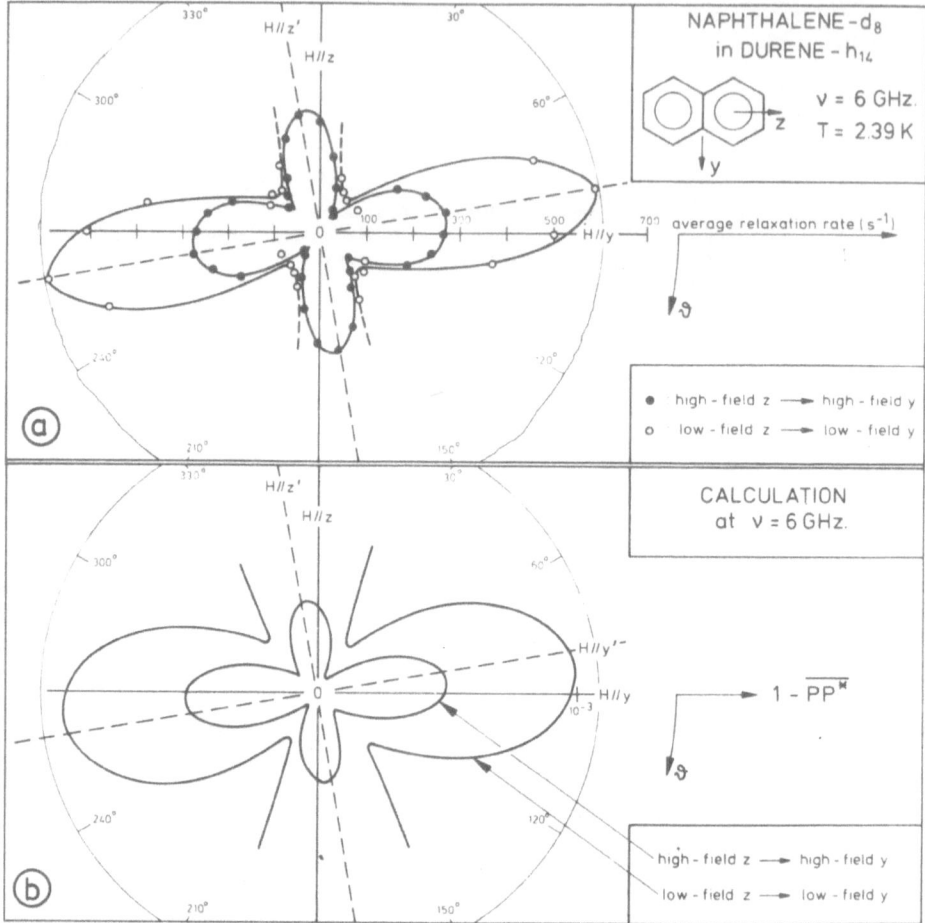

Fig. 14. (a) The average SLR rates as a function of the orientation of the magnetic field in the y-z plane of naphthalene in durene. The experimental conditions are the same as in Fig. 13(a). (b) The calculated average rotation $1 - \overline{|P|^2}$ of the spin eigenstates in T_e with respect to those in T_0 of naphthalene in durene as a function of the orientation of the magnetic field in the y-z plane. The conditions are the same as in Fig. 13(b)

where $P_{ii'}$ represents the scalar product of two corresponding eigenvectors in T_0 and T_e. The results of such calculations with the magnetic field in the x-y and y-z plane are presented in Fig. 13b and Fig. 14b. In these calculations the values were used for the angles between the spin axes in T_0 and T_e as derived from the temperature dependent EPR experiments (see Fig. 10 and equation 32). Further the zero-field splittings in T_0 and T_e were taken equal. In order to compare the results with the experimental findings the strength of the magnetic field was taken such that the energy difference of one of the $\Delta m = \pm 1$ transitions always is equal to 6 GHz, the frequency of the saturating microwave field.

From the results in Fig. 13b and Fig. 14b it appears that the magnetic field dependence of the factor given in the expression (36) indeed shows a remarkable resemblance with the orientational dependence of the measured SLR rates. Hence we conclude that in the presence of a magnetic field the SLR is also caused by a thermal excitation process to the higher lying librational state. In this context it should be mentioned that an analogous calculation of the orientational dependence of the SLR has been made by Vollmann [26]. In this perturbation treatment all three transitions are considered and a similar orientational pattern is found but with minimum and maximum values that are more extreme than the experimentally observed ones. The reason for this discrepancy is not clear.

To conclude this section on the system N in D we mention that according to our model presented in Section 2 we would expect an effect of the presence of T_e on the SSR rates in T_0. Unfortunately the dephasing of the triplet spins in N in D is dominated by the interaction with the nuclear spin system and the contribution of the thermal excitation to T_e is too small to be observable [20].

4.2. NAPHTHALENE-h_8 IN NAPHTHALENE-d_8

It is interesting to compare the results obtained on the system N in D with the SLR rates found in the triplet state of naphthalene-h_8 in naphthalene-d_8 (N-h_8 in N-d_8). In this system the triplet state of N-h_8 forms a shallow trap of only 100 cm^{-1} below the triplet excion band of the N-d_8 host material [47]. As a consequence phosphorescence from the N-h_8 impurities can only be observed below 9 K; at higher temperatures detrapping takes place. In Fig. 15 we show the phosphorescence spectrum of this system in the region of the 0—0 band. When comparing this spectrum with the one displayed in Fig. 9 for N-h_8 in D-h_{14} it is seen that in the isotopically mixed crystal the sidebands, caused by transitions from T_0 to (pseudo) localized phonon states in the ground state S_0, are absent. This is

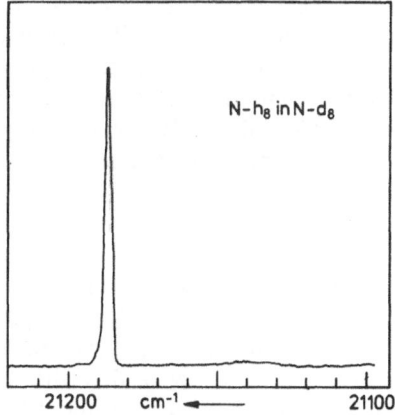

Fig. 15. The 0—0 band region of the phosphorescence spectrum of naphthalene-h_8 in naphthalene-d_8. $T = 5.6$ K.

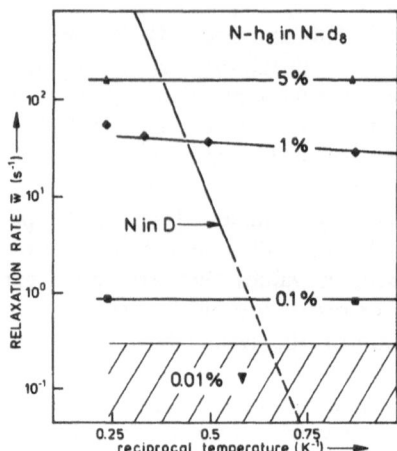

Fig. 16. The average SLR rate $\bar{w}$ in the lowest triplet state of naphthalene-h_8 in naphthalene-d_8 as a function of the temperature for different concentrations of N-h_8 in N-d_8. For comparison we have also drawn the behaviour of $\bar{w}$ in naphthalene-d_8 in durene-h_{14}.

reasonable because the N-h_8 guest molecule is a mild perturbation of the N-d_8 host crystal and localized phonon states are not expected to exist in contrast to chemically mixed crystals [48].

The behaviour of the SLR rate $\bar{w}$ in N-h_8 in N-d_8 also differs strongly from that in N in D. In Fig. 16 results on crystals with various concentrations of N-h_8 are presented. The SLR rate $\bar{w}$ depends strongly on the concentration but hardly on the temperature. Schwoerer et al. [11, 16] found similar results in their experiments in the presence of a magnetic field. At a concentration of 0.01 per cent the SLR rates in zero field are so slow that even at 4.2 K the spin levels are isolated from each other! For comparison we have also drawn the behaviour of $\bar{w}$ in N-d_8 in D-h_{14} and it is seen that in this crystal at 4.2 K the SLR rates are 4 orders of magnitude faster than in the diluted isotopically mixed crystal. Apparently the absence of a nearby local phonon state is the reason for the much slower SLR in N-h_8 in N-d_8. As a possible explanation for the strong concentration dependence it was suggested that in N-h_8 in N-d_8 the SLR is caused by magnetic dipole-dipole interactions between the N-h_8 monomers and the N-h_8 pairs [16].

4.3. QUINOXALINE IN DURENE

In Fig. 17 the behaviour is shown of the average relaxation rate $\bar{w}$ for quinoxaline-h_6 in durene-h_{14} (Q-h_6 in D-h_{14}) and in durene-d_{14} (Q-h_6 in D-d_{14}) as a function of the reciprocal temperature. It is clear that the curves must be described by a bi-exponential function. Further it is observed that the deuteration of the host material has a small but measurable effect on the SLR rates. The activation energies for Q-h_6 in D-h_{14} are (10.5 ± 1.5) cm^{-1} and (45 ± 4) cm^{-1}; for Q-h_6 in D-d_{14} (10.5 ± 1.5) cm^{-1} and (30 ± 3) cm^{-1}.

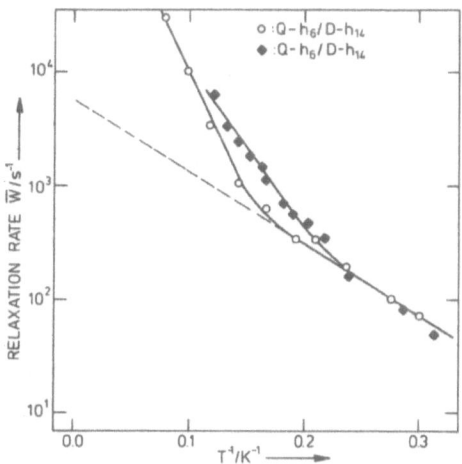

Fig. 17. The average SLR rates $\bar{w}$ in the triplet state of quinoxaline-h_6 in durene-h_{14} and in durene-d_{14} as a function of the inverse temperature. The curves indicate the presence of two activation energies. In Q-h_6 in D-h_{14}; 10.5 cm^{-1} and 45 cm^{-1}. In Q-h_6 in D-d_{14}; 10.5 cm^{-1} and 30 cm^{-1}.

In analogy with the system N in D hot band emission becomes observable in the phosphorescence spectrum at 10 K indicating the presence of a local phonon state T_e just above T_0. The exact position of the hot band is (12.5 ± 1) cm^{-1} for Q-h_6 in D-h_{14} and (10.6 ± 0.4) cm^{-1} for Q-h_6 in D-d_{14}. The linewidth of the hot band is 3 cm^{-1} corresponding with a lifetime of 2 ps. Since the results of the SLR suggest that another higher lying local phonon state is present about 30 cm^{-1} above T_0 a spectrum was taken at 25 K which revealed a weak band with a linewidth between 10—15 cm^{-1} at a distance of about 30 cm^{-1} from the 0—0 band [36].

The resonance frequencies of the zero-field transitions in Q in D shift with the temperature just like in N in D. The shifts for Q-h_6 in D-d_{14} are presented in Fig. 18. Below 50 K the curves show a bi-exponential temperature dependence similar to that of the SLR rates with activation energies of 10 cm^{-1} and 30 cm^{-1}. The $T_x - T_y$ transition can even be followed to 200 K and it is seen that above 50 K the transition frequency suddenly starts to move to lower values. This high temperature behaviour is described with an activation energy of 125 cm^{-1} and it is conceivable that we observe here the effect of thermal activation to a molecular vibration.

The directions of the spin axes of Q in D also show a dependence on the temperature. In Fig. 19 the results are presented of measurements of Q-d_6 in D-h_{14} performed in the same way as on N in D. From the semi-logarithmic plot in Fig. 19b an activation energy is derived of (47 ± 3) cm^{-1} for the results obtained above 15 K $(T^{-1} < 0.07$ K$^{-1})$. At lower temperatures the curves must be described with a lower activation energy but owing to the large uncertainty in the results it is not possible to derive an accurate value. In analogy with N in D we find

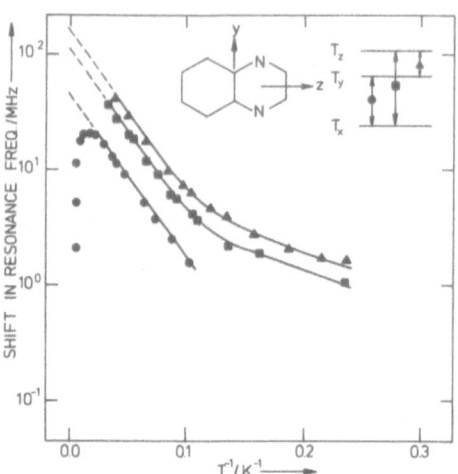

Fig. 18. The shifts of the zero-field resonance frequencies in the triplet state of Q-h$_6$ in D-d$_{14}$ as a function of the inverse temperature. The $T_z - T_x$ and $T_z - T_y$ transitions shift to lower frequencies. The $T_y - T_x$ transition, below 50 K, shifts to higher frequency but then reverses and shifts to lower frequency. Below 50 K the three curves can be described with the same activation energies as the curves for the average SLR rate $\bar{w}$.

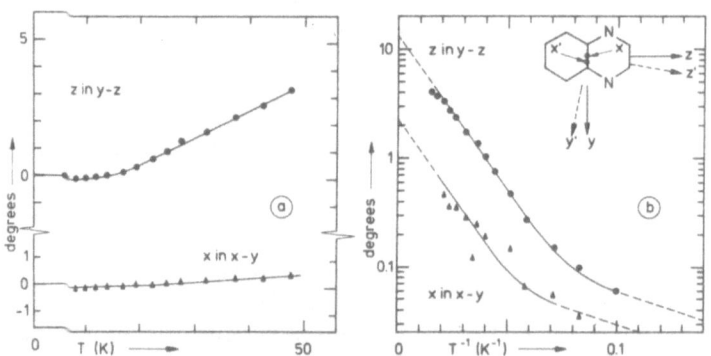

Fig. 19. (a) The change in direction of the spin axes in the principal planes of quinoxaline-d$_6$ in durene-h$_{14}$ as a function of the temperature. (b) The same results as in (a) as a function of the inverse temperature.

that the rotation of the spin axes is mainly in the plane of the molecule and that the out-of-plane x-axis rotates only very slightly in the x-y plane. It is concluded that in the 45 cm^{-1} local phonon state the molecule is rotated about its out-of-plane axis over 10° and about its long axis over 2°. These rotations are indicated in Fig. 19b.

The interesting conclusion is that in the system Q in D two local phonon states are involved in the SLR process between 1.2 K and 50 K. The low energy local phonon corresponding with a much smaller rotation of the spin axes than the high energy one. Moreover an effect of deuteration of the host on the SLR rates is observable. We think that the deuteration of the methyl groups of the neighbouring durene molecules is the main cause for the change in activation energy in analogy with our conclusion for N in D.

4.4. QUINOXALINE-h_6 IN NAPHTHALENE-d_8 (Q-h_6 IN N-d_8)

The triplet state of Q-h_6 forms a shallow trap about 90 cm^{-1} below the triplet exciton band of the host [49]. The SLR in this system has been studied extensively by Schwoerer and co-workers in the presence of a strong magnetic field [16–18, 27, 49, 50]. The authors concluded that the SLR is caused by direct processes in which the transition of an electron spin from one sublevel to the other is related with the emission or absorption of a phonon with the same energy. Here we shall show that, at least in zero-field, there is evidence that even in this system the SLR proceeds via two-phonon processes and that a local phonon state is involved.

In Fig. 20 the temperature dependence is shown of the average SLR rate $\bar{w}$ for Q-h_6 in N-d_8. The results suggest an activation energy of (22 ± 2) cm^{-1}. To confirm that this indeed corresponds with the presence of a local phonon state above T_0 we looked for hot band emission in the phosphorescence spectrum. In Fig. 21 a recording is shown at 4.2 K around the 0—0 band. A tiny peak is observed at the low energy side, 20 cm^{-1} from the 0—0 band, probably related to a local phonon state in the ground state S_0. Unfortunately a hot band could not be found because the phosphorescence already disappears at 8 K as a result of detrapping to the host exciton band. At this temperature a hot band at 20 cm^{-1} from the 0—0 band is too small to be observable.

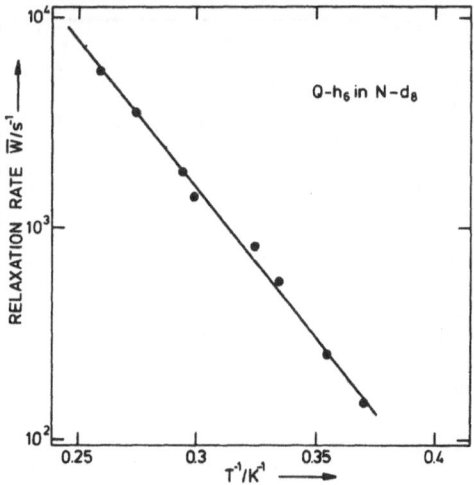

Fig. 20. The average SLR rate $\bar{w}$ in the triplet state of quinoxaline-h_6 in naphthalene-d_8 as a function of the inverse temperature in zero-magnetic field. The value of $\bar{w}$ has been obtained from the two time constants λ_1 and λ_2 in the recovery curve of the phosphorescence intensity upon a sweep through one of the zero-field transitions; $\bar{w} = (\lambda_1 + \lambda_2)/6$.

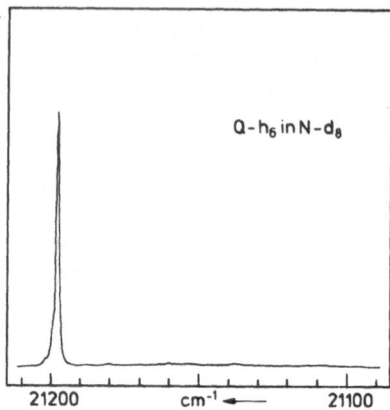

Fig. 21. The phosphorescence spectrum around the 0—0 band of quinoxaline-h_6 in naphthalene-d_8 at 4.2 K.

The temperature dependence of the zero-field transitions also points to the presence of a local phonon state above T_0. In Fig. 22 we see that the shift of the $T_z - T_x$ transition can be described with an activation energy of (28 ± 10) cm^{-1}. Although the uncertainty is rather large we exclude the possibility that an activation to the host exciton band is responsible for this effect.

Summarizing we can say that these results support the idea that a local phonon state, about 20 cm^{-1} above T_0, is involved in the SLR in Q-h_6 in N-d_8. It would be interesting to see whether a temperature dependence of the directions of the spin axes also occurs in this system.

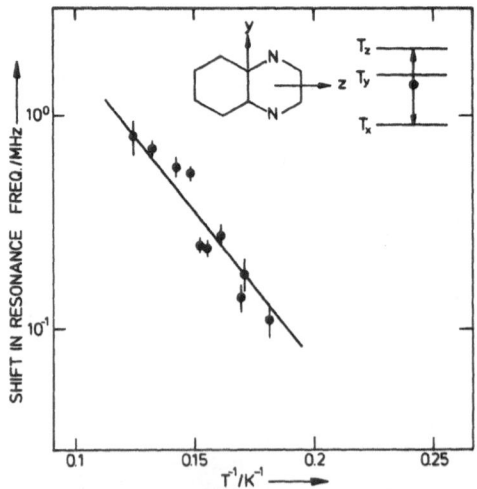

Fig. 22. The shift of the $T_z - T_x$ zero-field transition of quinoxaline-h_6 in naphthalene-d_8 as a function of the inverse temperature. The resonance frequency shifts to lower values with increasing temperature.

4.5. ANTHRAQUINONE-h_8 IN N-HEPTANE (AQ IN H) AND ANTHRAQUINONE-h_8 IN ANTHRAQUINONE-d_8 (AQ-h_8 IN AQ-d_8)

The average SLR rate $\bar{w}$ in AQ-h_8 in H has been determined by Avarmaa and Suisalu [51]. Their results in the region of dominant relaxation ($T > 4$ K) indicate an activation energy of (20 ± 2) cm^{-1}. When studying the phosphorescence spectrum as a function of temperature a hot band emission is found at (18 ± 1.5) cm^{-1} to the high energy side of the vibronic band at 20185 cm^{-1}. The intensity of this hot band grows in with the temperature according to an activation energy of (20.5 ± 1.5) cm^{-1}. From the linewidth a lifetime of 3 ps is derived. Since also a shift of the zero-field frequencies with the temperature is observed, that can be described with an activation energy of (19 ± 2) cm^{-1}, the conclusion seems justified that in this system the SLR is caused by a thermal activation to a nearby local phonon state at about 20 cm^{-1} [36].

This conclusion is further supported by the results of a study on the isotopically mixed crystal AQ-h_8 in AQ-d_8. Here it is found that below 2.3 K the SLR rate $\bar{w}$ is independent of the temperature but is strongly dependent on the concentration. In a sample with 0.2 per cent AQ-h_8 in AQ-d_8 the SLR is so slow that the spin levels even become isolated and only an upper limit for $\bar{w}$ can be given. This behaviour strongly resembles the observations in N-h_8 in N-d_8, where it was concluded that the isotopic impurity forms a very mild perturbation of the crystal and that as a consequence local phonon states do not occur. The rapid increase of $\bar{w}$ with the temperature above 2.3 K is attributed to detrapping to the host exciton band.

4.6. ANILINE-d_7 IN P-XYLENE-d_{10} (Ad_7 IN pX-d_{10})

An extensive study of the SLR and spin-spin relaxation (SSR) in the triplet state of aniline in p-xylene was performed by van Noort et al. [52]. This work was stimulated by an observation made during an investigation of the conformational instability of the metastable $^3\pi\pi^*$ state of this molecule [53]. The lowering of the symmetry of the electronic structure of A in pX was demonstrated by EPR experiments, which showed that the in-plane spin axes are no longer parallel to the molecular spin axes but rotated away over an angle of 74° (see Fig. 23). During this study it was observed that the spin axes not only deviate from the molecular axes but that their direction is also temperature dependent, indicating the presence

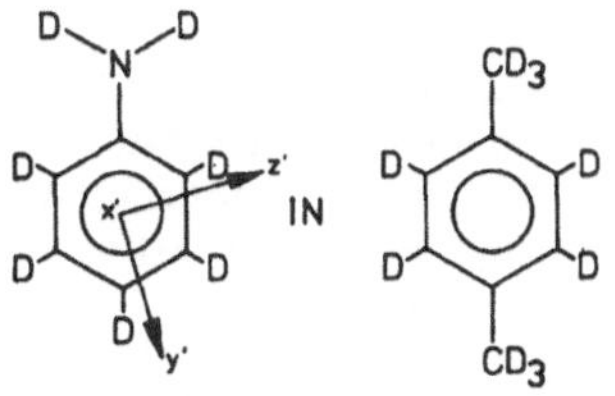

Fig. 23. The system aniline in p-xylene. The in-plane spin axes y' and z' deviate 74° from the molecular axes y and z as determined by EPR experiments in the presence of a magnetic field.

of a nearby local phonon state. In addition it was known from earlier work of van
't Hof [54] that the SSR rates in zero field and also the zero-field frequencies are
temperature dependent.

For us the system A-d_7 in pX-d_{10} is interesting for two reasons. First of all it is
the only chemically mixed crystal on which reliable measurements could be
performed of the individual SLR rates between the sublevels of the triplet state.
Secondly A-d_7 in pX-d_{10} is an example of a system where the effect of the thermal
excitation to the local phonon state T_e is also observable in the SSR.

In Fig. 24 the results are presented of the measurements on the temperature
dependence of the average SLR rate, the SSR rate, the direction of the in-plane

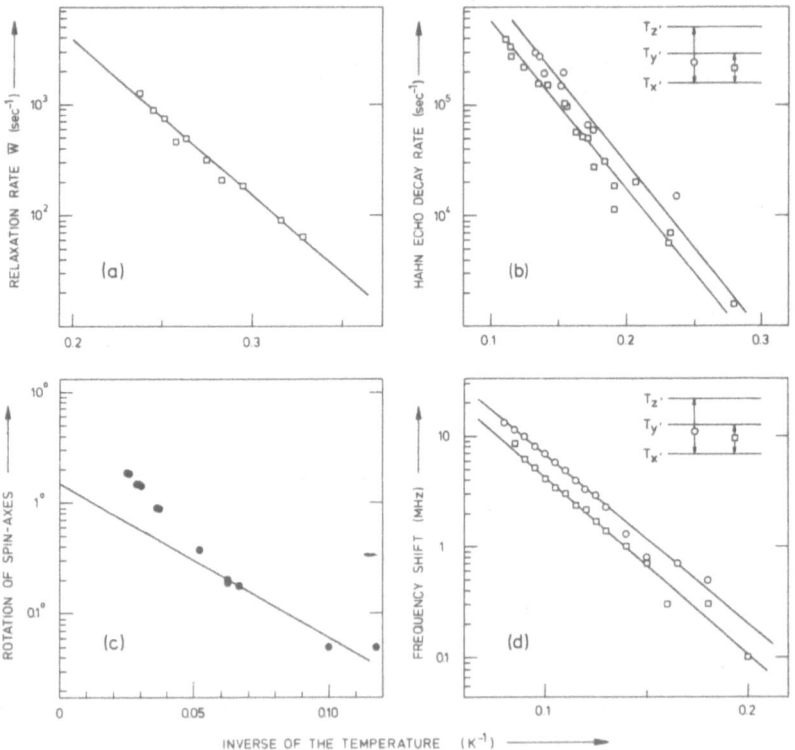

Fig. 24. The results of the different experiments performed on aniline-d_7 in p-xylene-d_{10}, (a) The
average SLR rate $\bar{w}$ as a function of the inverse temperature. The value of $\bar{w}$ was obtained from the
two time constants λ_1 and λ_2 in the recovery curve of the phosphorescence intensity upon a sweep
through one of the zero-field transitions; $\bar{w} = (\lambda_1 + \lambda_2)/6$. (b) The temperature dependent
part $T_2^{-1}(T)$ of the SSR rate for the $T_{z'} - T_{x'}$ and $T_{y'} - T_{x'}$ zero-field transitions as a function of the
inverse temperature. (c) The change in direction of the z' spin axis in the plane of the aniline
molecule as a function of the inverse temperature. (d) The shift of the resonance frequency of the
$T_{z'} - T_{x'}$ and the $T_{y'} - T_{x'}$ zero-field transitions as a function of the inverse temperature. x' is the
out-of-plane spin axis.

spin axes and the zero-field resonance frequencies in A-d_7 in pX-d_{10}. The most striking aspect is that they all show the same activation energy of about 25 cm^{-1}. More evidence that these four observations are related to the presence of a low lying local phonon state is provided by the appearance of a hot band in the phosphorescence spectrum, (24 ± 0.5) cm^{-1} above the 0–0 band.

The temperature dependence of the average SLR rate can be described by $\bar{w} = \bar{w}_\infty \exp\{-\Delta E/kT\}$. The values of $\bar{w}_\infty$ and ΔE are given in Table I.

The SSR rate T_2^{-1} was obtained from the decay of the microwave detected echo signal in two-pulse echo experiments on the $T_{z'} - T_{x'}$ and $T_{y'} - T_{x'}$ transitions. The temperature dependence of both curves can be described by an expression $T_2^{-1} = T_2^{-1}(T) + T_2^{-1}(0)$. Here the temperature independent part $T_2^{-1}(0)$ accounts for the contribution to the dephasing caused by the nuclear spin system. The temperature dependent part $T_2^{-1}(T)$ is shown in Fig. 24b and can be fitted by an expression of the form $T_2^{-1}(T) = T_2^{-1}(\infty) \exp\{-\Delta E/kT\}$. The values of the constants for the two transitions are given in Table I.

The directions of the in-plane spin axes as a function of the temperature were determined in a conventional EPR experiment following the same procedure as for N in D [40]. The variations obey the expression.

$$\theta(T) = \theta_\infty \exp\{-\Delta E/kT\} + \theta'_\infty \exp\{-\Delta E'/kT\}. \tag{37}$$

The values of the constants are given in Table I. No rotation with temperature of the out-of-plane axis was found.

The zero-field frequencies of the $T_{z'} - T_{x'}$ and $T_{y'} - T_{x'}$ transitions were measured in the range from 1.2 K to 12 K. With an increase in temperature the two lines shift in opposite directions; the $T_{z'} - T_{x'}$ to lower and the $T_{y'} - T_{x'}$ to higher frequency. The temperature dependence of these frequency shifts is given by

$$\varepsilon(T) = \varepsilon_\infty \exp\{-\Delta E/kT\}. \tag{38}$$

The two constants for the two transitions are presented in Table I.

TABLE I

The results of the different experiments performed on aniline-d_7 in p-xylene-d_{10}. The second column gives the activation energies ΔE, the third column the pre-exponential factors.

Experiment	ΔE (cm^{-1})	
$\bar{w}$	22.5 ± 15	$\bar{W}_\infty = (2.5 \pm 1) \times 10^6$ s^{-1}
$T_2^{-1}(T), (T_{z'} - T_{x'})$	27 ± 4	$T_2^{-1}(\infty) = (37 \pm 11) \times 10^6$ s^{-1}
$T_2^{-1}(T), (T_{y'} - T_{x'})$	24.5 ± 4	$T_2^{-1}(\infty) = (23 \pm 6) \times 10^6$ s^{-1}
$\theta(T), (z', y')$	21 ± 8	$\theta_\infty = (1.5 \pm 0.5)^0$
$\varepsilon(T), (T_{z'} - T_{x'})$	26 ± 3	$\varepsilon_\infty/2\pi = (-281 \pm 40)$ MHz
$\varepsilon(T), (T_{y'} - T_{x'})$	26 ± 2	$\varepsilon_\infty/2\pi = (+183 \pm 30)$ MHz
hot band	24 ± 0.5	

In addition to the system A-d$_7$ in pX-d$_{10}$ samples of A in pX were studied where only the guest or host molecules were deuterated or where the protons of the NH$_2$ group were replaced by deuterons. It appears that the position of the hot band depends slightly on the degree of deuteration. The SLR and SSR rates again show an exponential dependence on the temperature with activation energies slightly different from A-d$_7$ in pX-d$_{10}$ [52]. The temperature dependence of the rotation of the spin axes turned out to depend also somewhat on the deuteration of the guest and the host [52]. All these observations point to the conclusion that we are dealing with a librational motion of the aniline molecule in the force field of the surrounding host molecules and that this motion takes place predominantly in the plane of the molecule.

As mentioned before it was possible to obtain reliable values for the individual SLR rates in the system A in pX using the method discussed in Section 3.1c. In Fig. 3 an example is given of two optically detected recovery curves for A-d$_2$ in pX-h$_{10}$ at 3.2 K from which precise values could be obtained for the parameters λ_1, λ_2, C_1 and C_2. Since the radiative rate constants of the sublevels are well known [53] we derive; $w_{z'y'} = (270 \pm 10)$ s^{-1}, $w_{z'x'} = (5 \pm 5)$ s^{-1} and $w_{x'y'} = (15 \pm 5)$ s^{-1}. Thus SLR occurs predominantly between $T_{z'}$ and $T_{y'}$.

The results obtained on the mixed crystals of A in pX form a beautiful set to test the model for relaxation developed in Section 2 of this chapter. Here the approximation was made to consider only two sublevels in T_0 and the two related levels in T_e. This approximation applies very well to A in pX since the spin axes in T_e are only rotated in the plane of the molecule and the SLR relaxation occurs predominantly between the two sublevels related with the in-plane spin axes. So we have again

$$\bar{w} = (w_{z'y'} + w_{y'x'} + w_{z'x'})/3 \approx w_{z'y'}/3 = (T_1^{-1})/6. \qquad (39)$$

Since we know the angle of rotation θ of the spin axes in T_e and the zero-field splitting ω_{45} of the pair of levels in T_e we can determine the lifetime of this state with the help of (6). We find $\tau = (45 \pm 25)$ ps using $\theta_\infty = (1.5 \pm 0.5)°$ and $\omega_{45}/2\pi = 1746$ MHz. The large uncertainty in τ is caused by the uncertainty in θ_∞ derived from Fig. 24c.

When substituting these values for τ and ω_{45} in expression (7) for T_2^{-1} we find that the contribution of the first "dephasing" term is about 50 times larger than the second term. Unfortunately we have not been able to observe electron spin echo signals from the $T_{z'} - T_{y'}$ zero-field transition and hence we do not know the SSR rate T_2^{-1} of this particular transition. We do know however the SSR rates of the other two transitions. Here again the "dephasing" term is the dominant one. When substituting the appropriate values for the differences in resonance frequency we can fit the experimental findings for the $T_{y'} - T_{x'}$ and $T_{z'} - T_{x'}$ transition with lifetimes $\tau = (18 \pm 6)$ ps and $\tau = (12 \pm 4)$ ps respectively. In view of the large uncertainty of the value of τ derived from the T_1^{-1} measurements we consider the agreement as reasonable.

It is interesting to compare our results with those obtained in the system naphthalene in durene. There the rotation of the in-plane spin axes in T_e is about six times larger, the difference in resonance frequency $\delta/2\pi$ about five times smaller and the lifetime τ more than an order of magnitude shorter than in aniline. These differences explain why the thermal excitation in aniline in p-xylene has a much larger effect on the dephasing than in the case of naphthalene. Combined with the fact that in A-d$_7$ in pX-d$_{10}$ the dephasing caused by the nuclear spin system is small, this explains why the effect of thermal excitation is also visible in the SSR.

It is not so surprising that in aniline the zero-field frequencies in T_0 and T_e differ so much because it is known already from the work of Vergragt *et al.* [55] that the zero-field parameter E is very sensitive for perturbations. In the state T_0 the aromatic nucleus of aniline is distorted to a symmetry lower than C_{2v}, as evidenced by the skewness of the in-plane spin axes, and the observed rotation of 1.5° in T_e may partly or wholly correspond to a change in conformation with a simultaneous change in orientation of the spin axes relative to the molecular frame. This is in contrast with the situation in naphthalene where the rotation of 10° is expected to correspond with a rotation of the naphthalene molecule as a whole with a small change in the zero-field splitting frequency. The relative long lifetime $\tau = 20 - 40$ ps points to a much more localized motion of aniline in p-xylene than of naphthalene in durene.

5. Experimental Studies on Spin-Lattice and Spin-Spin Relaxation in the Triplet State of Naphthalene Dimers

A beautiful model system for studying the effect of a nearby excited state on the spin relaxation processes in the lowest triplet state of a phosphorescent molecule is provided by the dimers occurring in isotopically mixed crystals of naphthalene-h$_8$ in naphthalene-d$_8$ (N-h$_8$ in N-d$_8$). When doping a crystal of N-d$_8$ with a high concentration of N-h$_8$ a considerable portion of the guest molecules ends up as nearest neighbours. They may occupy either two adjacent translationally inequivalent positions in the unit cell (*AB*-pair) or two translationally equivalent ones (*AA*-pair).

The first excited triplet state excitation energy of N-h$_8$ is about 100 cm^{-1} smaller than that of N-h$_8$ and hence at low temperatures the N-h$_8$ molecules in their first excited triplet state are isolated from the host molecules. More important for us the near neighbour molecules of N-h$_8$ will be isolated dimers. Since the matrix element for excitation transfer between these two molecules ($|J_b| = 0.6$ cm^{-1} and $|J_{ab}| = 1.25$ cm^{-1}) is large compared to the zero-field splitting parameters for naphthalene (D ≈ 0.1 cm^{-1} and $|E| \approx 0.01$ cm^{-1}) the eigenstates of the pair in good approximation are given by

$$\phi \pm = \frac{1}{\sqrt{2}} \{|1^*2\rangle \pm |1\ 2^*\rangle\} \tag{40}$$

where $|1^*2\rangle$ is the state which describes a triplet excitation on molecule 1. The corresponding triplet energies are split by the amount $\Delta E = 2|J|$ (the Davydov splitting). From optical spectroscopy it was established that the symmetric and antisymmetric states of the AA- and AB-dimer are located almost symmetrically around the triplet state of the monomer. From polarization experiments it was confirmed that in the AA-pair ϕ_+ is the lowest and that in the AB-pair ϕ_- is the lowest state [56—58].

In this section we shall show that the SLR and SSR in the lowest triplet state of the AB-dimer and in that of the AA-dimer is caused by a two-phonon, Orbach-type process to the higher Davydov component. The experiments allow us to derive information about the properties of the excited triplet state that cannot be obtained via other experimental means.

The experiments on the AB-dimers were performed on a crystal of N-d_8 grown from the melt doped with 0.6 or 2.0 per cent of N-h_8. The naphthalene crystal structure is monoclinic and contains two translationally inequivalent molecules A and B per unit cell. It is known that at these high concentrations one observes the EPR signals of monomer N-h_8 guest molecules and of AB pairs [59]. In Fig. 25 we show a stereographic projection of the experimentally observed directions of the principal axes of the two monomers A and B together with those of the AB-pair, which are labelled by x^*, y^* and z^*. The ac-plane is a mirror plane between the two molecules. The z^* axis of the pair is perpendicular to this plane whereas the x^* and y^* axes are located in the ac-plane.

In the first experiment the temperature dependence was measured of the

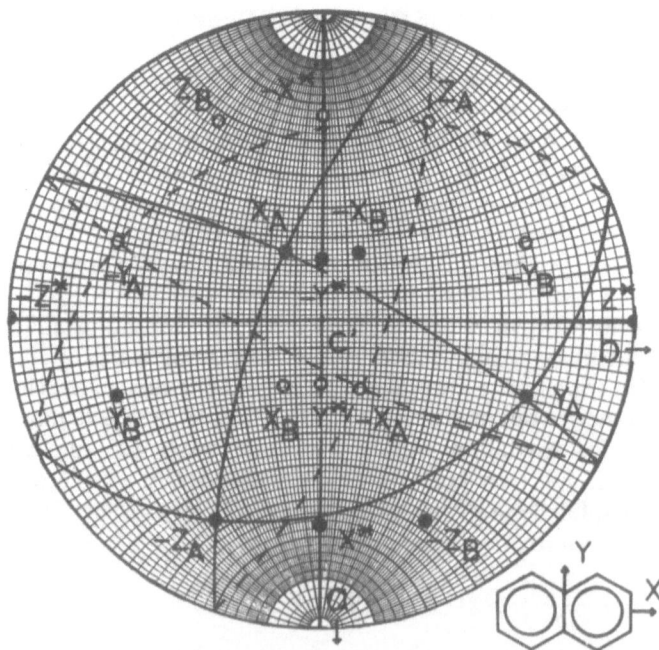

Fig. 25. A stereographic projection of the principal axes directions for the two translationally inequivalent molecules A and B, and for the AB-pair (labelled with *) with respect to the crystallographic axes a, b and c' in the isotopically mixed crystal of naphthalene-h_8 in naphthalene-d_8.

resonance frequencies of the zero-field transitions in the lowest Davydov component of the AB-pair [60]. An example of the optically detected $T_{y^*} - T_{x^*}$ transition is given in Fig. 26 at 1.18 K and 4.2 K, in which the shift in resonance frequency can clearly be observed. In Fig. 27 we have plotted the shift in resonance frequency of the $T_{y^*} - T_{z^*}$ transition as a function of the temperature. It is seen that on a logarithmic scale the shift can be fitted to a straight line with a slope of 2.44 cm^{-1}.

In the second experiment electron spin-echo signals were generated by a sequence of a $\pi/2$ and π microwave pulse resonant with the $T_{y^*} - T_{x^*}$ and $T_{x^*} - T_{z^*}$ zero-field transition of the AB-pair [61]. The pulses are separated by a time τ and follow a laser flash after a delay time t_d. This laser flash at a wavelength of 31526 cm^{-1}, with a bandwidth of 0.5 cm^{-1} and an energy of 200 µJ $-$ 7 mJ, selectively excites the triplet state of the AB-pair via the 0—0 band of the lowest AB^+ Davydov component in the $S_1 \leftarrow S_0$ absorption (see Fig. 28). The phase memory time T_2 (or rather T_M) was measured from the decay of the echo signal as a function of the time τ while keeping t_d constant. The temperature dependence of T_2^{-1} as obtained for these two transitions is represented in Fig. 29 for the 2 per cent crystal. Identical results were obtained for the 0.6 per cent crystal. The signals of the third $T_{y^*} - T_{z^*}$ transition were so weak that no reliable measurements could be performed. The delay time $t_d = 10$ ms was chosen to allow the crystal to thermalize with the ^{3}He bath after the laser flash.

As can be seen from the results in Fig. 29, below 0.4 K, T_2^{-1} reaches a constant value $T_2^{-1}(0) = 0.35 \times 10^6$ s^{-1} for the $T_{x^*} - T_{z^*}$ as well as the $T_{x^*} - T_{y^*}$ transition. As the temperature is increased a drastic shortening of T_2 sets in. Assuming that the behaviour of T_2^{-1} can be described as the sum of two contributions

$$T_2^{-1} = T_2^{-1}(T) + T_2^{-1}(0) \tag{41}$$

and subtracting $T_2^{-1}(0) = 0.35 \times 10^6$ s^{-1} from the best fit to the experimental

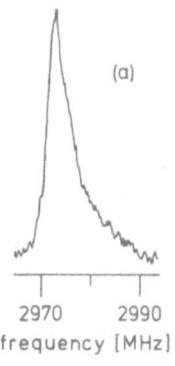

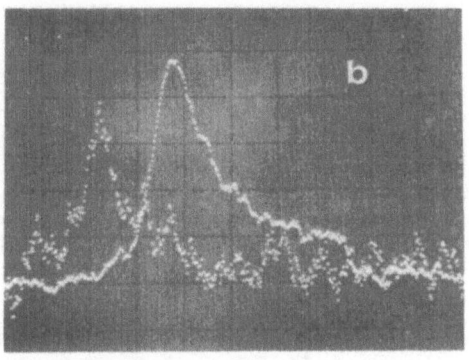

Fig. 26. (a) The optically detected $T_{x^*} - T_{y^*}$ zero-field transition of the naphthalene AB-pair. $T = 1.18$ K. (b) The frequency shift of the $T_{x^*} - T_{y^*}$ transition which occurs by raising the temperature from 1.18 K (right) to 4.22 K (left). The scale is 5 MHz per division.

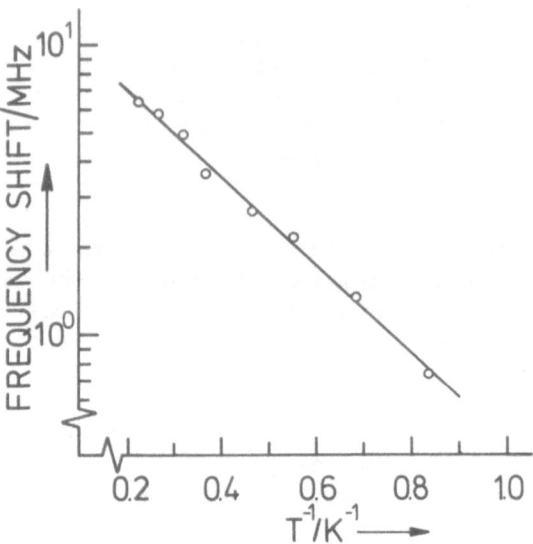

Fig. 27. The frequency shift of the $T_{y^*} - T_{z^*}$ transition of the naphthalene AB-pair as a function of the inverse of the temperature.

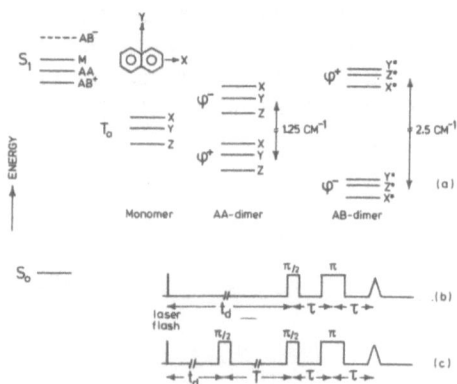

Fig. 28. (a) A simplified energy level scheme of the first excited singlet and triplet states of the naphthalene-h_8 monomer, AA- and AB-dimer in the N-h_8 in N-d_8 isotopically mixed crystal. The splittings of the two Davydov components in the triplet state of the two dimers are represented on an expanded scale. (b) and (c) A schematic drawing of the pulse sequences in the two-pulse and three-pulse ESE experiments following the laser flash.

values of T_2^{-1} we obtain the temperature dependent contribution $T_2^{-1}(T)$ as given by the broken lines in Fig. 29a and b. These two lines are described by

$$T_2^{-1}(T) = T_2^{-1}(\infty) \exp\{-\Delta E/kT\}. \tag{42}$$

The average value of $T_2^{-1}(\infty) = (1.4 \pm 0.5) \times 10^7 \text{ s}^{-1}$ and the average value of $\Delta E = (2.6 \pm 0.5) \text{ cm}^{-1}$.

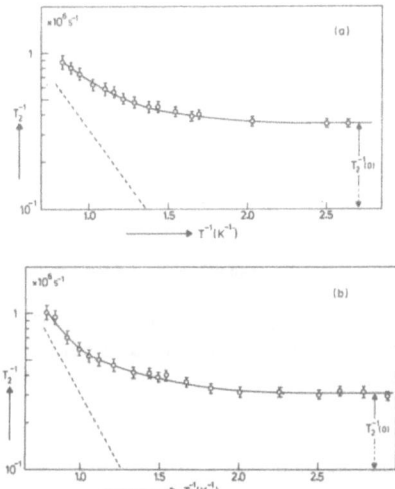

Fig. 29. (a) The temperature dependence of the spin-spin relaxation rate T_2^{-1} as measured from the decay of the two-pulse echo signal on the $T_{x^*} - T_{y^*}$ zero-field transition of the AB-dimer at 2972 MHz. (b) The results of a similar experiment on the $T_{x^*} - T_{z^*}$ zero-field transition at 1659 MHz.

The measurement of T_1 was performed via a three pulse sequence in which the $\pi/2$ and π pulse combination was preceded by a saturating $\pi/2$ pulse (see Fig. 28). It turned out to be very difficult to obtain reliable data for the temperature dependence of T_1^{-1} because of long term instabilities in the spectrometer. Nevertheless in the $T_{x^*} - T_{y^*}$ transition values for T_1 at 0.92 K and 0.4 K of (1.5 ± 0.5) μs and (370 ± 100) μs respectively were obtained.

To investigate the effects of energy transfer on the T_1 and T_2 measurements the $S_1 \leftarrow S_0$ excitation spectrum was recorded by varying the wavelength of the laser flash, while detecting the ESE signal of one of the zero-field transitions of either the AB-pair, the AA-pair or the monomer at a delay of $t_d = 50$ ms after the laser flash. It appears that an ESE signal of the triplet state of the AB-dimer can only be observed upon excitation in the $AB+$ or $AB-$ components of the $S_1 \leftarrow S_0$ absorption. Thus it seems that communication between monomers, AA- and AB-dimers can be excluded during a time $t_d < 50$ ms [61, 62].

In addition to the zero-field measurements on the AB-pair, ESE experiments were also performed in the presence of a magnetic field at X-band frequencies with continuous illumination from a high pressure mercury arc. In Fig. 30 the peak at the right is the high field AB-pair line with $B // x^*$. In the same scan we observe the two coinciding resonances of the monomers, which occur at slightly lower field. An orientation diagram near x^* is shown in Fig. 31 for this high field transition. The resonance of the pair line does not fall exactly halfway between the monomer lines because the x^*-axis does not coincide exactly with the projection of the z-axes on the mirror plane (see Fig. 25).

The temperature dependence of T_2^{-1} was measured with the magnetic field turned 4° away from the x^* direction in order to avoid complications arising from cross-relaxation processes with the monomers. In this orientation the pair line is

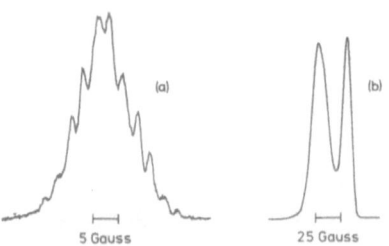

Fig. 30. (a) The high-field **B** // **z** EPR transition of the naphthalene monomer detected with the X-band ESE spectrometer via the intensity of the two-pulse ESE signal while scanning the applied magnetic field. $T = 1.18$ K. (b) A similar spectrum taken with the magnetic field parallel to the x^* axis of the AB-pair. The line at the right is the high-field x^* transition of the pair. The line at the left is a superposition of the signals from monomers of type A and B.

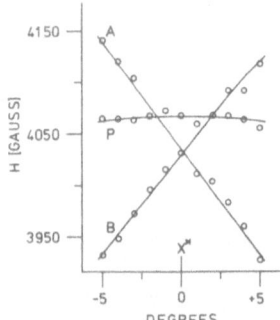

Fig. 31. The positions of the high field EPR lines of the monomers A and B and of the AB-pair as a function of the orientation of the magnetic field in the $z_A - x^*$ plane in the vicinity of the x^* axis.

well separated from the two monomer lines. In Fig. 32 the behaviour of T_2^{-1} versus T^{-1} is shown between 1.18 K and 2.17 K (the lambda point of liquid helium). Above this temperature helium bubbling causes sudden jumps in the resonance frequency of the cavity and it was impossible to perform reliable measurements. At low temperatures T_2^{-1} can be described by

$$T_2^{-1} = T_2^{-1}(\infty) \exp\{-\Delta E/kT\} \tag{43}$$

with $T_2^{-1}(\infty) = (2.9 \pm 0.7) \times 10^6$ s^{-1} and $\Delta E = (2.5 \pm 0.1)$ cm^{-1}.

The experimental results clearly show that thermal activation from the lower to the upper Davydov component of the triplet state of the AB-pair is responsible for the temperature dependence of the resonance frequencies and the phase memory times in zero-field as well as in the presence of a magnetic field. Here we shall show that the mathematical model developed in Section 2 can be used to explain the observed effects and to obtain information about the properties of the upper Davydov component.

We first consider the Hamiltonian describing the triplet state of the dimer

$$\mathcal{H} = \varepsilon_0(b_A^+ b_A + b_B^+ b_B) + J_{AB}(b_B^+ b_A + b_A^+ b_B) + $$
$$+ \mathbf{S} \cdot \overline{\overline{T}}_A \cdot \mathbf{S} + \mathbf{S} \cdot \overline{\overline{T}}_B \cdot \mathbf{S} \tag{44}$$

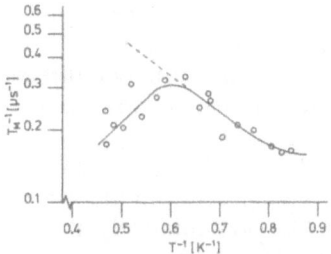

Fig. 32. The decay rate $T_M^{-1}(T_2^{-1})$ of the naphthalene AB-pair as a function of the inverse temperature for the high field EPR transition with the magnetic field making an angle of $-4°$ with the x^* axis (see Fig. 31).

where J_{AB} is the exchange integral, ε_0 the excitation energy of the isolated molecule, b_A^+ and $\underline{b}_A$ the creation and annihilation operators of the triplet state at molecule A and $\overline{\overline{T}}_A$ and $\overline{\overline{T}}_B$ are the fine structure tensors of the two differently oriented molecules. Since J_{AB} is much larger than $\overline{\overline{T}}$ it is reasonable to represent the Hamiltonian (44) on a basis formed by the symmetric and antisymmetric combinations of product states as given in (40), using the triplet spin eigenfunctions of the pair. Hence we have a six dimensional basis

$$\phi_\pm(\mu), \mu = x^*, y^*, z^* \tag{45}$$

where $\mu = x^*$ means that the spin is in the eigenstate $|x^*\rangle$ of the average spin Hamiltonian of molecules A and B [59, 63]. On this basis $\mathcal{H}$ is represented by

		Φ_+			Φ_-		
		$\|x^*\rangle$	$\|y^*\rangle$	$\|z^*\rangle$	$\|x^*\rangle$	$\|y^*\rangle$	$\|z^*\rangle$
Φ_+	$\|x^*\rangle$	$-J_{AB}-T_{xx}^*$					
	$\|y^*\rangle$	$-T_{xy}^*$	$-J_{AB}-T_{yy}^*$				
	$\|z^*\rangle$	0	0	$-J_{AB}-T_{zz}^*$			
Φ_-	$\|x^*\rangle$	0	0	$-T_{xz}^*$	$J_{AB}-T_{xx}^*$		
	$\|y^*\rangle$	0	0	$-T_{yz}^*$	$-T_{xy}^*$	$J_{AB}-T_{yy}^*$	
	$\|z^*\rangle$	$-T_{xz}^*$	$-T_{yz}^*$	0	0	0	$J_{AB}-T_{zz}^*$

$$\tag{46}$$

where

$$T_{\lambda\mu}^* = -(\lambda \cdot x_A)(\mu \cdot x_A)X - (\lambda \cdot y_A)(\mu \cdot y_A)Y - (\lambda \cdot z_A)(\mu \cdot z_A)Z;$$
$$\lambda, \mu = x^*, y^*, z^* \tag{47}$$

so that $(\lambda \cdot z_A)$ represents the projection of the z-axis of monomer A onto the principal axes of the pair. Further X, Y and Z are the zero-field energies of the monomer molecules.

For very large values of J_{AB} the off-diagonal elements are negligible and the zero-field splittings in the lower and upper states are identical. However in the case of the AB-pair the off-diagonal elements cannot be neglected. As a result not only the zero-field frequencies in the two Davydov components differ slightly but also the principal axes have slightly different directions. By carrying out the diagonalization of the Hamiltonian matrix (46) one finds for the difference $\delta/2\pi$ of the three zero-field frequencies $(T_{x^*} - T_{y^*})$, $(T_{x^*} - T_{z^*})$ and $(T_{y^*} - T_{z^*})$ in Φ_+ and Φ_-, the values -24.3 MHz, 1.0 MHz and -25.3 MHz respectively. Further it is found that the x^*_+ and z^*_+ principal axes in Φ_+ are rotated over an angle of $1.5°$ with respect to x^*_- and z^*_- in Φ_-, predominantly in the $x^* - z^*$ plane, whereas the y^*_+ axis makes an angle of only $0.3°$ with y^*_-.

The three quantities measured in zero-field and in the presence of a magnetic field show an exponential dependence on the temperature with an activation energy that is in surprisingly good agreement with the Davydov splitting of 2.5 cm^{-1} observed in the optical experiments [56, 64]. When calculating the differences in resonance frequency $\delta/2\pi$ from the measurements of the shifts ε, assuming $(\delta/\Gamma')^2 \ll 1$ and assuming that the second term between the brackets in (8) is negligible (this condition will be discussed below) we find the results of Table II. It is seen that there is a nice agreement with the results of the diagonalization of (46). The only appreciable deviation occurs for the $T_{x^*} - T_{z^*}$ transition where the difference in resonance frequency $\delta/2\pi$ is predicted to be small. Here the effect of the rotation of the spin axes, as expressed by the second term between the brackets in (8), is of comparable magnitude as the effect of the difference in resonance frequency as expressed by the first term.

When considering the expression (7) for the variation of T_2^{-1} with the temperature and taking the values of δ derived above it is possible to exclude that the observed temperature dependence of T_2^{-1} in zero-field is caused by the difference in resonance frequency $\delta/2\pi$. If this were true than one would predict that the effect of thermal excitation on T_2^{-1} for the two transitions studied would differ by a factor of 25, whereas in actual practice hardly a difference is observed.

An important experimental finding is that at 0.92 K $T_2 = 1/2\, T_1$. This means

TABLE II

The zero-field splittings of the two Davydov components ϕ_+ and ϕ_- of the AB-pair and their differences $\delta/2\pi$. All values are given in MHz. The columns indicated by "exp." show the results derived from the experiments. The frequencies for ϕ_- are the experimental values extrapolated to $T = 0$. The columns indicated by "theor." give the values derived from the diagonalization of the Hamiltonian matrix (47).

Transition	ϕ_-		ϕ_+		$\delta/2\pi$	
	exp.	theor.	exp.	theor.	exp.	theor.
$T_{x^*} - T_{y^*}$	2974.0	2965.3	2951.2	2941	-22.8 ± 2.0	-24.3
$T_{y^*} - T_{z^*}$	1317.2	1318.0	1302.7	1292.7	-14.5 ± 1.0	-25.3
$T_{x^*} - T_{z^*}$	1656.0	1647.3	1651.9	1648.3	-4.1 ± 0.7	$+1.0$

that the observed SSR rate is caused by a T_1-type process resulting from the rotation of the spin axes. In order to calculate the average SLR rate we must use expression (6) and use the average value of $\sin^2 2\theta$,

$$\sin^2 2\theta = 1 - \frac{1}{3} \sum_u \langle u_+ | u_- \rangle^2; (u_\pm = x_\pm^*, y_\pm^*, z_\pm^*). \tag{48}$$

For the AB-pair the average spin axis rotation is 1.2°. By taking $T_2^{-1}(\infty) = 1/2\, T_1^{-1}(\infty)$ and using the average value of $T_2^{-1}(\infty) = (1.4 \pm 0.5) \times 10^7$ s^{-1} we can then derive the lifetime $\tau = (\Gamma')^{-1}$ of the upper Davydov component. Since the reliability of this ascertainment is rather limited, as will be clear from the spin-spin relaxation measurements presented in Fig. 29, we can only give an estimate of this lifetime; $\tau = (\Gamma')^{-1} = (1 - 4) \times 10^{-10}$ s. When using the parameters so obtained we also find that the second "T_1-type" term in expression (7) for T_2^{-1} gives a contribution to T_2 that is of the same order of magnitude as the contribution of $1/2\, T_1$.

It is evident that at 0.4 K the thermally activated process does not contribute significantly to the value of T_2. Here T_2 is determined by the fluctuations in the local field of the triplet spins caused by random flip-flop motions in the nuclear spin system [20].

The spin relaxation behaviour in the presence of a magnetic field, as studied with the X-band ESE spectrometer operating at 0.3 T, is somewhat different from that in zero-field. In the presence of this magnetic field the spin eigenfunctions of Φ_+ and Φ_- in good approximation become eigenstates of the Zeeman interaction and the angle θ between corresponding spin eigenvectors becomes so small that the "T_1" contribution to T_2 becomes negligible. This is the reason why it is possible to observe ESE signals in the X-band spectrometer at 1.2 K whereas in zero-field at the same temperature the relaxation is so fast that this is no longer possible. In the presence of the magnetic field the first "dephasing" term in expression (7) dominates and one can explain the observations on the basis of the difference in resonance frequency $\delta/2\pi$. This was done in the paper by Botter et al. [60] where the significance of the rotation of the spin axes for the relaxation behaviour was not yet realized. So we see the interesting effect that by varying the magnetic field we can either make the "T_1" effect or the "dephasing" effect the dominating factor in the spin-spin relaxation in the lowest Davydov component of the AB-pair.

An observation that we like to comment on is the lengthening of T_2^{-1} at temperatures higher than 1.7 K ($T^{-1} = 0.6$ K^{-1}) in the magnetic field experiment (see Fig. 32). The maximum in T_2^{-1} corresponds with an effect known as motional narrowing and it occurs when $|\delta\Gamma^{-1}| \approx 0.5$ [65]. Here the triplet spins jump up and down so fast that the loss in phase begins to average out. The reason that this effect is not observed in zero-field is related to the fact that here the "dephasing" term in the expression (7) for T_2^{-1} is overshadowed by the "T_1" term.

The low temperature approximation, that was used to derive the formulae (6), (7) and (8) in Section 2, strictly speaking is not valid in the whole temperature range where the measurements were performed. In particular this means that the lifetime of the upper Davydov component can not be assumed to be temperature independent as remarked already by Reineker and co-workers [29, 30] and Silbey [28]. This may explain the spread in the values for the lifetime of the upper Davydov component derived from the experiments.

In addition to the AB-pair an experimental study was made by Botter et al. [66] on the lowest triplet component of the AA-dimer. Here the zero field splittings and the directions of the spin axes in Φ_+ and Φ_- are equal to each other and to those of the monomer as a result of the translationally equivalence of the two partner molecules. Hence one does not expect that phonon induced transitions between Φ_+ and Φ_- would result in a temperature dependence of T_2^{-1}, T_1^{-1}, or of the resonance frequency of the three zero-field transitions in the lowest state Φ_+.

Surprisingly enough a temperature dependence of T_2^{-1} does occur, suggesting that small differences in resonance frequency exist between corresponding transitions in the two Davydov components. This effect was measured on the $T_z - T_x$ zero-field transition of the AA-dimer which is observable as a satellite of the $T_z - T_x$ transition of the monomer in a crystal of N-d_8 doped with 5 per cent N-h_8 (see Fig. 33b). In Fig. 34 we show the decay of the two-pulse ESE signal with the microwave frequency tuned to this satellite. The slow component of the bi-exponential decay is related to the monomer, because its decay rate is almost equal to the one found with the spectrometer tuned to the main peak. The fast component however is attributed to the AA-pair. The temperature dependence of its decay rate T_2^{-1}, between 1.18 K and 4.2 K, is presented in Fig. 35. At low temperatures T_2^{-1} can be described by

$$T_2^{-1}(T) = T_2^{-1}(\infty)\exp\{-\Delta E/kT\} \tag{49}$$

with $T_2^{-1}(\infty) = (13.0 \pm 0.5) \times 10^5 \text{ s}^{-1}$ and $\Delta E = (1.0 \pm 0.1) \text{ cm}^{-1}$. The latter value is very close to the spectroscopically determined value of the Davydov splitting $2|J_b|$ of the AA-pair.

A temperature dependence of the resonance frequency of the $T_z - T_x$ transition

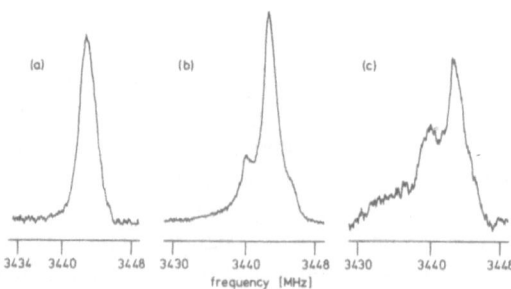

Fig. 33. (a) The optically detected $T_x - T_z$ zero-field transition of naphthalene-h_8 in a single crystal of naphthalene-d_8. The concentration of the guest is 0.1 per cent. Microwave power 10^{-2} W. The line is recorded by modulating the amplitude of the microwave power at 230 Hz and detecting synchronously in the phosphorescence intensity. $T = 1.18$ K. (b) The same transition detected in a crystal with 5 per cent guest concentration. Microwave power 1 W. Modulation frequency 230 Hz. $T = 1.18$ K. (c) The same transition detected in the 5 per cent crystal but now with a modulation frequency of 60 kHz. $T = 1.18$ K.

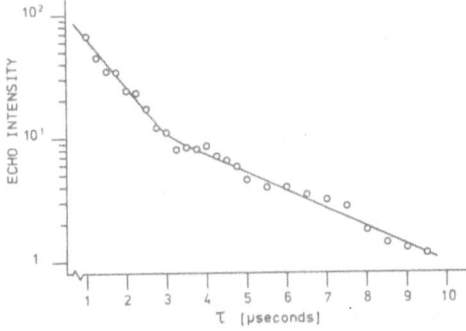

Fig. 34. The intensity of the two-pulse ESE signal, as a function of the time τ between the microwave pulses, of the $T_x - T_z$ transition of the AA-pair at 3440 MHz. $T = 2$ K.

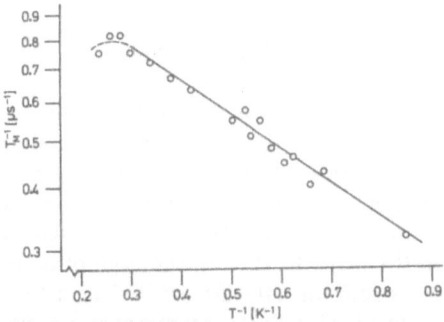

Fig. 35. The decay rate $T_M^{-1}(T_2^{-1})$ of the two-pulse ESE signal, obtained on the $T_x - T_z$ transition of the AA-pair at 3440 MHz, as a function of the inverse temperature.

of the AA-pair could not be observed within the experimental accuracy of 0.2 MHz.

The SLR rates among the spin levels of the AA-pair could only be estimated from the dependence of the intensity of the optically detected zero-field transition on the modulation frequency of the applied microwave field. As is observed in Fig. 33c, at high modulation frequencies, the relative intensity of the AA-pair signal has grown relative to the monomer signal. From the observed dependence we estimate the SLR rate at 1.18 K to be in the order of 10^4 s^{-1}.

The experimental results presented above support the idea that in the case of the AA-pair the T_2 and T_1 relaxation rates also are determined by a thermal activation process in which the higher lying Davydov component is involved. Since only a limited amount of experimental data is available we are not able to derive a picture as detailed as for the AB-pair. Nevertheless we can draw the following conclusions. First from the fact that no shift in resonance frequency is observed we derive that the difference in resonance frequency $|\delta/2\pi| < 1$ MHz. Secondly from the observation that the SLR relaxation rates are considerably slower than the spin-spin relaxation rates we conclude that the rotation θ of the spin axes in the upper state is extremely small and that "T_1" effects on T_2 may be neglected. On the basis of the upper limit for the difference in resonance frequency $\delta/2\pi$ Botter et al. [66] derived that $\tau = (\Gamma')^{-1} > 3 \times 10^{-8}$ s at 1.18 K. An independent piece of information is supplied by the bending of the curve of T_2^{-1} versus the temperature in Fig. 35 at 3.56 K. We think that here again the effect of motional averaging sets in which occurs when $|\delta\Gamma^{-1}| \approx 0.5$ [65]. When using this condition it is found that at 3.56 K $|\delta/2\pi| = 0.7$ MHz and $\tau = (\Gamma')^{-1} = 7 \times 10^{-8}$ s [66].

A remarkable result of this study is that the lifetime $(\Gamma')^{-1}$ of the upper Davydov component of the AA-pair is so much longer than for the AB-pair. This question and the question of the temperature dependence of the lifetime of the upper Davydov component in the AB- and AA-pair has been discussed by Reineker and Silbey in particular in relation with the effect of motional averaging observed in the two systems. For a detailed discussion of this problem the reader is referred to the chapter of Silbey.

6. Conclusions

From the results presented in Section 4 it has become clear that two-phonon processes reminiscent of Orbach-processes in ionic crystals, play an important role in the SLR in photo-excited triplet state molecules present as a guest in a foreign host crystal. The close-lying states involved in this process are so-called "local or pseudo-local phonon states". They correspond with localized librational motions of the guest molecule in the force field of the surrounding host molecules and have energies typically in the order of $10-20$ cm^{-1}. The SLR arises because in the excited state the molecule librates in an asymmetric potential well so that at the average the axes of quantization are no longer parallel to those in the lowest triplet state.

The energy and lifetime of the excited librational state, and the rotation of the spin axes vary considerably from one system to the other. This expresses the fact that these quantities depend on the subtleties of the intermolecular forces determining the shape of the potential well and the coupling of the librational motion to the lattice phonons. As a result the absolute values of the SLR rates differ strongly between different combinations of guest and host molecules. Moreover the orientational dependence of the SLR rates, in the presence of a magnetic field, does not reflect the symmetry of the guest molecule but rather that of guest and surrounding host molecules.

The SSR in the triplet state is mostly determined by fluctuations in the local field of the electron spins, caused by flip-flop motions in the nuclear spin system. However in systems where the proton spins are replaced by deuterons this contribution to the phase memory time of the triplet spins may become so small that the effect of the presence of a nearby local phonon state becomes visible as a shortening of the dephasing with increasing temperature.

The AB- and AA-dimers occurring in isotopically mixed crystals form beautiful examples to illustrate the effect of thermal excitation to a nearby excited triplet state on the dynamic properties of the triplet spins. From the temperature dependence of the SLR and SSR rates, and the resonance frequency we have not only been able to test the theoretical model but the results also show that the description of the dimers at low temperatures has to be in terms of superposition states and not in terms of random hopping between the two partner molecules.

7. Acknowledgement

An important part of this work was performed as part of the research program of the "Stichting voor Fundamenteel Onderzoek der Materie (FOM)" with financial support from the "Nederlandse Organisatie voor Wetenschappelijk Onderzoek (NWO)".

References

1. C. J. Gorter, *Paramagnetic Relaxation*, Elsevier, Amsterdam (1947).
2. A. A. Manenkov and R. Orbach, *Spin-Lattice Relaxation in Ionic Solids*, Harper and Row, New York, Evanston and London (1966).
3. A. Abragam and B. Bleaney, *Electron Paramagnetic Resonance of Transition Ions*, Clarendon Press, Oxford (1970).
4. I. Waller, *Z. Physik* **79** (1923), 370.
5. W. Heitler and E. Teller, *Proc. Roy. Soc. (London)* **155** (1936), 629.
6. J. H. van Vleck, *Phys. Rev.* **57** (1940), 426.
7. R. de L. Kronig, *Physica* **6** (1939), 33.
8. R. Orbach, *Proc. Roy. Soc. (London)* **A264** (1961), 458; **A264** (1961), 485.
9. W. B. Mims, 'Electron Spin Echoes', in *Electron Paramagnetic Resonance*, editor S. Geschwind, Plenum Press, New York, London (1972), 263.
10. P. H. H. Fischer and A. B. Denison, *Mol. Phys.* **17** (1969), 297.
11. M. Schwoerer and H. Sixl, *Z. Naturforsch.* **24a** (1969), 952.
12. J. P. Wolfe, *Chem. Phys. Lett.* **10** (1971), 212.
13. L. H. Hall and M. A. El-Sayed, *J. Chem. Phys.* **54** (1971), 4958.
14. J. Zuclich, J. U. von Schütz, and A. H. Maki, *Mol. Phys.* **28** (1974), 33.
15. D. A. Antheunis, B. J. Botter, J. Schmidt, P. J. F. Verbeek, and J. H. van der Waals, *Chem. Phys. Lett.* **36** (1975), 225.
16. M. Schwoerer, U. Konzelmann, and D. Kilpper, *Chem. Phys. Lett.* **13** (1972), 272.
17. U. Konzelmann, D. Kilpper, and M. Schwoerer, *Chem. Phys. Lett.*, *Z. Naturforsch.* **30a** (1975), 754.
18. F. Dietz, H. Port, and M. Schwoerer, *Chem. Phys. Lett.* **50** (1977), 26.
19. D. E. Kaplan, M. E. Browne, and J. A. Cowen, *Rev. Sci. Instr.* **32** (1961), 1182.
20. C. A. van 't Hof and J. Schmidt, *Mol. Phys.* **38** (1979), 309.
21. R. J. Silbey, in *Organic Molecular Aggregates*, editors P. Reineker, H. Haken, and H. C. Wolf, Springer Verlag, Heidelberg (1983), 67.
22. H. S. Gutowsky, D. W. McCall, and C. P. Schlichter, *J. Chem. Phys.* **21** (1953), 279.
23. H. M. McConnell, *J. Chem. Phys.* **28** (1958), 430.
24. H. Levinsky and H. Brenner, *Chem. Phys.* **40** (1979), 111.
25. P. J. F. Verbeek and J. Schmidt, *Chem. Phys. Lett.* **63** (1979), 384.
26. W. Vollmann, *Chem. Phys.* **57** (1981), 157.
27. F. Dietz, U. Konzelmann, H. Port, and M. Schwoerer, *Chem. Phys. Lett.* **58** (1978), 565.

28. R. J. Silbey, Chapter 4 of this book.
29. P. Reineker, J. Köhler, U. Schmid and R. Silbey, *J. Chem. Phys.* **83** (1985), 623.
30. U. Schmid and P. Reineker, *Mol. Phys.* **55** (1985), 77.
31. H. Sixl, Thesis, University of Stuttgart (1971).
32. D. A. Antheunis, Thesis, University of Leiden, (1974).
33. H. M. van Noort, B. Wirnitzer, and J. Schmidt, *Chem. Phys. Lett.* **85** (1982), 359.
34. S. Vega, *J. Chem. Phys.* **61** (1974), 1093.
35. J. F. C. van Kooten, Thesis, University of Leiden (1984).
36. P. J. F. Verbeek, Thesis, University of Leiden (1979).
37. A. J. van Strien, Thesis, University of Leiden (1982).
38. J. A. Kooter, Thesis, University of Leiden (1978).
39. P. J. F. Verbeek, C. A. van 't Hof, and J. Schmidt, *Chem. Phys. Lett.* **51** (1977), 292.
40. P. J. F. Verbeek, A. I. M. Dicker, and J. Schmidt, *Chem. Phys. Lett.* **56** (1978), 585.
41. A. I. Kitaigorodski, *Tetrahedron* **14** (1961), 230.
42. D. E. Williams, *J. Chem. Phys.* **45** (1966), 3770.
43. T. R. Koehler and J. Schmidt, *Chem. Phys. Lett.* **75** (1980), 38.
44. T. R. Koehler, H. M. van Noort, B. Wirnitzer, and J. Schmidt, *Chem. Phys. Lett.* **102** (1983), 95.
45. D. E. Truhlar, *J. Comput. Phys.* **10** (1972), 123.
46. P. J. F. Verbeek, H. J. Blanken, and J. Schmidt, *Chem. Phys. Lett.* **60** (1979), 358.
47. H. Port and H. C. Wolf, *Z. Naturforsch.* **23a** (1968), 315.
48. P. H. Chereson, P. S. Friedman, and R. Kopelman, *J. Chem. Phys.* **56** (1972), 3716.
49. U. Konzelmann and M. Schwoerer, *Chem. Phys. Lett.* **18** (1973), 143.
50. K. F. Renk, H. Sixl, and H. Wolfrum, *Chem. Phys. Lett.* **52** (1977), 98.
51. R. Avarmaa and A. Suisalu, *Chem. Phys. Lett.* **52** (1977), 567.
52. H. M. van Noort, C. A. van 't Hof, B. Wirnitzer, and J. Schmidt, *Chem. Phys. Lett.* **81** (1981), 351.
53. H. M. van Noort, Ph. J. Vergragt, J. Herbich, and J. H. van der Waals, *Chem. Phys. Lett.* **71** (1980), 5.
54. C. A. van 't Hof, Thesis, University of Leiden (1977).
55. Ph. J. Vergragt, J. A. Kooter, and J. H. van der Waals, *Mol. Phys.* **33** (1977), 1523.
56. C. L. Braun and H. C. Wolf, *Chem. Phys. Lett.* **9** (1971), 260.
57. H. Port and H. C. Wolf, *Z. Naturforsch.* **30a** (1975), 1290.
58. F. Dupuy, P. Pee, R. Lallane, J. P. Lemaistre, C. Vaucamps, H. Port, and Ph. Kottis, *Mol. Phys.* **35** (1978), 595.
59. M. Schwoerer and H. C. Wolf, *Mol. Cryst.* **3** (1967), 177.
60. B. J. Botter, C. J. Nonhof, J. Schmidt, and J. H. van der Waals, *Chem. Phys. Lett.* **43** (1976), 210.
61. J. F. C. van Kooten and J. Schmidt, *Mol. Phys.* **55** (1985), 351.
62. H. Port, D. Vogel, and H. C. Wolf, *Chem. Phys. Lett.* **34** (1975), 23.
63. J. S. King, Thesis, University of Chicago (1973).
64. D. M. Hanson, *J. Chem. Phys.* **52** (1970), 3409.
65. P. W. Anderson, *J. Phys. Soc. Japan* **9** (1954), 316.
66. B. J. Botter, A. J. van Strien, and J. Schmidt, *Chem. Phys. Lett.* **49** (1977), 39.

THE DYNAMICS OF ONE-DIMENSIONAL TRIPLET EXCITONS IN MOLECULAR CRYSTALS

J. SCHMIDT

Centre for the Study of Excited States of Molecules, Huygens Laboratory, University of Leiden, P.O. Box 9504, 2300 RA Leiden, The Netherlands.

1. Introduction

The present chapter is devoted to the study of the dynamic properties of photo-excited triplet excitons in molecular crystals. We will focus our attention mainly on the system 1,2,4,5-tetrachlorobenzene (TCB). This molecular crystal has several important features which makes it a prototype for investigating the properties of tightly bound excitons with well-defined intermolecular interactions. First of all the intermolecular interaction, in good approximation, is one-dimensional and consequently the band structure is relatively simple. Second, the triplet excitons not only can be studied by optical means but also by magnetic resonance methods. The latter technique will play an important role in our study. In addition to TCB we will also discuss results obtained on the related one-dimensional system para-dichlorobenzene (DCB), which confirm a number of our conclusions concerning the properties of one-dimensional triplet excitons.

Since the early work of Frenkel in 1931 [1] and Davydov [2] in 1962 on band structures in molecular crystals many publications have appeared dealing with collective excitations of such systems. Within the exciton model, the process of absorption of light is envisioned to occur in the following way. The photon incident on the crystal has a well-defined energy and momentum $\hbar Q$ where Q is the photon wave vector. In the process of absorption an exciton wave packet with the same energy is created. This exciton has a quasi-momentum $\hbar k$ and for momentum to be conserved k must be equal to Q.

In a perfect crystal the exciton retains its initial momentum. However, anything that disturbs the translational symmetry of the perfect lattice, such as impurities, lattice defects and phonons, will cause a scattering among the various k states and will lead to a thermalization of the exciton system. This model is valid only at low temperatures where the interaction of the excitons and the phonons is weak. At high temperatures exciton-phonon interactions will become so strong that this description is no longer valid. The excitation then is localized on a single molecule in the crystal and moves stochastically between the lattice sites. The former case is generally referred to as the wave-like or "coherent" exciton description, whereas the latter is known as the "incoherent" or diffusive model.

The experimental and theoretical studies to be presented in this chapter apply to the low-temperature regime where the "coherent" picture is valid. We shall show that by the application of modern, pulsed, narrow-band, tunable dye lasers it

J. Fünfschilling (ed.), Relaxation Processes in Molecular Excited States, 51–111.
© *1989 by Kluwer Academic Publishers.*

is possible to prepare the triplet excitons in a non-equilibrium situation. The subsequent evolution towards thermodynamical equilibrium then is observed by pulsed electron spin echo methods enabling us for the first time to study the scattering among the exciton **k** states in great detail. The experiments allow us to draw conclusions about the scattering mechanisms and about the spatial delocalization of the excitons in the crystal. In particular it will appear that in the TCB crystal studied by us the scattering is an impurity "assisted" process. This means that the exciton-phonon scattering can be explained by only considering the conservation of energy. This is in contrast with earlier theoretical models [3, 4] in which one always assumed that both energy and quasi-momentum must be conserved in the exciton-phonon scattering process.

Before presenting the experiments and their results, we shall first present a brief review of the recent literature on triplet excitons in molecular crystals. For a more comprehensive discussion of this subject the reader is referred to the works of Davydov [4], Craig and Walmsley [5], Knox [6], Burland and Zewail [7] who also consider the case of dimers and the more recent book by Broude *et al* [8]. Further we review briefly the properties of photo-excited triplet molecules and the various aspects of electron spin echo techniques. For a thorough discussion of time domain EPR experiments on photo-excited triplet state molecules we refer to the review paper of Schmidt and van der Waals [9].

2. Review of the Recent Literature

Until the end of the sixties investigations on singlet and triplet excitons have been performed exclusively via optical spectroscopy using conventional light sources. The first successful EPR experiments on triplet excitons were reported in 1967 by Haarer *et al.* [10], and Haarer and Wolf [11] on single crystals of anthracene and naphthalene at room temperature in the presence of an external magnetic field. It was concluded from an analysis of the EPR line shapes that at this temperature the electronic excitation moves stochastically in the ab plane of the crystal, and therefore that it can be described by a two-dimensional diffusive model. Shortly afterwards, similar EPR experiments were reported on tetracene [12] and pyrene [13]. A quantitative analysis of the observed EPR linewidth and spin-lattice relaxation (SLR) was developed by Reineker [14], based on the Haken and Strobl [15] stochastic model for exciton motion. Although one agrees on the diffusive nature of the motion of the triplet excitation in these crystals at room temperature the fit of the theory to the experimentally observed orientation dependence of the EPR linewidths was not satisfactory. Later it was shown by Rosenthal *et al.* [16] that orientational disorder contributes significantly to the observed linewidths. We shall see that inhomogeneous contributions to the linewidths of EPR and optical transitions of triplet excitons are very common, making it difficult to draw conclusions about the details of the exciton motion from an analysis of the lineshapes.

In the systems discussed above the major intermolecular interaction is between the two inequivalent molecules in the unit cell. This leads to a splitting of the exciton band in two components, the Davydov components [4], and to a relatively complicated band structure. The two components can be observed via optical spectroscopy at liquid helium temperatures. For instance Port et al. [17, 18] have obtained beautiful absorption spectra of the two 0—0 Davydov components in anthracene and napthalene. The optical linewidth depends strongly on the crystal quality, and in favorable cases it even proved possible to observe the individual magnetic substates in low magnetic fields as well as in zero field [19]. Unfortunately it has not been possible yet to observe EPR signals of the triplet excitons under these conditions. This is probably related to the fact that in these two-dimensional exciton systems the trapping of the exciton on chemical or structural impurities is very fast.

The only triplet exciton systems on which EPR signals have been observed at liquid helium temperatures are 1,4-dibromonaphthalene (DBN) [20] and 1,2,4,5-tetrachlorobenzene (TCB) [21]. The attraction of the two systems is that their intermolecular interaction in good approximation is one-dimensional and that the band structure is simple. In DBN the EPR signals have been observed by Schmidberger and Wolf [20] in the presence of a relatively high magnetic field via optical detection on one of the Zeeman components and via conventional EPR techniques. The results confirmed the earlier conclusions of Hochstrasser and Whiteman [22] that the excitons in DBN are mainly one-dimensional with a bandwidth of about 30 cm^{-1}. From a study of the temperature dependence of the EPR lineshapes it was concluded that above about 16 K the exciton becomes incoherent. Further, from a study of the orientational dependence of the SLR, it was established that this is caused by interchain motion.

The optical experiments on DBN have resulted in conclusions about the correlation times for exciton scattering that differ considerably from the results of Schmidberger and Wolf. For instance Burland et al. [23, 24] measured the origin of the optical absorption of the exciton in DBN at low temperatures and found, like Hochstrasser and Whiteman [22], an asymmetric lineshape. This asymmetry was explained by assuming that the optically accessible $k = 0$ state is at the bottom of the band. The width of the absorption line was attributed either to ^{13}C impurities [24] or stuctural disorder [25]. From an analysis of the optical lineshape [24] the correlation time for exciton scattering was estimated to be about 60 ps at 2 K, orders of magnitude shorter than the value derived by Schmidberger and Wolf at the same temperature.

In the analysis of the optical and EPR lineshapes in the case of DBN the conclusions depend strongly on the assumption whether we are dealing with inhomogeneous or homogeneous lineshapes. We believe that inhomogeneous broadening related to structural disorder contributes considerably to the observed lineshapes, making it difficult to obtain reliable results on correlation times for exciton scattering. The influence of structural disorder on optical lineshapes is demonstrated very dramatically by the experiments of Port et al. [18, 19] on anthracene and naphthalene. It was shown that in carefully treated anthracene crystals grown from the vapor linewidths as narrow as 0.01 cm^{-1} could be observed in the $T_0 \leftarrow S_0$ absorption, an order of magntiude smaller than those observed in melt-grown crystals. The authors however, still hesitate as to whether they are dealing with homogeneous lineshapes. More recently they have also obtained high resolution spectra in DBN with much narrower lineshapes than ever before [26].

The attraction of the triplet exciton system of TCB is that again we are dealing with a simple one-dimensional system. In addition to the optical absorption and emission lines of the excitons one can also observer EPR and optically detected EPR (ODMR) transitions in zero field as well as in the presence of a magnetic field. An important feature of this system is that the various exciton **k** states correspond with slightly different zero-field frequencies. This property enables us to study the evolution of an exciton **k** state by means of the intensity of its corresponding zero-field transition. In other words, we have a system where the exciton scattering can be studied in real time.

The main conclusions derived in this chapter pertain to experiments performed on TCB at low temperatures and for that reason we shall discuss its history in some detail. The interest in the triplet exciton system of TCB started with the work of Francis and Harris [21] who observed for the first time the zero-field transitions of the triplet excitons in this system by means of optical detection. The peculiar lineshape of these transitions was explained by considering the spin-orbit coupling (SOC) between the singlet and triplet exciton band. Here we shall not discuss the details of their model (this will be postponed till Section 4) but only give the important conclusion that as a result of this interaction the zero-field frequencies of the triplet excitons become slightly dependent on the wave vector **k**. In fact, they could derive a dispersion relation for this **k** dependence of the zero-field frequencies which has the same functional dependence as the dispersion relation for the energy $E(\mathbf{k})$ of the triplet excitons.

It was realized by Harris and his group that this microwave dispersion opens a unique possibility to study the triplet excitons via magnetic resonance methods. In fact, since the paper of Francis and Harris a large number of publications have appeared from this group and from others dealing with problems concerning the dynamics of the transfer of energy in this exciton system. In these papers one tries to ascertain the average lifetime of the **k** states, the number of molecules participating in the excitation, the probability of the **k** to **k**′ scattering, and the temperature above which the excitons become incoherent. Let us briefly review the main conclusions and the main points of controversy in the publications in the period 1971—1979. It soon appeared possible to establish a reliable value for the exciton bandwidth from an analysis of the steady state ODMR zero-field lineshapes. This bandwidth turned out to be 1.36 cm^{-1} with **k** = 0 at the top of the band, 374.8 nm (26680 cm^{-1}) above the ground state [27, 28]. The time evolution of the phosphorescence of the exciton and the dominant X-trap, which lies about 17 cm^{-1} below the exciton band, upon selective laser flash excitation into the exciton band was measured in detail by Shelby *et al.* [29] and Güttler *et al.* [30]. Their conclusions are roughly the same, i.e. below 2 K the trapping rate of the exciton is 10^2–10^3 s^{-1}, its lifetime about 1 ms and that of the X-trap 35 ms. At higher temperatures the communication between trap and exciton becomes so fast that they decay with the same rate making it impossible to study them seperately.

The problem of the scattering of the triplet excitons and their delocalization was studied via various methods, but the experiments often led to rather conflicting results. This must be attributed to the fact that the conclusions are based on line fitting procedures of the steady state EPR and optical lineshapes in combination with various theoretical models [21, 27, 31—35]. Just as in the case of DBN one is never certain of the inhomogeneous contributions to the lineshape. For instance, from an analysis of the optical absorption lineshape of the exciton band Burland *et al.* [32] concluded that the optical dephasing time, which should be related to the characteristic scattering time of the excitons, is about 20 ps. This must be compared with estimates of the dephasing time of the triplet exciton spins, obtained from line fitting procedures of the ODMR transitions, of the order of about 1 μs [27, 35]. Various suggestions were made to account for this difference [32, 34]. In a very detailed study of the ODMR lineshapes as a function of the temperature Breiland and Saylor [35], applying the exchange formalism, not only derived conclusions regarding the triplet dephasing time but also concluded that $k - k'$ scattering should be restricted to about one third of the total bandwidth. Since in TCB the dispersion of the triplet excitons is considerably smaller than the dispersion of the acoustic phonons, this restricted scattering seemed to be in agreement with predictions based on theoretical models for exciton-phonon coupling. These theories [3, 4] are based on the idea that in the exciton-phonon coupling the rules for conservation of energy and quasi-momentum apply simultaneously thereby reducing considerably the number of phonons that can scatter the excitons at such low temperatures.

In a series of papers Dlott, Fayer and Wieting [36, 37] investigated the effects of impurity scattering and transport topology on spatial exciton migration and trapping. According to their arguments exciton trapping in TCB is strongly influenced by impurities having their excited state just above that of the exciton band. This conclusion was supported by a study of the time evolution of the X-trap phosphorescence as a function of the concentration of deuterated TCB molecules that had been incorporated in the crystal. A similar macroscopic model was developed by Zwemer and Harris [38] who argued that it is necessary to include tunneling and thermal promotion mechanisms in order to understand energy transfer in one-dimensional systems with traps. The quantitative results of these studies however should be considered with care since in the experimental verification one does not observe directly the behaviour of the various exciton states.

In 1978 the first time resolved electron spin echo experiments on triplet excitons in TCB were reported by Botter *et al.* [39]. They observed these signals upon excitation with a pulsed N_2 laser directly in the $T_0 \leftarrow S_0$ absorption about 3000 cm^{-1} above the lowest triplet exciton band. The echo signals could be detected 10 μs after the laser flash. The lineshape of the zero-field transition so observed proved that at this delay a Boltzmann distribution exists over the exciton band. This experimental result made it possible for the first time to establish the triplet spin dephasing time, but it was not yet possible to follow directly the scattering among the exciton k-states. A very remarkable shortening of the phase memory time was observed with increasing temperature, an effect attributed to phonon induced scattering among the exciton **k**-states. From a study of the stimulated spin echo signals it was concluded that the **k** − **k**′ scattering is restricted to small intervals in **k**-space, a conclusion that was shown later to be erroneous.

In 1980 Van Strien *et al.* [40] presented the results of a similar study where now the optical excitation source was a pulsed tunable dye laser allowing for excitation directly into the 0–0 band of the $T_0 \leftarrow S_0$ absorption. In this experiment it appeared that this excitation leads to a strong excess population in the **k** ≈ 0 region of the exciton band. Moreover it proved possible to follow directly the evolution of this non-thermal population distribution and thus to study for the first time the scattering among the exciton **k** states. This work triggered a series of experimental and theoretical papers which led to some important conclusions concerning the mechanism of the scattering and trapping of the triplet excitons in this one-dimensional system. The contents of these papers will be discussed in detail in this chapter. In addition we shall consider more recent results on the related one-dimensional system p-DCB which suggest that here the weak inter-stack interaction leads to an extremely rapid spin-lattice relaxation of the triplet exciton spins and also to a very rapid trapping of the excitons.

3. Triplet Excitons

The concept of excitons was first introduced by Frenkel [1] in 1931 to describe the collective electronic excitations of pure molecular crystals. The Frenkel exciton is usually designated as the "tight-binding" exciton. This expresses the fact that the ground state and the electronically excited state are strongly localized on individual molecules. The eigenstates of the crystal are linear combinations of these localized excitations, each belonging to a wave vector **k** of the reciprocal lattice. Frenkel called these states "excitons".

It was shown by Wannier [41] that in semiconductors and dielectrics with a high dielectric constant an exciton may be viewed alternatively as a conduction-band electron and a valence-band hole, bound together but with a considerable separation. The two pictures are not so unrelated as it might seem. Since an excited state of an atom can be described as an electron bound closely to an ion by the Coulomb interaction the Frenkel and the Wannier excitons differ physically merely by their "radii", that is the degree of separation of the "electron" and the "hole".

Frenkel excitons and Wannier excitons are the two limiting models of collective currentless excitations of dielectrics in which the excitation is distributed throughout a large region of the crystal. Since we will be dealing with molecular crystals we shall talk exclusively about Frenkel excitons. First we shall discuss some general aspects of the Frenkel model based on the description developed by Davydov [4]. Then we restrict ourselves to triplet excitons and in particular to one-dimensional or linear chain triplet excitons.

Let us consider a large crystal with one molecule in the unit cell. We also assume that the molecules are rigidly fixed in the equilibrium positions, i.e. we neglect the presence of phonons. A unit cell is a parallelepiped with three non-coplanar basis vectors $\mathbf{a}_1$, $\mathbf{a}_2$, $\mathbf{a}_3$. The position of the lattice points where the molecules are fixed is determined by the lattice vector

$$\mathbf{n} = \sum_{i=1}^{3} n_i \mathbf{a}_i, \tag{1}$$

where n_i are integers. If the length of the crystal edges is given by $N_i \mathbf{a}_i$ ($i = 1, 2, 3,$), we find for the total number of molecules in the crystal $N = N_1 \times N_2 \times N_3$. The total Hamiltonian in this approximation is

$$\mathscr{H} = \sum_{\mathbf{n}} \mathscr{H}_{\mathbf{n}} + \frac{1}{2} \sum_{\mathbf{n,m}}' V_{\mathbf{nm}}, \tag{2}$$

where $\mathscr{H}_{\mathbf{n}}$ is the Hamiltonian for a molecule at the $\mathbf{n}$-th site, and $V_{\mathbf{nm}}$ is the interaction energy operator of two molecules $\mathbf{n}$ and $\mathbf{m}$. The prime in the second summation indicates that terms with $\mathbf{n} = \mathbf{m}$ are absent in the sum. If for a moment we neglect $V_{\mathbf{nm}}$, we find for the total wavefunction describing the ground state of the crystal

$$\psi^0 = \Pi_{\mathbf{m}} \phi_{\mathbf{m}}^0 \tag{3}$$

with $\phi_{\mathbf{m}}^0$ the wave function of an isolated molecule. The excited states corresponding to the f-th excitation of the molecule at position $\mathbf{n}$ of the crystal are given by

$$\psi_{\mathbf{n}}^f = \phi_{\mathbf{n}}^f \Pi_{\mathbf{m} \neq \mathbf{n}} \phi_{\mathbf{m}}^0 \tag{4}$$

The functions ψ^0 and $\psi_{\mathbf{n}}^f$ must be antisymmetrized with respect to all electrons resulting in additional energy terms that contain the overlap integrals of the wavefunction of adjacent molecules. For the sake of simplicity we shall ignore these terms in this general discussion, but we shall show their importance in the specific case of triplet excitons.

The states $\psi_{\mathbf{n}}^f$ are N-fold degenerate because any of the N molecules of the crystal can be in the f-th excited state. Since all molecules are the same, instead of specifying the site of the excited molecule in the crystal, it is customary to introduce N new orthonormalized functions

$$\psi^f(\mathbf{k}) = N^{-1/2} \sum_{\mathbf{n}} \psi_{\mathbf{n}}^f e^{i\mathbf{k} \cdot \mathbf{n}} \tag{5}$$

These functions are known as Bloch functions and differ in the wave vector $\mathbf{k}$, which is defined by

$$\mathbf{k} = \sum_{i=1}^{3} \frac{2\pi}{N_i} \ell_i \mathbf{b}_i; -\frac{1}{2} N_i < \ell_i \leq \frac{1}{2} N_i \tag{6}$$

The vectors $\mathbf{b}_i$ are the basis vectors of the reciprocal lattice and are related to the basis vectors **a** of the direct lattice by $\mathbf{b}_i \cdot \mathbf{a}_j = \delta_{ij}$. The set of all **k** vectors is often referred to as the first Brillouin zone.

The N-fold degeneracy of the states $\psi^f(\mathbf{k})$ is removed under the influence of the intermolecular interaction V_{nm}. Since in molecular crystals V_{nm} is small compared with the intramolecular interactions, we may still use, in first approximation, the wavefunctions ψ^0 and ψ^f_n to calculate the energy of the ground state and excited state of the crystal in the presence of the intermolecular interaction V_{nm}. If we then take the difference between the energy of the ground state and the excited states corresponding with $\psi^f(\mathbf{k})$, we obtain the excitation energy

$$E_f(\mathbf{k}) = \Delta\varepsilon_f + D_f + L_f(\mathbf{k}). \tag{7}$$

Here, $\Delta\varepsilon_f$ is the excitation energy of a free molecule, D_f is the change in the interaction energy of one molecule with all the surrounding molecules in its transition to the f-th excited state, and $L_f(\mathbf{k})$ is an addition to the excitation energy which is a function of the wave vector **k**,

$$L_f(\mathbf{k}) = {\sum_{m}}' \beta^f_{nm}\, e^{i\mathbf{k}\cdot(\mathbf{n}-\mathbf{m})} \tag{8}$$

In this case the matrix element

$$\beta_{nm} = \langle \phi^0_n\, \phi^f_m |\, V_{nm}\, | \phi^0_m\, \phi^f_n \rangle \tag{9}$$

determines the transition of the excitation f from molecule **n** to molecule **m**.

In large crystals adjacent values of **k** differ very little from one another, so the N values of the excitation energy $E_f(\mathbf{k})$ form a quasi continuous band of excited states. Each of the excited states, labeled with **k**, is a collective excited state of the entire crystal. These elementary excitations, or excitons, are characterized by the quasi-momentum $\hbar\mathbf{k}$ and energy $E_f(\mathbf{k})$ and are usually referred to as quasi-particles. Since each molecule plays an identical role in the formation of an exciton state, the associated excitation is distributed throughout the crystal and is not concentrated on one molecule. The states associated with excitations of a limited region of the crystal are represented by wave packets, i.e. linear combinations of $\psi^f(\mathbf{k})$, for example

$$\psi^f = \sum_{k} w(\mathbf{k})\, \psi^f(\mathbf{k}); \tag{10}$$

where $w(\mathbf{k})$ differs from zero only in a small region of **k** space. The excitations ψ^f do not have a definite energy or a definite wave vector value. The smaller the region of excitation (Δx), the greater the uncertainty in the wave vector (Δk_x)

$$\Delta x\, \Delta k_x = 2\pi \tag{11}$$

When the crystal is subjected to electromagnetic radiation, the excited region has linear dimensions at least in the order of magnitude of the wavelength of the radiation. For visible and ultraviolet light the wavelength is considerably greater than the lattice constant, therefore $\Delta x \gg a_x$ and consequently $\Delta k_x \ll 2\pi/a_x$.

Furthermore, since the law of conservation of quasi-momentum must be satisfied, $\mathbf{k} = \mathbf{Q}$. Here $\mathbf{Q}$ is the wave vector of the incident light and since $\mathbf{Q} \cdot \mathbf{a} \ll 1$ we see that only excitons with $\mathbf{k} \approx 0$ can be created by light.

The main difference between singlet and triplet excitons is in the matrix element for excitation transfer β_{nm}^f. It involves the intersite interaction operator V_{nm} which characterizes the Coulomb interaction of electrons and nuclei of both molecules. In the calculation of β_{nm}^f in neutral molecules with a center of symmetry, one first considers electric dipole-dipole interactions of the dipole moment of the transition to the f-th excited state. Clearly the transition dipole moment for the allowed singlet-singlet transitions are much larger than those for the spin-forbidden triplet-singlet transitions. In general, in an aromatic hydrocarbon crystal, the calculation gives singlet exciton bandwidths of several hundreds cm^{-1}. For triplet excitons however bandwidths no greater than 10^{-3} cm^{-1} are predicted, whereas in experimental practice values in the order of $1-10$ cm^{-1} are found. The observed bandwidths in triplet excitons can be explained by considering the intermolecular exchange interaction which depends on the overlap of the wave functions of adjacent molecules. It can be taken into account by appropriate antisymmetrization of the wave functions (3) and (4).

If there is more than one molecule in the unit cell the above description becomes slightly more complicated. We then start by taking Bloch functions $\psi_\alpha^f(\mathbf{k})$ similar to (5) for every molecule α within the unit cell. If there is also an interaction between the different molecules α it is necessary to construct linear combinations of $\psi_\alpha^f(\mathbf{k})$ each of which describes a different branch of the exciton band. The number of these branches is equal to the number of molecules in the unit cell, and the energy splitting between the $\mathbf{k} = 0$ states of these branches is usually called Davydov splitting [4]. The absolute value of this splitting is proportional to the interactions between the molecules within the cell. Therefore in a crystal with more than one molecule in the unit cell one can only distinguish different branches if the intermolecular interaction between those sites is non-negligible. If on the other hand this interaction is so small that no Davydov splitting is observed it is allowed to apply the description for one molecule in the unit cell.

The above considerations apply to an ideal crystal without vibrations and lattice defects. In this situation the excitations can be considered as ideal exciton states corresponding to a definite $\mathbf{k}$ value, that is, the range of $\mathbf{k}$ values within an exciton wave packet (10) is very small and does not change in time. However when translational symmetry is destroyed, for instance in an imperfect crystal or by lattice vibrations, this is no longer true. The coupling of the exciton states to the phonon bath leads to a scattering of the $\mathbf{k}$ states, the rate of which increases with increasing temperature. As a result of this scattering not only the mean $\mathbf{k}$ value of a wave packet changes in time but also its width $\Delta \mathbf{k}$. As a consequence of the uncertainty relation (11), the degree of delocalization decreases with increasing width of the exciton wave packet. When this degree of delocalization becomes of the order of the lattice constant, the excitation becomes localized and its motion can be described by a random walk on the lattice. Although strictly speaking we then are no longer dealing with excitons in the sense of the definition by Davydov, one usually refers to it as an "incoherent exciton" in contrast to the "coherent exciton" described above. The experimental results to be discussed in this chapter will be described in terms of the "coherent exciton" model.

4. The Systems TCB and p-DCB

4.1. THE ENERGY LEVELS OF ISOLATED TCB AND p-DCB MOLECULES

In Figure 1 the energy level scheme is presented of TCB diluted in a single crystal of durene. The labels on the left side of the levels give the orbital symmetry in D_{2h} (the symmetry group of TCB). For the sublevels of the triplet states the total symmetries are indicated by the labels at the right.

The radiative properties of the individual spin components of T_0 are determined by the spin-orbit coupling (SOC) in the molecule. This interaction gives matrix elements between singlet and triplet states of the same total symmetry and thus allows electric dipole transitions to occur between particular triplet spin levels and the singlet ground state. When considering Figure 1 it is clear that in TCB the spin component T_y is allowed to interact via SOC with the $^1B_{3u}$ excited singlet state. Similarly the component T_x will interact with the first excited singlet state S_1 of species B_{2u}. However when investigating the size of these coupling constants it turns out that the interaction between the T_y component of the lowest triplet state and the $n\pi^*$ state S_i exceeds the other one by at least an order of magnitude [42]. So one predicts dominant x-polarized phosphorescence from T_y and this is confirmed by measurements of the decay rates of the spin components; $k_y = 30$ s^{-1} and $k_x = 1.4$ s^{-1}. The third spin component T_z also acquires singlet character but via indirect vibronic spin-orbit coupling. In Figure 1 we have indicated a possible route for this interaction via an excited triplet state of species B_{3u} which couples via SOC with the first excited $^1B_{2u}$ state. The emission from T_z to S_0 however is symmetry forbidden and proceeds to a b_{2g}vibrational state of S_0 via y-polarized light. Although this transition is vibrationally induced it appears that the decay rate $k_z = 27$ s^{-1} is almost equal to k_y [42]. The reason for this is twofold; first the admixture of $^1B_{2u}$ character into the excited $^3B_{3u}$ state is relatively large because of the proximity of these two interacting levels and second the oscillator strength of the $S_1 \leftarrow S_0$ transition, which corresponds with a $\pi \rightarrow \pi^*$ excitation, is considerably larger than the $S_i \leftarrow S_0$ transition which corresponds with a $n \rightarrow \pi^*$ excitation.

Since the intersystem crossing (ISC) process from $S_1 \rightarrow T_0$ also is determined by SOC the populating rates P_u of the spin components of T_0 differ greatly. In TCB it appears that $P_z \approx P_y \gg P_x$ [42] and thus upon excitation into the singlet manifold we preferentially populate the T_z and T_y sublevels.

When using pulsed lasers it is possible to excite an appreciable number of TCB molecules directly from the singlet ground state S_0 into the triplet state T_0. Since the transition to the T_y sublevel is the only one that is optically accessible from S_0 one thus creates a transient overpopulation in this particular sublevel. This

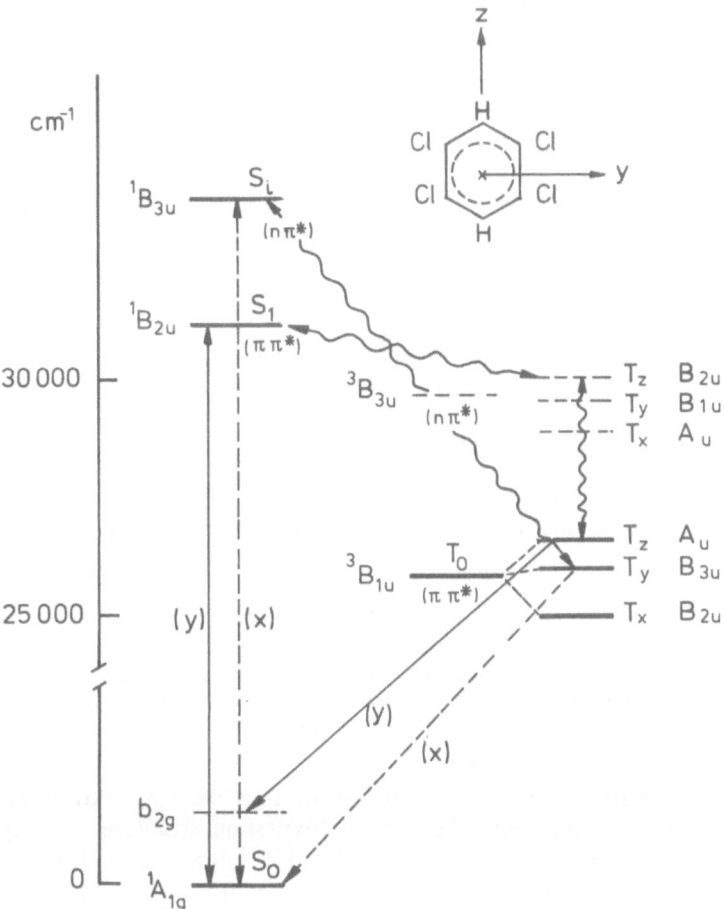

Fig. 1. The lower electronic states of TCB. The labels at the left of the energy levels are orbital symmetries in D_{2h}; for the singlets they also represent the total symmetries. For the sublevels of the triplet state the total symmetries are indicated by the labels at the right. The routes for SOC (or vibronic SOC) are indicated by the wavy lines. The sublevel T_y of T_0 acquires singlet character via direct SOC with a $n\pi^*$ excited singlet state of the same symmetry. It emits phosphorescence to the singlet ground state and the totally symmetric vibrational states of S_0. The sublevel T_z of T_0 acquires singlet character via indirect vibronic SOC coupling with the first excited $\pi\pi^*$ singlet state S_1 with symmetry $^1B_{2u}$. Its phosphorescence proceeds to a b_{2g} vibrational state of the singlet ground state [42]. The labels $n\pi^*$ and $\pi\pi^*$ refer to the two singly occupied orbitals in a simple molecular orbital picture.

property will be used to our advantage in the Electron Spin Echo (ESE) experiments on the triplet excitons in TCB.

The strongly differing radiative decay rates of the triplet sublevels allow for optical detection of magnetic reasonance transitions between the triplet sublevels in zero field as well as in the presence of a magnetic field. In zero field however no

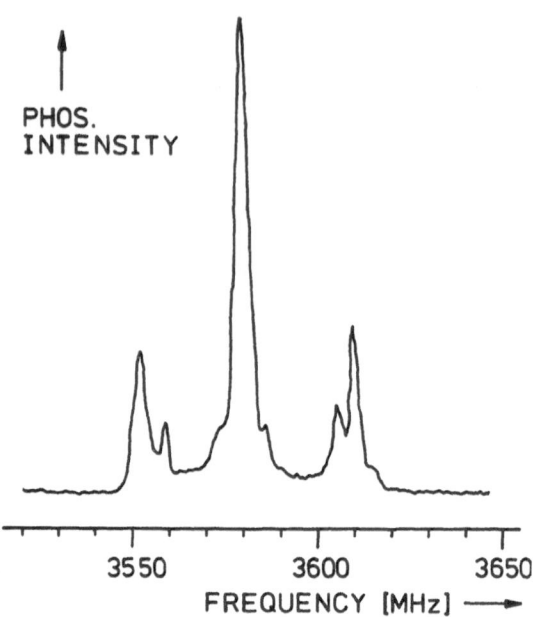

Fig. 2. The optically detected $T_y - T_x$ zero-field transition of an isolated TCB molecule (TCH-h_2 in TCB-d_2). The central peak is the allowed transition broadened by second order hyperfine interaction. The satellites are "forbidden" transitions and correspond to simultaneous flips of the ^{35}Cl and ^{37}Cl nuclei [43, 44].

first order hyperfine splitting will occur in the electron spin levels and the resonance lines do not show the familiar hyperfine structure of ordinary EPR spectra. In the case of TCB and p-DCB this HF interaction indirectly leads to a very characteristic power dependent structure of the zero-field spectra because these molecules contain chlorine nuclei with $I = 3/2$ having a sizable quadrupole splitting. By virtue of the HF interaction "forbidden" transitions will occur separated from the "allowed" one by the chlorine quadrupole splitting and corresponding to a nuclear spin "flip" in addition to that of the electron spin [43, 44]. In Figure 2 we show the $T_y - T_x$ zero-field transition of TCB-h_2 present as a trap in a single crystal of TCB-d_2 where such satellites appear. In the zero-field transitions of the triplet excitons in TCB these satellites cannot be observed. Here the excitation is "diluted" over a large number of molecules and consequently the hyperfine interaction is so small that the probability for these forbidden transitions vanishes. This property provides an important argument to distinguish between zero-field transitions of excitons and traps in pure TCB crystals [21].

4.2. ONE-DIMENSIONAL TRIPLET EXCITONS IN TCB

The attraction of TCB is that in the neat crystal the interaction energy β_{nm} (9) has an appreciable value only in one direction. As discussed before for triplet excitons this interaction energy depends on the overlap of the molecular wave functions and consequently is non-zero only between neighbouring molecules. In Figure 3 we show the crystal structure of TCB which is monoclinic at room temperature with two molecules in the unit cell. As can be inferred from the crystal structure the intermolecular exchange interaction is predominantly along the short **a** axis. The generally accepted value is $\beta = +0.34$ cm^{-1}. The intermolecular interaction in the other directions is estimated to be at least four orders of magnitude smaller [45]. Consequently there are two types of exciton chains (A and B) between which only a very weak interchain interaction exists. Obviously the A and B chains are indistinguishable in zero-magnetic field.

The energy dispersion for a one-dimensional exciton system like TCB with nearest neighbour interaction β can be calculated easily. We find (omitting vector signs in this linear case)

$$E(k) = E_0 + 2\beta \cos ka, \tag{12}$$

where E_0 is the energy of the center of the band and a the intermolecular distance. Since β is positive in TCB the $k = 0$ states are at the top of the band. Each of the exciton k states however is split into three spin components and since we are dealing with translationally equivalent molecules we would expect that this splitting is the same as for isolated TCB molecules. However in 1971 Francis and Harris [21] showed that the zero-field splitting of the triplet excitons is slightly k

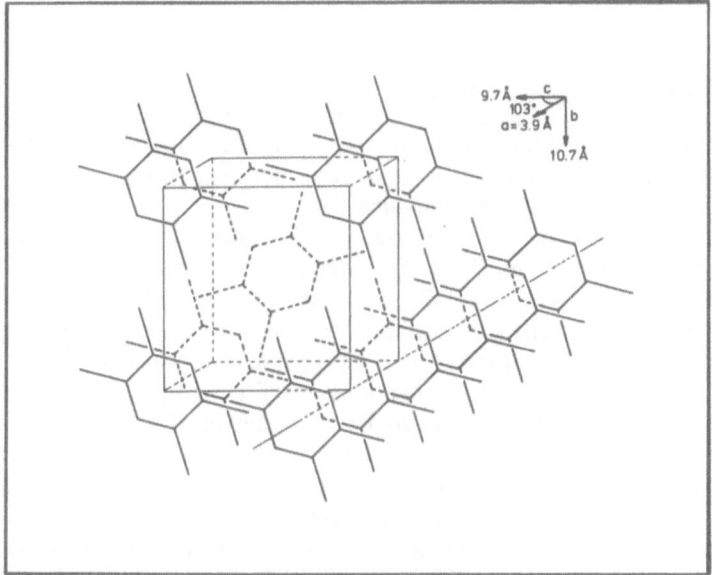

Fig. 3. The room temperature crystal structure of TCB. The unit cell of the monoclinic crystal structure has been drawn schematically. One edge of the unit cell is extended along the short **a** axis to illustrate the direction of the largest intermolecular interaction.

dependent as a result of spin-orbit coupling (SOC) between triplet and singlet excitons with the same value of k. This dispersion of the microwave transition energy is of great importance for our experiments because it allows us to observe the behaviour of the various k states. For this reason we shall discuss it here in some detail.

As explained in Section 4.1 in TCB singlet character is admixed into the T_y sublevel via direct SOC, into T_z via vibronic SOC whereas the singlet admixture in T_x in first approximation may be neglected. Let us first consider the effect of this SOC on the spin component T_y. The corresponding wave function can be written,

$$T_y(k) = T_y^0(k) + \sum_i \frac{\langle S_i(k)|\mathcal{H}_{soc}|T_y^0(k)\rangle}{\Delta E_i(k)} S_i(k) \qquad (13)$$

Here $T_y^0(k)$ stands for the pure triplet spin component without SOC and with an energy dispersion as given in (12). The $S_i(k)$ are the admixed singlet states and $\Delta E_i(k) = E_{s_i}(k) - E_{T_y^0}(k)$. Since the singlet exciton bandwidth is significantly larger than the triplet exciton bandwidth, $\Delta E_i(k)$ is k dependent. As a result the singlet admixture into T_y is k dependent and consequently its energy also varies with k. When neglecting the singlet admixture in T_x one finds via a straightforward perturbation calculation the following dispersion relation for the resonance frequency of the $T_y - T_x$ transition [21],

$$\nu^{yx}(k) = \nu_0^{yx} + 2\mu^{yx} \cos ka \qquad (14)$$

with

$$\mu^{yx} = \zeta_y^2(\beta_S - \beta_T)/(E_{ST}^0)^2. \qquad (15)$$

Here ζ_y is the SOC constant for T_y, β_S and β_T are the singlet and triplet molecular interaction energies and E_{ST}^0 is the separation between triplet and singlet states in the absence of intermolecular interactions.

Using reasonable values for the various parameters in (15) we can obtain an estimate of the width of the $T_y - T_x$ transition. For instance taking $(\beta_S - \beta_T) = 200 \text{ cm}^{-1}$, $E_{ST}^0 = 10^4 \text{ cm}^{-1}$ and $\zeta_y = 10 \text{ cm}^{-1}$ we get $\mu^{yx} \approx 2 \times 10^{-4} \text{ cm}^{-1} = 6 \text{ MHz}$. We shall see furtheron that this result is very close to the experimentally observed broadening of the exciton $T_y - T_x$ zero-field transition.

Following similar arguments we can derive a dispersion relation for the $T_z - T_x$ zero-field transition. Although the singlet character in T_z is acquired via vibronic SOC it appears that a similar perturbation calculation can be performed leading to an expression for $\nu^{zx}(k)$ analogous to (14). In Fig. 4 we give a schematic depiction of the resulting variation of the zero-field splitting with k.

The steady-state lineshapes of the $T_y - T_x$ zero-field transitions can be calculated using expression (14) for the k dependence of the zero-field frequencies and assuming a Boltzmann population distribution over the k states. In Fig. 5 we show results of optically detected magnetic resonance (ODMR) experiments on the $T_y - T_x$ and $T_z - T_x$ transitions taken from the original paper of Francis and Harris

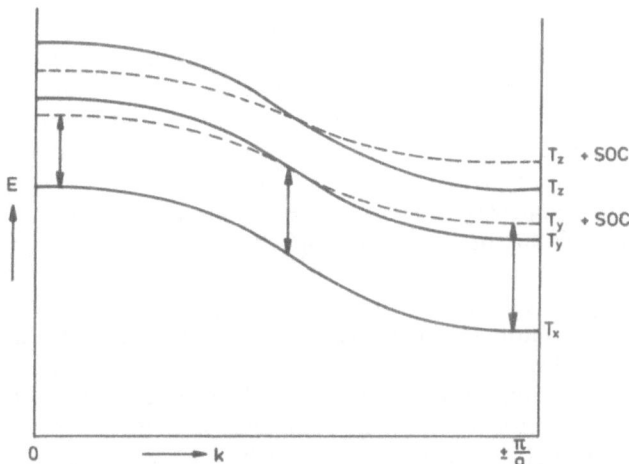

Fig. 4. A schematic depiction of the k dependence of the zero-field splitting in the linear chain triplet exciton system of TCB as a result of selective SOC. The positions of the triplet sublevels in the absence of SOC are indicated by the full lines. The dashed lines indicate their positions including SOC. Note that the zero-field splittings and the exciton dispersion are not drawn to scale for illustrative purposes.

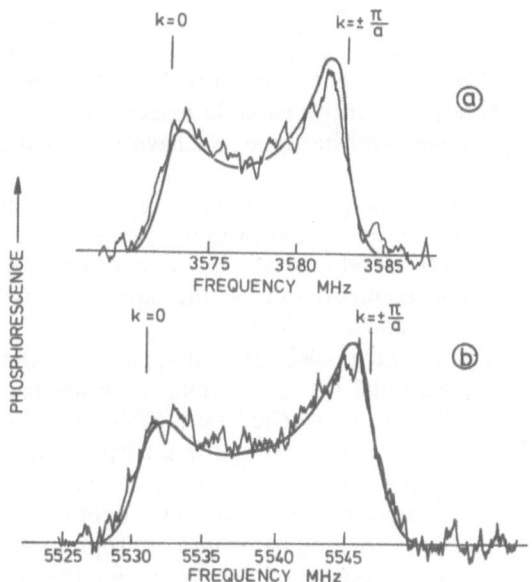

Fig. 5. (a) The optically detected $T_y - T_x$ zero-field transition of the triplet excitons in TCB. The solid line is the theoretically predicted curve at $T = 3.2$ K. (b) The optically detected $T_z - T_x$ zero-field transition of the triplet exciton in TCB. The solid line again is the theoretically predicted curve at $T = 3.2$ K. After reference [21].

together with the theoretically predicted lineshapes. The theoretical curves have been derived using $4\beta = 1.36$ cm^{-1} for the triplet exciton bandwidth, $\mu^{yx} = -2.5$ MHz for the $T_y - T_x$ and $\mu^{zx} = -3.9$ MHz for the $T_z - T_x$ transition respectively. The spectra show two peaks of different intensity which correspond with the maxima in the density of states function at the top and the bottom of the exciton band. The difference in height of the two maxima is related to the Boltzmann factor across the band [21]. Since the $k = 0$ state is at the top of the band the corresponding peak in the microwave spectra becomes smaller when going to lower temperature.

The beautiful correspondence between the optically detected and theoretical lineshapes of the exciton zero-field transitions confirms that in the liquid helium temperature range the "coherent" picture applies for the triplet excitons in TCB. The important consequence of the model of Francis and Harris is that every k state has its own zero-field splitting parameter and therefore can uniquely be defined by its zero-field resonance frequency. It is this particular property that will allow us to study the dynamic properties of the exciton wave packets.

5. Experimental

The Electron Spin Echo (ESE) experiments to be described in the next sections have been performed in the presence as well as in the absence of an external magnetic field. In the first kind of experiment the microwave frequency is tuned into resonance with one of the zero-field splittings of the photo-excited triplet state. The second kind of experiment is performed at a fixed microwave frequency of about 9 GHz (X-band) and with a variable external magnetic field. Therefore, although the basic principles are the same, we have used two distinctly different ESE spectrometers.

The zero-field ESE spectrometer is in principle identical to the one described in [9, 46]. Since then however an improvement in the performance has been achieved by introducing more recently developed microwave components. In Fig. 6 we show schematically the present configuration of the apparatus which covers three frequency octaves 1—2, 2—4 and 4—8 GHz.

First we discuss the microwave pulse generating part. In order to create a phase shift in the microwaves, essential for spin-locking experiments, the output of the CW signal oscillator (SO) can be switched by a SPDT PIN switch between two parallel coaxial lines. A variable phase shifter is incorporated in one of the arms, while in the other an adjustable attenuator compensates for the power loss in the phase shifter. The parallel branches are joined in a hybrid junction and the CW signal is then converted by a PIN modulator into appropriate pulses. These pulses are amplified to a maximum power of 10 Watts by the traveling wave tube amplifier (TWTA). Owing to the high noise figure of this power amplifier a third PIN switch is incorporated in the system to attenuate the noise at the moment that the echo appears. At A the frequency of the microwaves is measured by an electronic counter, and at D the microwave pulses are monitored. At B and C we

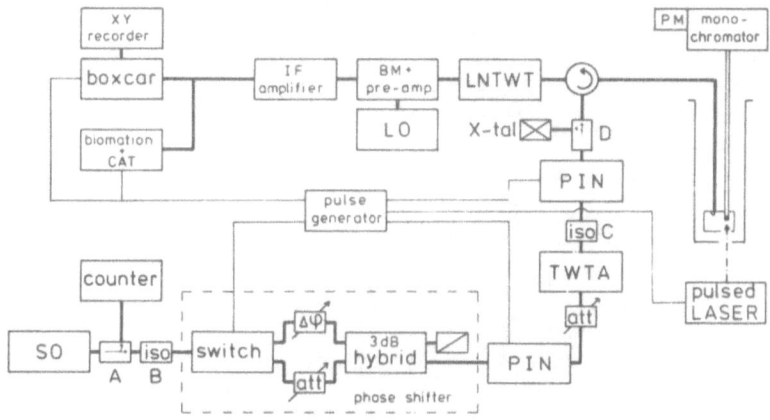

Fig. 6. The schematic diagram of the zero-field electron spin echo spectrometer.

have introduced isolators to absorb the microwaves reflected fom the switches. The microwave pulses are guided through a circulator to the cavity.

The other part of the circuit serves to detect the echoes generated in the cavity. The echoes, and also the reflected driving pulses, pass via the circulator through a low-noise traveling wave tube amplifier (LNTWTA) which acts as an amplifier for the echo signal but also as a limiter for the reflected driving pulses. The signals are transferred to an intermediate frequency (IF) of 30 MHz in a balanced mixer (BM) with the help of a local oscillator (LO). After amplification in the IF amplifier the signals are rectified in a video output. The echoes can either be observed on an oscilloscope or to improve the signal-to-noise ratio can be fed into a fast transient recorder connected to a signal averager. Another means of detection is the boxcar integrator which can be used in combination with a scan delay generator and XY-recorder. The apparatus has a threshold sensitivity of about 10^{-12} W with a bandwidth of 8 MHz.

The crystals are placed in a tunable re-entrant cavity of the type described by Erickson [47]. In Fig. 7 we show such a cavity together with the approximate electric and magnetic field configuration. By changing the position of the plunger the resonance frequency can be varied over one octave. The quality factor Q of the cavity is about 500 depending somewhat on the resonance frequency. With a microwave power of 10 W incident on this cavity the maximum amplitude of the microwave magnetic field B_1 is equivalent to a precession frequency $\gamma_e B_1$ of about 3 MHz. The recovery time of the receiver after a 10 W pulse is about 800 ns enabling the observation of a two-pulse echo 1.6 μs after the beginning of the first microwave pulse. The sample is mounted at the end of a quartz light pipe in a small teflon container with a quartz bottom. It is positioned in the upper part of the cavity in a region where the microwave magnetic field has its maximum value and it is irradiated through a hole in the bottom. The light pipe can be turned around its axis and moved up and down slightly so that the orientation of the sample with respect to the direction of $\mathbf{B}_1$ can be varied.

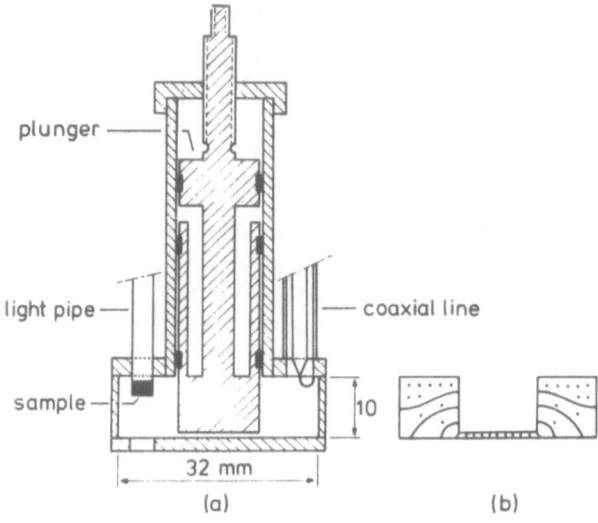

Fig. 7. (a) The tunable microwave cavity. (b) A schematic drawing of the electric and magnetic field configurations. The electric fields are indicated by lines and the circumferential magnetic field lines by dots.

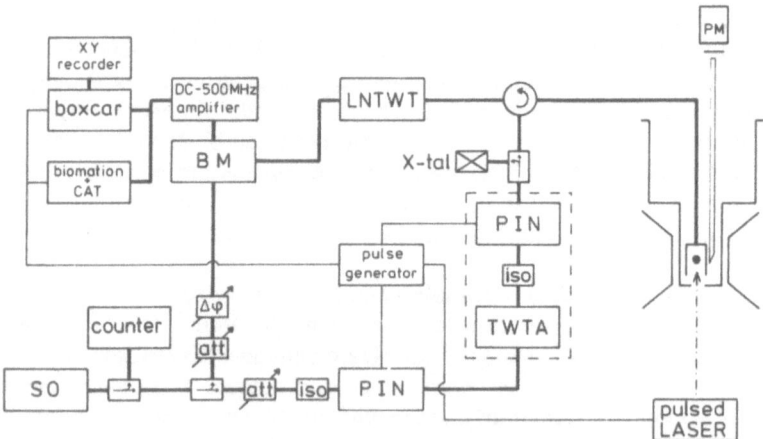

Fig. 8. The schematic diagram of the X-band electron spin echo spectrometer.

In Fig. 8 a schematic diagram is shown of the X-band ESE spectrometer. Apart from the lack of a microwave phase switching assembly the main difference with the zero-field set-up is that it is capable of homodyne or phase sensitive detection. In the homodyne scheme a small fraction of the CW microwaves generated by the SO is guided via an adjustable phase shifter into the LO input of a double balanced mixer. The LO and SO signals are mixed to a DC signal with a bandwidth of 500 MHz. As a result of this mixing scheme there is a well-defined phase relation between the LO and SO inputs of the BM and it is possible to obtain information from the phase of the echo signal (microwave absorption or emission). For instance from a comparison with the free radical signal which occurs around $g = 2$ and always has an absorptive phase one can ascertain whether the triplet echo signal corresponds with absorption or emission.

The cavity is a slotted tube resonator built according to a design by Mehring *et al.* [48] and is shown in Fig. 9. Its great advantages are its open structure allowing easy optical access and its relatively high filling factor. The quality factor Q is about 1000 and the recovery time about 400 ns compared with 800 ns for the zero-field spectrometer. To obtain even more intense microwave pulses the TWTA, PIN switch combination can be replaced by a Litton pulsed amplifier capable of delivering 1 kW pulses with a minimum width of 100 ns. This increases the microwave B_1 field inside the cavity to an equivalent precession frequency $\gamma_e B_1 \approx 8$ MHz.

The sample tube is mounted in a gear arrangement on the body of the cavity allowing for rotation of the sample about a horizontal axis. The static magnetic field can be rotated about a vertical axis. In this way any of the crystal axes can be aligned parallel to the magnetic field. The reproducibility of this system is 0.2°. Magnetic field values are measured by means of proton resonance Gauss meter.

Most experiments to be described in this chapter were performed in the liquid ^{4}He temperature range between 4.2 K and 1.2 K in conventional bath cryostats; a four window cryostat for the experiments in zero-magnetic field and a cryostat with a window in the bottom for the experiments in the presence of a magnetic field. However in one of the zero-field experiments we used a ^{3}He cryostat inserted in the ^{4}He dewar system to reach temperatures lower than 1.2 K. The attraction of liquid ^{3}He is that its vapour pressure at 1.2 K is about 30 times higher than that of liquid ^{4}He. Consequently it is relatively easy to obtain temperatures as low as 0.3 K by pumping on a bath of liquid ^{3}He. For a detailed description of this equipment and its operation we refer to [49].

For excitation we have used an excimer pumped dye laser system which delivers light pulses with a duration of 15 ns and with high energy in the wavelength region of the $T_0 \leftarrow S_0$ absorption of TCB and p-DCB. The pump laser (Lambda Physik EMG 102) uses a XeCl excimer as the laser medium and produces pulses with an energy of 150 mJ per pulse at a wavelength of 308 nm. In the second stage a dye laser (Lambda Physik FL 2000) converts the output of the excimer laser into a longer wavelength. At 375 nm (the wavelength of the $T_0 \leftarrow S_0$ 0−0 absorption in TCB) pulses with an energy of 5−10 mJ and a spectral width of 0.7 cm^{-1} are obtained. An intra-cavity etalon can be introduced in the dye laser to reduce the bandwidth to ≈ 0.2 cm^{-1}.

Some of the experiments have been performed by exciting the sample with a mercury arc (Osram HBO 200 W). The light beam is then filtered by a water solution of Ni- and Co-sulphate and a UG-5 glass filter. The phosphorescence or fluorescence of the sample is either passed through a suitable combination of high- and low-pass filters or through a monochromator. In zero field we used a Spex 0.5 m or a Spex 0.85 m double monochromator. In the X-band experiments the

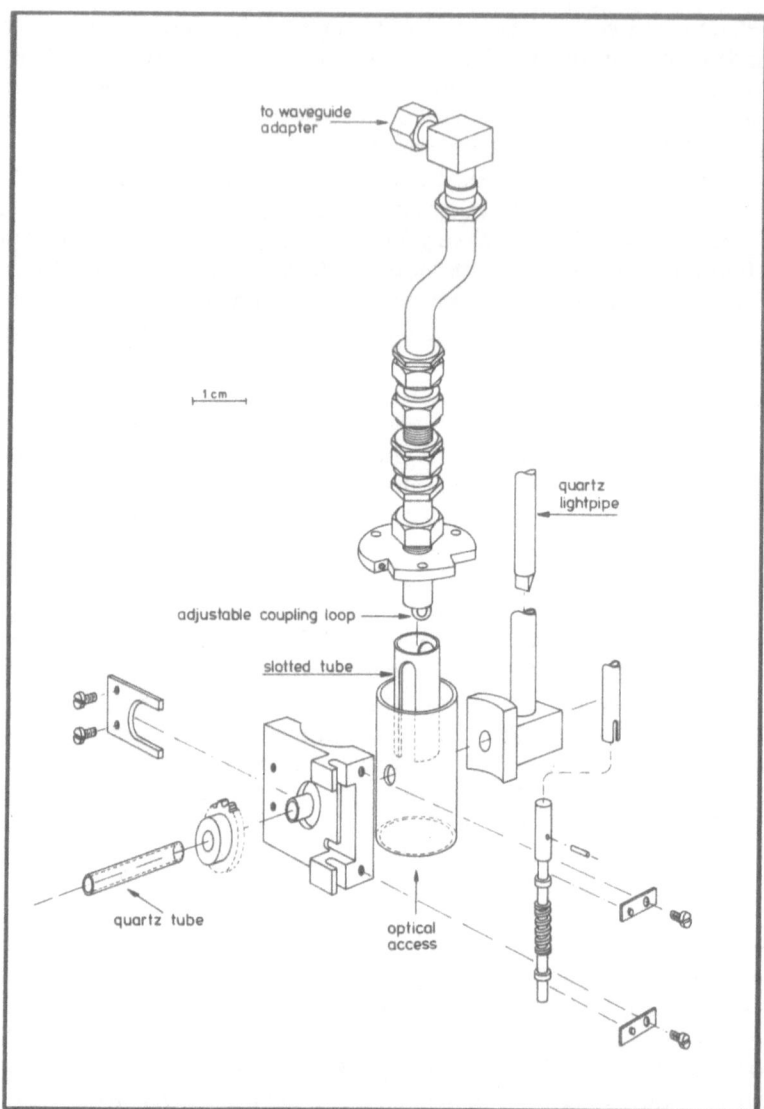

Fig. 9. An exploded view of the slotted tube resonator.

phosphorescence was observed via a Spex 0.25 m double monochromator. As photomultipliers we used EMI 9524 or EMI 9658 tubes.

The crystals of TCB were prepared from two different starting materials. For the first we used TCB from Fluka. This material was zone refined for 140 passes and TCB crystals were grown from the melt by the Bridgman technique. From an analysis via a gas chromatograph coupled to a mass spectrometer it was established that the dominant chemical impurity in the sample was one-bromotrichlorobenzene. Its concentration was about 0.1% and could not be lowered by further zone refining. The only other detectable impurity was 1,2,3,5-tetrachlorobenzene with a concentration of less then 10 ppm. In a later stage we were able to obtain crystals with a concentration of chemical impurities below 20 ppm by using as starting material TCB obtained from Aldrich. These crystals after growing from the melt were annealed to reduce structural imperfections. In this process the crystal was first cooled very slowly to room temperature and then reheated and kept at a temperature 2° below the melting point for 60 hours.

p-DCB purchased from BDH was zone refined for 100 passes. The resulting product was analyzed by gas chromatography and found to contain less than 1 ppm of p-DCB isomers or other impurities. Neat crystals of p-DCB were also grown from the melt by the Bridgman technique and subsequently annealed to reduce structural imperfections.

6. The Dynamic Properties of Triplet Excitons in TCB

6.1. THE SCATTERING PROCESS AMONG THE EXCITON k STATES

The first time-resolved electron spin echo (ESE) experiments on triplet exitons in TCB were performed in 1978 by Botter et al. [39]. They observed these signals on the $T_y - T_x$ zero-field transition upon excitation of the sample with the 337 nm line of a pulsed (10 ns) N_2 laser directly in the singlet-triplet absorption about 3000 cm^{-1} above the $T_0 \leftarrow S_0$ 0—0 band. The echo signals were generated by two microwave pulses resonant with the $T_y - T_x$ zero-field transition with an interval τ of about 1 μs. The delay between the laser flash and the first microwave pulse had to be set at 10 μs because the electromagnetic interferences of the spark-gap triggered N_2 laser prevented the use of shorter delays. In Fig. 10 we show a typical echo signal and in Fig. 11 the $T_y - T_x$ zero-field lineshape obtained from a measurement of the echo height as a function of the microwave frequency at 1.13 K. This lineshape can be fitted nicely to the theoretical curve derived with the aid of expression (14) for the microwave dispersion of the exciton k states and assuming a Boltzmann population distribution over the k states.

The nice fit of the experimental and theoretical curve in Fig. 11 indicates that any overpopulation in the $k \approx 0$ region of the exciton band that might have been established by the optical excitation has disappeared within 10 μs by vibrational relaxation and/or $k - k'$ scattering. It is important to note however that the ESE signals correspond with a strong overpopulation of the T_y spin level as compared with T_z and T_x. This spin alignment is of great practical importance because it leads to very large ESE signals after laser excitation. It can be understood by inspecting the SOC routes in TCB. As illustrated in Fig. 1 the T_y and T_z spin levels of TCB are contaminated with singlet character but only the transition $T_y \leftarrow S_0$ is symmetry allowed. The excess population of T_y with respect to T_z and T_x

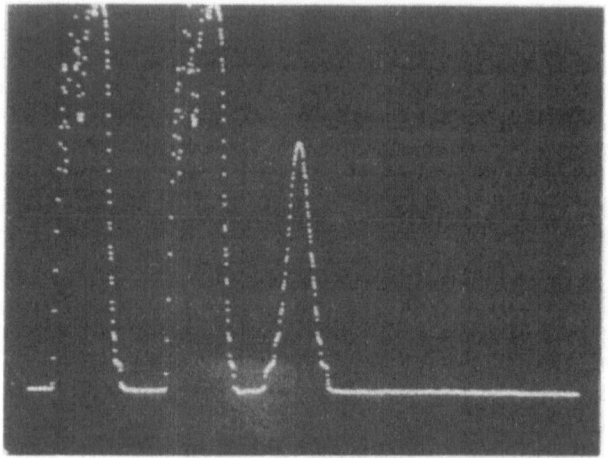

Fig. 10. Two-pulse echoes of the triplet excitons in TCB obtained with two microwave pulses with an equal duration of 300 ns and a power of 1 W. The delay after the laser flash is 10 μs and the interval τ between the pulses is 2 μs. $T = 1.2$ K. Horizontal 1 μs per division. The signal was averaged 64 times.

disappears in about 100—150 μs by spin-lattice relaxation (SLR) processes (see Section 6.4).

The phase memory time T_M of the triplet spins is obtained from a measurement of the two pulse echo signal by varying the interval τ between the microwave pulses. Reliable values could only be obtained for the $k \approx \pm \pi/a$ excitons which give relatively strong ESE signals. Between 1.13 K and 1.8 K T_M first remains constant but then shortens upon further increase of the temperature. The variation of T_M^{-1} with the temperature is shown in Fig. 12 and can be described by,

$$T_M^{-1} = T_M^{-1}(0) + T_M^{-1}(T) \tag{16}$$

In addition to the ESE experiments on the triplet excitons measurements were also performed on the dominant X-traps which lie about 17 cm^{-1} below the exciton band and which correspond with TCB molecules present as structural defects in the crystal lattice (see Section 6.3). The temperature dependence of T_M^{-1} of these traps is also shown in Fig. 12 and can be described by the same expression (16).

From the two-pulse echo experiments we derive a homogeneous linewidth of the triplet exciton spin packets $(\pi T_M)^{-1}$ of the order of a few hundred kHz i.e. only a small fraction of the total inhomogeneous linewidth of 10 MHz. Above 1.6 K the spin packet linewidth increases presumably as a result of increased scattering among the exciton k states. Since different k states have slightly different zero-field frequencies this scattering results in instabilities of the resonance frequency and thus to a shortening of the phase memory time.

It was argued by Botter et al. [39] that impurities play an important part in this scattering process because it appeared that T_M depended on the choice and

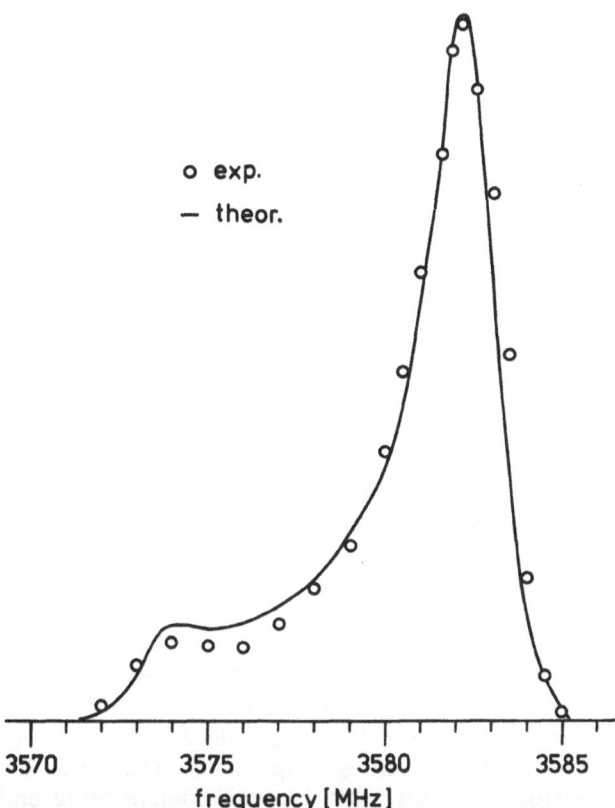

Fig. 11. The two-pulse echo detected lineshape of the triplet exciton $T_y - T_x$ zero-field transition upon excitation with the pulsed N_2 laser at 337 nm, 3000 cm^{-1} above the origin of the $T_0 \leftarrow S_0$ absorption. The open circles indicate the intensity of the ESE signals. The laser flash has a duration of 10 ns. The delay between the laser flash and the first microwave pulse is 10 μs. The time interval between the microwave pulses is 1 μs. $T = 1.13$ K. The drawn line indicates the theoretical lineshape assuming a Boltzmann distribution over the k states.

"history" of the sample. For instance the temperature independent part $T_M(0)$ of the excitons could be shortened from almost 3 to 1.8 μs by allowing the sample to warm to room temperature and cooling it again. Since TCB exhibits a phase transition from a high temperature monoclinic to a low temperature triclinic structure at 188 K [50, 51] it was thought that in this way additional structural defects were introduced in the sample.

The strongest argument that impurities are involved in the scattering of the exciton k states is provided by the observation that the temperature dependence of T_M^{-1} of exciton and X-trap behave in about the same way. Above about 1.6 K T_M of exciton and X-trap starts to shorten and although this variation with the temperature depends on the crystal quality, exciton and trap always behave similarly.

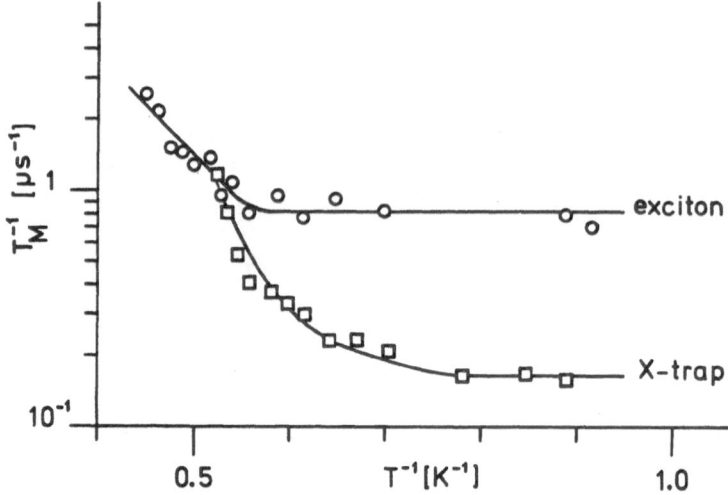

Fig. 12. The decay rate T_M^{-1} for the $T_y - T_x$ zero-field transition of the excitons and X-traps as a function of the inverse temperature. The measurements on the excitons were performed at 3582.5 MHz, i.e. the resonance frequency of the $k = \pm \pi/a$ excitons.

The dead time of the ESE spectrometer did not allow observations of echo signals above 2.2 K but it is likely that the exciton scattering continues to increase above this temperature until finally the scattering rate will be in the order of magnitude of the total linewidth of 10 MHz. Indeed it was observed that above 5 K the ODMR linewidth shows a strong temperature dependence. The two peaks in the line move towards each other with increasing temperature until at 8 K they coincide at the center of the line (see Fig. 13). Beyond this temperature the linewidth starts to broaden again and above 15 K the ODMR signal can no longer be observed. In Fig. 14 we have plotted the ODMR linewidth and the homogeneous linewidth $(\pi T_M)^{-1}$ of the spin packets versus the inverse of the temperature.

It was proposed by Botter et al. that the narrowing of the ODMR line is an example of motional narrowing caused by scattering of the exciton k states at a rate faster than the zero-field linewidth of about 10 MHz. At 5 K the situation of motional narrowing is reached and the average scattering rate must be in the order of 10^8 s^{-1} [52]. At higher temperatures the scattering takes place so rapidly that finally at 8 K we observe the motionally narrowed line exactly at the position of the $k = \pm \pi/2a$ excitons (see Fig. 13).

The broadening of the ODMR line above 8 K was explained by assuming that in this region the average scattering rate approaches the magnitude of the intermolecular interaction β. We rather think that this rate approaches the resonance frequency of the zero-field transition of 3.5 GHz. This would mean that at 8 K the scattering rate is of the order of 10^{10} s^{-1}.

The results obtained in these experiments left little doubt that the shortening of T_M and the narrowing of the resonance line is caused by increased scattering of the packets of exciton k states with the temperature and that impurities play a role in the scattering mechanism. Several questions however remained unanswered. For instance what is causing the temperature independent part of T_M^{-1} at low temperatures and how is the homogeneous linewidth $(\pi T_M)^{-1}$ related to the width Δk of the exciton wave packet? Further can anything be said about the scattering probability function? The latter question was studied by Botter *et al.* via measurements of the decay of the stimulated echo signals. It was concluded that at low temperatures (1.2 K) there is a preference for scattering to nearby k states. We shall not discuss these experiments any further because the results obtained by van Strien *et al.* presented hereafter allow us to study the scattering among the k states in great detail. They moreover show that the conclusions derived from the stimulated echo experiments are not correct.

A remarkable result of the above experiment is that no indication is found of a preferential population of the $k \approx 0$ excitons following the laser excitation. In the next experiment performed by van Strien *et al.* [40] it was shown that the optical selection rule does apply. In this experiment the exciton system is now excited

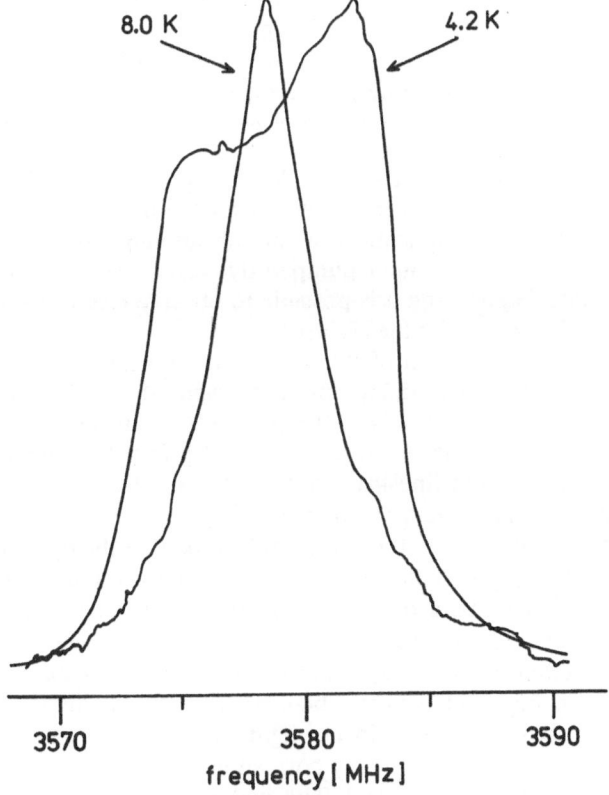

Fig. 13. The optically detected lineshape of the triplet exciton $T_y - T_x$ zero-field transition at 4.2 K and at 8 K.

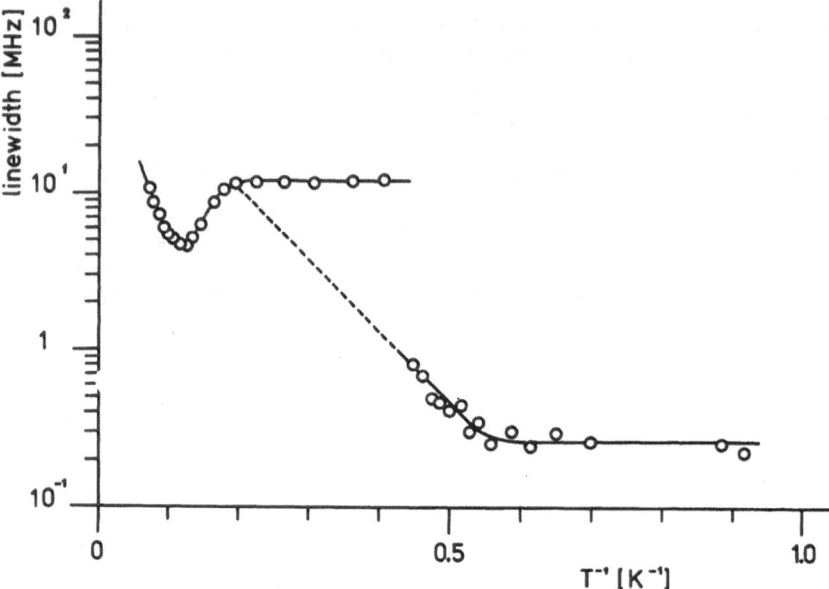

Fig. 14. The ODMR linewidth (upper curve) and $(\pi T_M)^{-1}$ (lower curve) for the triplet exciton $T_y -$ T_x zero-field transition as a function of the inverse of the temperature.

selectively in the 0—0 band of the $T_0 \leftarrow S_0$ absorption by means of an excimer pumped dye laser tuned to 374.8 nm. The advantage of this excitation is twofold. First we circumvent the complication of the vibrational relaxation following the laser excitation. Second the excimer pumped dye laser system produces very little electromagnetic interference and it is possible to observe electron spin echo signals at much shorter delays than with the N_2 laser.

In Fig. 15 the result is shown of this experiment on the $T_y - T_x$ exciton zero-field transition at 1.15 K. The delay time t_d between the exciting laser pulse and the first microwave pulse now only is 0.5 μs. Care has been taken to reduce the microwave power as much as possible to prevent power broadening effects. It is seen that this ESE detected lineshape differs strongly from the one obtained by Botter *et al.* (cf. Fig. 11). A very strong signal is present around 3572.5 MHz the resonance frequency of the $k \approx 0$ excitons indicating that the optical selection rule indeed leads to an overpopulation of the $k \approx 0$ excitons. On the other hand we also observe a weaker signal around 3582.5 MHz the resonance frequency of the $k \approx \pm \pi/a$ excitons. We shall see hereafter that the corresponding population of the $k \approx \pm \pi/a$ excitons is caused by scattering processes in the exciton band.

The lineshape in Fig. 15 has been obtained with an exciting laser beam with a spectral width of about 0.2 cm^{-1}. In the figure we have indicated by the broken line the spread in resonance frequency corresponding to a spread of 0.2 cm^{-1} on the optical scale of the exciton band. The width of 1.1 MHz (full width at half maximum) has been derived with the help of the dispersion relations (12) and (14) for the exciton band and the zero-field frequencies. This result suggests that the

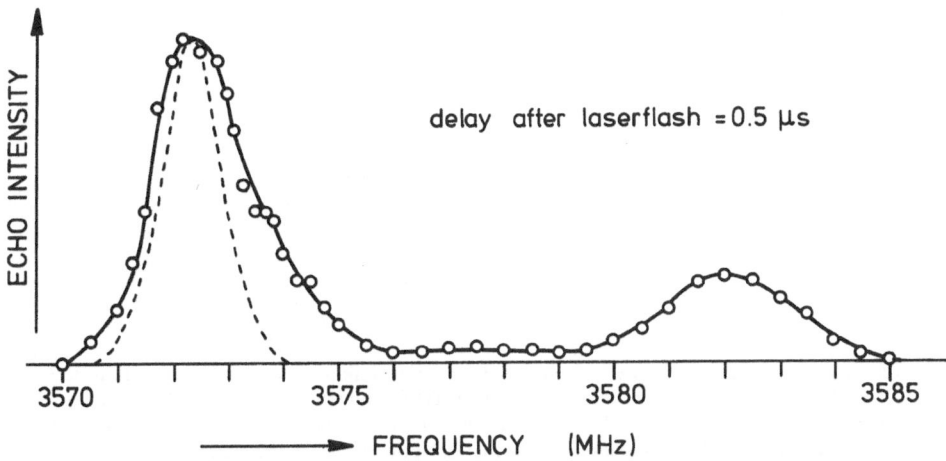

Fig. 15. The ESE detected lineshape of the $T_y - T_x$ triplet exciton zero-field transition in TCB following laser flash excitation in the $T_0 \leftarrow S_0$ absorption origin at 374.8 nm. The open circles indicate the echo intensity. The laser bandwidth is 0.2 cm^{-1}, the pulse duration is 15 ns and the pulse energy 3 mJ. The microwave pulses have a peak power of 0.25 W and the pulse length is 0.5 μs. The dashed line indicates the spread in microwave frequency corresponding to the bandwidth of 0.2 cm^{-1} of the laser. Since the delay time is not infinitely short one observes already the effect of scattering in the exciton band via the presence of a signal corresponding to the $k \approx \pm \pi/a$ excitons.

spread in k states around $k \approx 0$ is determined by the bandwidth of the laser. Indeed when increasing this bandwidth to about 0.7 cm^{-1} the ESE detected lineshape again at $t_d = 0.5$ μs exhibits a width corresponding almost exactly to a spread of 0.7 cm^{-1} on the optical scale of the exciton band. We think that this broadening effect and apparent relaxation of the optical selection rule is related to structural defects in the crystal which cause a spread in the optical energies of the triplet excitons. This conclusion was supported by later experiments on crystals that were subjected to an annealing treatment and where it appeared that this broadening effect was no longer present. We shall come back to this problem further on in this section.

In Fig. 16 we show in a three-dimensional representation the evolution of the non-Boltzmann population distribution created with the laser flash. The curves represent the intensity of the echo signal at a given resonance frequency as a function of the delay t_d between the laser flash and the first microwave pulse with a fixed interval $\tau = 1.5$ μs between the microwave pulses. It is seen that the overpopulation of the $k \approx 0$ excitons has reached its thermal equilibrium value in a few μs and that the $\pm \pi/a$ excitons grow in roughly in the same time. The curve also shows why in Fig. 15 we observe a signal at 3582.5 MHz corresponding to the $\pm \pi/a$ excitons. Even at this short delay scattering has already established an appreciable population of the $k \approx \pm \pi/a$ excitons.

When studying the $T_z - T_y$ transition a three-dimensional picture similar to Fig. 16 is obtained. As explained in Section 4 the T_y sublevel is the only one accessible

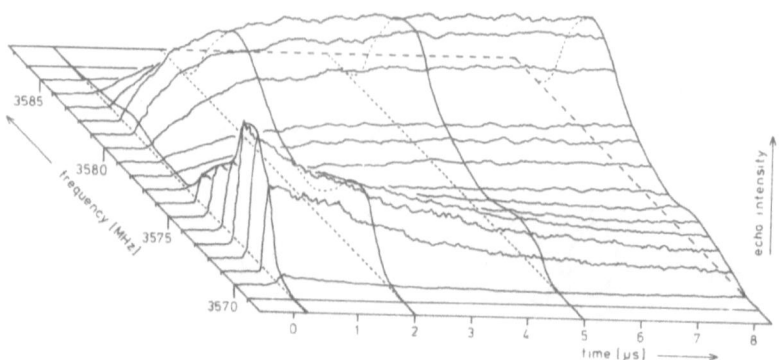

Fig. 16. A three dimensional representation of the time evolution of the ESE detected linshape of the $T_y - T_x$ triplet exciton zero-field transition in TCB following laser flash excitation into the absorption origin at 374.8 nm. The laser bandwidth is 0.2 cm^{-1}, the pulse duration is 15 ns and the pulse energy is 3 mJ. The microwave pulses have a peak power of 1 W and a duration of 250 ns. The time is the time interval t_d between the laser flash and the first microwave pulse. The interval between the microwave pulses is 1.5 μs.

from the S_0 ground state and strong ESE signals can be observed for this transition. The echo intensity at 1957.5 MHz corresponding to the $k \approx 0$ excitons also shows a strong overpopulation and reaches its equilibrium value in about 8—10 μs. The signal at 1963 MHz corresponding to the $k \approx \pm \pi/a$ excitons simultaneously grows in with about the same time constant.

Before discussing the results of Fig. 16 in more detail it is good to make already a short comment. It will be clear that the observed scattering pattern can not be explained by the theory of exciton-phonon coupling where it is assumed that at low temperatures one phonon processes occur and that quasi-momentum and energy are conserved simultaneously [4—7]. Since the energy dispersion of acoustical phonons is so much larger than that of the triplet excitons one would expect that scattering then can only occur in a very limited region of k space around $k \approx 0$. This is in disagreement with the observed scattering pattern which suggests, at first glance, an unrestricted scattering probability in k space.

To obtain more information on the scattering of the excitons in TCB we next measured the temperature dependence of the decay rate $T_s^{-1}(k \approx 0)$ of the initial excess population of the $k \approx 0$ excitons following the laser flash. This rate represents the total scattering probability from $k \approx 0$ to all the other exciton k states. In Fig. 17 we show the results from 1.15 K to 2.5 K. At higher temperatures the phase memory time T_M becomes so short that the echo signals can no longer be observed. It is seen that the scattering rate first increases slowly with increasing temperature but that beyond 1.9 K it increases much more rapidly. It is interesting to note that this variation looks very similar to the temperature dependence of T_M^{-1} observed by Botter et al. [39] (see Fig. 12).

The scattering rates from other k states of the exciton band cannot be found in this way. However with the help of spin-locking experiments we can obtain in principle the scattering rates out of any k state in the band by tuning the frequency

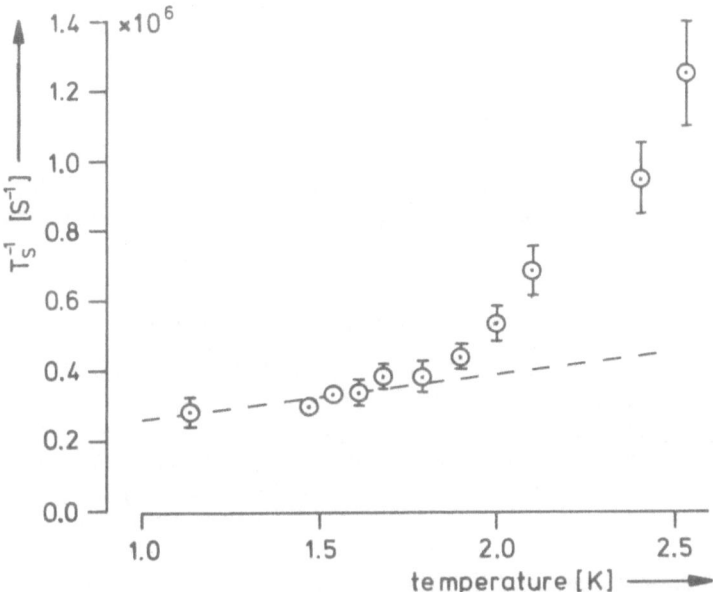

Fig. 17. The experimentally determined temperature dependence of the scattering rate $T_s^{-1}(k \approx 0)$, as obtained from the decay of the excess population of the $k \approx 0$ excitons at various temperatures. The dashed line indicates the calculated temperature dependence of $T_s^{-1}(k \approx 0)$ using the scattering probability given in expression (19) and using the model described in the text.

of the ESE spectrometer to the corresponding resonance frequency in the zero-field transition. In such an experiment first a $\pi/2$-pulse is given followed immediately by a long pulse shifted $\pi/2$ in phase. This phase shifted microwave field "locks" the spins in a frequency interval of about $2\gamma_e B_1$ MHz around the applied microwave frequency. By keeping $\gamma_e B_1$ smaller than the zero-field linewidth of about 10 MHz we can measure the probability for scattering away from a well-defined region of the band via the decay of the spin-locked signal as a function of the lock time. This decay time is usually indicated by $T_{1\rho}$. By tuning the spectrometer to 3572.5 MHz (the frequency in the $T_y - T_x$ transition corresponding to the $k \approx 0$ excitons) we find at 1.15 K $T_{1\rho}(k \approx 0) = (2.2 \pm 0.2)$ μs. At 3582.5 MHz we find $T_{1r}(k \approx \pm \pi/a) = (10 \pm 1)$ μs.

The fact that $T_{1\rho}(k \approx 0)$ is almost equal to the scattering time $T_s(k \approx 0) = (2.7 \pm 0.2)$ μs indicates that the spin-locking experiment indeed measures the scattering rate from $k \approx 0$ to the other k states. This supports the idea that $T_{1\rho}(k \approx \pm \pi/a)$ is a good measure of the scattering time of the $k \approx \pm \pi/a$ excitons to the other k states. In particular the ratio $T_{1\rho}(k \approx \pm \pi/a)/T_{1\rho}(k \approx 0)$ = 4.6 is close to the Boltzmann factor $\exp(4\beta/k_B T) = 5.5$ ($4\beta = 1.35$ cm^{-1} and T = 1.15 K). This finding shows that the principle of detailed balance applies to the scattering process in the band.

The effect of the scattering of the triplet excitons is also observable in the microwave induced changes in the phosphorescence emission. For instance it appears that a microwave pulse resonant with the zero-field frequency of the $k \approx 0$ excitons in the $T_y - T_x$ transition leads to an immediate change in the intensity of the 0—0 emission band whereas a pulse resonant with the $k \approx \pm \pi/a$ excitons results in the build-up of the optical signal with a time constant of about $10 \ \mu s$. This result can be explained by assuming that the optical selection rule applies for the 0—0 emission band and that therefore saturation of the zero-field transition of the $k \approx \pm \pi/a$ excitons can be observed only after scattering towards $k \approx 0$ [40].

6.2. THE THEORETICAL MODEL FOR THE EXCITON SCATTERING

The triplet exciton bandwidth in TCB is considerably smaller than the energy dispersion of the acoustic phonons. Thus in the limit of pure exciton systems one would expect that one-phonon scattering processes (in which the change of the exciton state is accompanied by the creation or the annihilation of a phonon with simultaneous conservation of energy and quasi-momentum) are only possible around $k \approx 0$. In contrast the experimental observations clearly indicate that such a restriction does not exist in the scattering process.

In an attempt to account for the observed scattering pattern a model was developed by van Strien et al. [53] in which it was assumed that one-phonon processes are induced by the presence of impurities or structural dislocations causing local perturbations of the exciton-phonon coupling. It appeared that this model could give a satisfactory account of the observed scattering pattern displayed in Fig. 16 and the temperature dependence of the scattering rate shown in Fig. 17 at low temperatures i.e. where $k_B T$ is small compared with the exciton bandwidth.

In an effort to obtain more insight into the problem of impurity induced exciton-phonon scattering Benk and Silbey [54] made a detailed theoretical study of the effect of several types of impurities present in a linear chain of molecules. First they considered the model treated by van Strien et al. [53], which they called the interstitial model and in which it was assumed that an impurity molecule on a neighbouring stack interacts with a molecule on the stack under consideration. This breaks the translational invariance of the exciton-phonon interaction and allows one-phonon processes to occur. Since the translational invariance of the excitation energy and the exchange interaction is conserved the exciton states are the unperturbed states as given in expression (5). In the second model a finite chain was considered and the impurities are thought to be disruptive preventing the excitation of getting past. In this case the exciton states are,

$$\psi(k) = \{2/(N+1)\}^{1/2} \sum_n \psi_n \sin kn, \tag{17}$$

with

$$k = \pi v/(N+1); v = 1, 2, 3, \ldots, N \tag{18}$$

In the third case, which was named the substitutional model, an infinite chain was considered incorporating one molecule with a slightly lower energy. Here we no longer deal with unperturbed exciton states but instead have perturbed band excitons (which for the sake of simplicity we will still indicate by the symbol k) and quasi-localized trap states (x). A consequence of this third model is that trapping can be viewed simply as a form of scattering between two eigenstates of a single Hamiltonian describing band and trap states together. An important aspect of the latter two models is that the impurities correspond with disruptions of the translational symmetry of the excitation energy while the exciton-phonon coupling retains it translational symmetry.

The expression derived by Benk and Silbey for the scattering probability A_{kk}, in the case of exciton-phonon coupling due to a non-translational exciton-phonon interaction (the interstitial model) reads as follows,

$$A_{kk'} = \frac{1}{\pi} \, C_{ii}^2 \left(\frac{2\beta}{c_\perp} \right)^2 \left(\frac{2\beta}{c_\parallel} \right) (|\cos k - \cos k'|)^3 \tag{19}$$

$$\times \, [\tilde{\chi}^{ac} + 2 \cos k' \, \tilde{F}_1^{ac}]^2 \, f_T(\Delta E_{kk'}).$$

Here C_{ii} is the concentration of interstitial impurities that break the translational invariance of the exciton-phonon interaction. The parameters $C_\parallel$ and $C_\perp$ determine the dispersion relation of the acoustic phonons parallel and perpendicular to the exciton chain,

$$\omega(q) = \{ C_\parallel^2 \, q_\parallel^2 + C_\perp \, q_\perp^2 \}^{1/2}. \tag{20}$$

Where $\omega(q)$ indicates the frequency of a phonon with wave vector q. Further the parameters $\tilde{\chi}^{ac}$ and $\tilde{F}_{ac}^1$ determine the q dependence of the local and non-local exciton-(acoustic) phonon coupling. Finally the factor $f_T(\Delta E_{kk'})$ is equal to

$$f_T(\Delta E_{kk'}) = \exp\{\Delta E_{kk'}/k_B T\}/[\exp\{\Delta E_{kk'}/k_B T\} - 1] \tag{21}$$

when $\Delta E_{kk'} > 0$ and

$$f_T(\Delta E_{kk'}) = [\exp\{\Delta E_{kk'}/k_B T\} - 1]^{-1} \tag{22}$$

when $\Delta E_{kk'} < 0$.

For the case of a finite chain we find for the scattering probability the expression

$$A_{kk'} = \frac{2}{3\pi} \, C_{di}^2 \left(\frac{2\beta}{C_\perp} \right)^2 \left(\frac{2\beta}{C_\parallel} \right)^3 \sin^2 k \sin^2 k'$$

$$\times \, |\cos k - \cos k'| \, (\chi^{ac} + 2 \cos k \, F_1^{ac})^2 \, f_T(\Delta E_{kk'}) \tag{23}$$

where C_{di} is the average concentration of impurities which limit our molecular chains.

Finally in the third case a linear chain of 2N molecules is considered with a substitutional impurity at one site. The site energy of the guest is lower than that of the host molecules by the amount ΔU_0 and the nearest neighbour interaction between guest and host is β' rather than β. For the case of purely diagonal disorder i.e. $(\Delta U_0)/2\beta \gg 1$ and $\beta'/\beta = 1$ it is found that,

$$A_{kk'} = C_{ti}^2(\chi^2/5\pi)\left(\frac{2\beta}{C_\perp}\right)^2\left(\frac{2\beta}{C_\parallel}\right)^5\left(\frac{2\beta}{\Delta U_0}\right)^2 \sin^2 k \sin^2 k'$$

$$\times \left(|\cos k - \cos k'|\right)^3 (\cos k + \cos k')^2 f_T(\Delta E_{kk'}). \tag{24}$$

Where C_{ti} is the average concentration of trapping impurities.

When comparing the expressions (19), (23) and (24) it is seen that the scattering rate for the case of an interstitial impurity breaking the translational symmetry of the exciton-phonon interaction is considerably faster than those for the other two cases. The ratios of the scattering rates being given by

$$A_{kk'}(\text{interstitial})/A_{kk'}(\text{finite chain}) = (C_\parallel/2\beta)^2 \gg 1 \tag{25}$$

and

$$A_{kk'}(\text{interstitial})/A_{kk'}(\text{substitutional}) = (C_\parallel/2\beta)^4 \gg 1. \tag{26}$$

To test whether we can simulate the observed scattering pattern presented in Fig. 16 with the expressions (19), (23) or (24) for the scattering rates we have performed computer calculations of the evolution of the population distribution over the exciton k states for the three different cases. In this numerical calculation we consider a finite chain and it is convenient to adopt the notation for k given in (18). The dispersion relation for the energy E_k of exciton state k in a chain of N molecules then is $E_{(k)} = E_0 + 2\beta \cos k$ with k defined according to equation (18). The optically accessible state $k = \pi/(N+1)$ ($v = 1$) is at the top of the band.

The Pauli master equation [55] governing the population $P_k(t)$ of exciton state k is given by,

$$\frac{dP_k(t)}{dt} = \sum_{k'} P_{k'}(t) A_{kk'} - P_k(t) \sum_{k'} A_{k'k} \tag{27}$$

where $A_{kk'}$ is the probability for scattering from k' to k. We assume that the Markovian approximation is valid, i.e. each scattering event is independent of the preceding ones, which implies that the $A_{kk'}$ are time independent [56]. The solution of (27) is found with the help of laplacian transformation techniques,

$$P_k(t) = P_k(0)e^{-A_k t} + \sum_{k'} A_{kk'} \int_0^t e^{A_k(t'-t)} P_{k'}(t') \, dt', \tag{28}$$

where $A_k = \sum_{k'} A_{k'k}$ is the total scattering rate from state k to all the other k' states. Since the convolution integral in (28) cannot be solved analytically, we have used a numerical method to calculate $P_k(t)$. To this end we have developed an iterative computer program using the linear approximation of (28)

$$P_k(t + \delta t) = P_k(t)\, e^{-A_k \delta t} + \frac{1}{A_k}(1 - e^{-A_k \delta t}) \sum_{k'} A_{kk'} P_{k'}(t). \tag{29}$$

This approximation is valid only if δt is so small that $P_{k'}(t')$ can be considered a constant and consequently taken out of the integral in (28). To solve the $P_k(t)$ we have taken for $A_{kk'}$ the expression (19), (23) and (24) for the three cases considered by Benk and Silbey.

In Fig. 18 we show the scattering patterns calculated according to the three microscopic scattering probabilities. To relate a given exciton state k to a particular zero-field frequency we have used the dispersion relation $v(k) = v_0 + 2\mu\cos v\pi/(N+1)$ $(v = 1, 2, 3, \ldots, N)$ with $\mu = 2.5$ MHz and $v_0 = 3577.5$ MHz. In addition we have used $N = 50$ for the number of molecules in the chain and a

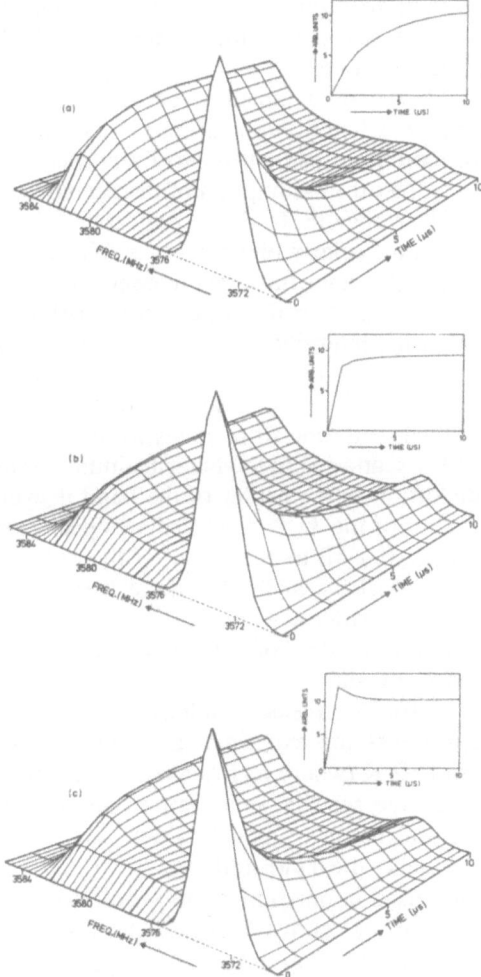

Fig. 18. The scattering patterns calculated according to the three different impurity models for the scattering probabilities. (a) Interstitial impurity, (b) Finite chain, (c) substitutional impurity. The time evolution curves of the $\pm \pi/2a$ states are shown in the insets for clarity.

gaussian inhomogeneous broadening of 1.5 MHz for the spin packet related to a particular exciton state. The value of 50 has been chosen for practical reasons. (An estimate of the average chain length is obtained from the dependence of the phase memory time on the zero-field frequency i.e. on the particular value of k (see Section 6.5).) The results of the computer simulations however turn out to be independent of the chain length since A_k becomes independent of N for $N > 15$. The inhomogeneous width of 1.5 MHz is larger than the homogeneous spin packet linewidth of about 0.1 MHz as derived from the spin echo experiments. It has been estimated from the overall zero-field lineshape and from the variation of the initial ESE detected lineshape with the bandwidth of the exciting laser. The only adjustable parameter in the fit is the constant factor in the expressions (19), (23) and (24). It is given a value such that the computed decay rate $T_S^{-1}(k \approx 0)$ is the same as the experimentally observed one.

It is seen from Fig. 18a that of the three cases the interstitial model already considered by van Strien $et\ al.$ [53] (who assumed that a local perturbation exists of the exciton-phonon coupling) can satisfactorily account for all regions of the observed pattern. By contrast the assumption of disruptive impurities confining the excitation to limited linear regions leads to noticably changes in the predicted scattering (see Fig. 18b). Particularly obvious is the failure to establish Boltzmann equilibrium within 10 μs. An important factor contributing to this behavior is a more rapid population of states near the band center, with $k = \pm \pi/2a$, acquiring within 1 μs 85 percent of the intensity it has after 10 μs. Accompanying this quick filling-in of the center is a pronounced decrease in buildup near the edge at $k \approx \pm \pi/a$. Also the decay of $k \approx 0$ is no longer exponential as a slow component resulting from backscattering from states in the center becomes apparent. Much the same is true for scattering involving substitutional impurities as depicted in Fig. 18c, but the distortion in the center of the band is even more pronounced. In this case the population of the $k \approx \pm \pi/2a$ states peaks after a short time and then decreases gradually. After 10 μs the height of the electron spin echo signal is predicted to be 80 percent of its value at 1 μs, a trend clearly opposed to that displayed in the experimental pattern.

As a further test of the validity of the interstitial model we calculated the temperature dependence of the scattering pattern to obtain the temperature dependence of $T_S^{-1}(k \approx 0)$. The result is indicated in Fig. 17 by the dashed line which very nicely reproduces the experimentally observed temperature dependence of $T_S^{-1}(k \approx 0)$ up to 2 K (where $k_B T$ is almost equal to the bandwidth 4β of the triplet exciton band). We see this as an important support for the validity of our one-phonon interstitial impurity model, which seems to apply up to a temperature where $k_B T$ is equal to the bandwidth.

The scattering models used in our simulation procedures cannot account for the temperature dependence of T_S^{-1} above $T = 2$ K. This is not very surprising since above this temperature two-phonon Raman-type processes are possible. Such processes are expected to induce a uniform scattering probability in the first Brillouin zone. Further a strong temperature dependence is predicted, usually in the order of $T^5 - T^7$. In the limited temperature range between 2 K and 2.5 K it is difficult to ascertain the exact temperature dependence. However we do have an important additional piece of information from the work of Botter et al. [39]. As shown in Fig. 13 at 5 K a motional averaging of the whole $T_y - T_x$ zero-field lineshape sets in. As explained before at this temperature the average scattering rate A must be of the order of the total linewidth, i.e. $A \approx 10^8 \, s^{-1}$. Combining this value with the estimated contribution of two-phonon processes to the scattering rate T_s^{-1} at 2.5 K we derive a temperature dependence between these two temperatures of $T^{(6.4 \pm 0.6)}$. We therefore are led to believe that Raman-type processes dominate the scattering between 2.5 K and 5 K. Whether impurities also play a role here remains an open question. In principle Raman-type processes without involving impurities can fulfill both the condition for the conservation of energy and quasi-momentum. However impurities again relax the second condition and this may facilitate this scattering process.

Another two-phonon process is the Orbach process where a third intermediate level is involved. At low temperatures any of the exciton states in principle can act as an intermediate thus facilitating the scattering to nearby k states for which the one-phonon process only gives a very low probability. Recent work of Jackson and Silbey [57] shows that the Orbach type scattering is included in the theory presented above. They also discuss other two-phonon processes in both pure and impure crystals.

To conclude this section we can say that we have found a satisfactory model for the scattering at low temperatures $(kT \leqslant 4\beta)$ by assuming one-phonon processes induced by the presence of impurities. Of the three models considered the interstitial model, in which it is assumed that an impurity breaks the translational symmetry of the exciton-phonon coupling, not only is the most efficient one but it also gives the best fit to the observed scattering pattern. It is interesting to note that in the other two models the impurities only lead to mild perturbations of the exciton states and that the preference for scattering around $k \approx 0$ as it applies for pure exciton systems is only partly relaxed.

6.3. IMPURITIES AND TRAPS AND THEIR RELATION TO THE SCATTERING MECHANISM

In the previous section we have supplied evidence that the observed scattering in the triplet exciton band of TCB is caused by the presence of impurities. The question then arises what kind of impurities are involved in the scattering process. We do not have a complete answer to this problem but from the observations of Botter et al. we have an indication that the X-traps which lie about 18 cm^{-1} below the triplet exciton band and which correspond with slightly distorted TCB molecules (see Section 6.4) might be involved. They found that the phase memory time T_M of the X-traps varies with the temperature in a way almost similar to that of the triplet excitons (see Fig. 12). This effect can not be explained by trapping or detrapping mechanisms because the related rate constants are in the order of $10^3 \, s^{-1}$ [29, 30], i.e. much slower than the observed values of T_M. Moreover phonon assisted excitation exchange between trap and band can be excluded. A possible explanation is that in the exciton scattering process this X-trap is excited by a (pseudo)-localized phonon resulting in a modulation of its zero-field frequency. This modulation causes a shortening of T_M and it is expected that this effect becomes more pronounced if the scattering rate increases with the temperature.

In an effort to obtain more information on the nature of the impurities responsible for the scattering in TCB a number of experiments were performed by van Kooten et al. [58]. First the effects of chemical impurities on scattering were evaluated through the use of samples of very high and very low purity. Second the nature of various crystal imperfections and their influence on scattering was tested in a comparison of results from carefully grown, annealed crystals and crystals subjected to severe thermal stress. Further, trapping dynamics below 1 K were investigated to test for possible k-dependence in a temperature regime where the equilibrium band populations could be controlled easily.

By a careful analysis of the chemical composition of the samples used by van Strien et al. [53] to study the scattering in TCB, it was found that the crystals contained one principal chemical impurity, 1-bromo,2,4,5-trichlorobenzene, at a level of 0.1 per cent. This chemical impurity is responsible for the presence of a so-called Y-trap 50 cm^{-1} below the origin of the exciton emission. Reduction of the concentration of the bromo impurity to 20 ppm is accompanied by an almost complete disappearance of emission from the Y-trap at 4.2 K and a 15 fold increase in the ratio of exciton-to-X-trap emission at 1.2 K. Despite the virtual elimination of the Y-trap there is no change in the $k-k'$ scattering as measured in the ESE detected method. The time constants associated with decay and buildup of the $k \approx 0$ and $k \approx \pm \pi/a$ excitons remain identical within the experimental error to those measured on the original crystal. Moreover, intentional incorporation of chemical impurities to a level of 0.4 per cent also does not alter the basic scattering behaviour. No significant differences could be discerned in crystals grown from impure starting material containing high concentrations of dichloro- and tetrachlorobenzene isomers, as well as one-bromo-trichlorobenzene.

Isotopic impurities present in otherwise pure crystals might be expected to influence scattering. Molecules of TCB-h_2 form traps with a depth of 20 cm^{-1} when introduced as guests into a host crystal of TCB-d_2 [59]. The situation is reversed when TCB-h_2 is the host, since the deuterated molecules act as antitraps, or barriers, with a height of 20.9 cm^{-1}. Barriers of this sort have been proposed as scattering centres in one-dimensional systems. However the presence of a 2.3 per cent concentration of TCB-d_2 in TCH-h_2 still produced no significant changes in the scattering rates or overall appearance of the scattering pattern.

The influence of structural imperfections on the scattering was investigated by measuring the effect of a careful annealing procedure on the scattering rates. During this annealing process the crystal, after growing from the melt, was first cooled very slowly to room temperature and then reheated and kept at a temperature 2° below the melting point for 60 hours. It was found that this procedure did not affect the scattering rates nor did it change the emisssion pattern from excitons and traps. However the width of the $k \approx 0$ peak in the ESE line shape detected immediately following the laser flash is reduced by 75 per cent to 1.25 MHz. In an effort to measure the influence of the cooling of the sample through the phase transition (which occurs at 188 K [51]), several experiments were performed in which the cooling rate was reduced to 1° per 20 minutes in the temperature region of this transition. Again no measurable effect on the scattering rates was observed. Even the harsh treatment of rapidly warming up from 4.2 K to room temperature and then immediately reintroducing it into liquid helium did not change the scattering rates.

The trapping dynamics were investigated in the temperature range between 1.25 K and 0.35 K. In Fig. 19 the changes are illustrated observed in the phosphorescence spectrum of a chemically pure, annealed crystal of TCB. At 1.25 K the emission from the X-trap is visible as a strong peak. However as the temperature decreases the intensity of the small peak 8 cm^{-1} below the energy of the triplet exciton band begins to grow and it becomes the dominant one below 0.8 K. This proves that in addition to our X-trap at 18 cm$^-$ below the exciton band, another structural trap is present with a comparable concentration. We have indicated these two traps as X_1 and X_2 in the figure. The changes in equilibrium trap populations are accompanied by a steady decrease in the ratio of exciton-to-trap phosphorescence, with the exciton signal hardly visible at 0.35 K. The weakness of the exciton 0—0 emission at this low temperature arises from the equilibrium population distribution, which favours the $\pm \pi/a$ levels at the bottom of the band. The signal observed at 0.35 K in the region between the exciton and X_1 indicates that additional shallow trap states, previously unknown, are populated at very low temperatures.

The phosphorescence from each trap builds up and decays as a biexponential,

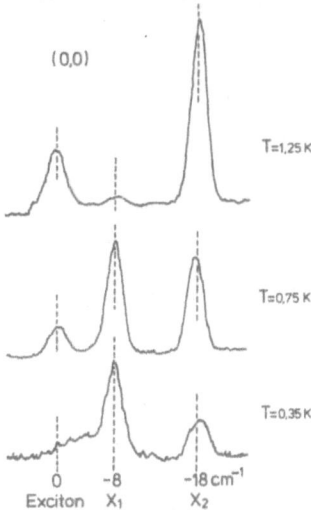

Fig. 19. Steady-state phosphorescence spectra of the 0—0 transition in TCB at three temperatures.

on a millisecond time scale. The rate constants characteristic of the buildup and decay of both traps are represented in Fig. 20. The estimated experimental errors are 10 per cent for the times and 0.04 K for the temperatures. Within these limits, it is apparent that phosphorescence from the shallow X_1 trap builds up independently of temperature below 0.80 K, with a time constant of (9.0 ± 1.0) ms. In this range of temperatures, the trap lifetime is also constant at (47 ± 3) ms. The trends observed for the deeper X_2 trap over the same range of temperature are similar to those of X_1. The decay of the exciton is nearly constant down to 0.60 K, although its time constant increases from 25 to 36 ms as the temperature is further lowered to 0.35 K.

From the results obtained so far it is clear that the evolution of the electronic excitation initially prepared in the $k \approx 0$ level of the lowest TCB triplet exciton band proceeds on two time scales. First $k - k'$ scattering promotes thermal equilibrium across the band within microseconds after the creation of the original non-equilibrium distribution. Second the exciton decays over a period of milliseconds, owing to radiative and non-radiative decay processes and trapping. Scattering and trapping are not unrelated, however, because the structural and chemical impurities that give rise to energy traps may also facilitate exciton-phonon scattering by breaking the translational symmetry of the one-dimensional chain.

Comparison of emission spectra from crystals containing high and low concentrations of one-bromo-trichlorobenzene allows responsibility for the Y-trap to be assigned to this particular inpurity. It becomes equally clear that the Y-trap is not an important scattering centre. That $k - k'$ scattering is determined by factors other than either the level or identity of chemical impurities present is suggested

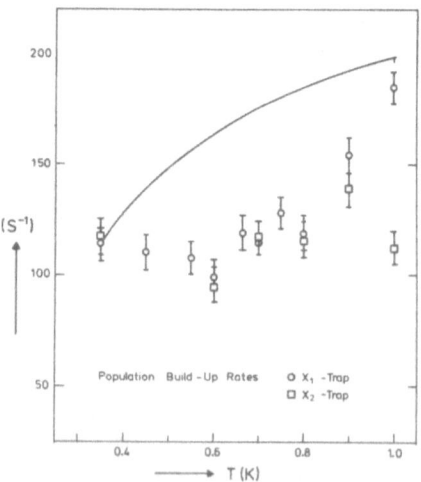

Fig. 20. Trapping rates versus temperature for the X_1 traps (circles) and the X_2 traps (squares), as obtained from the measured buildup times. The solid curve depicts the temperature dependence predicted for substitutional trapping centers (see text).

strongly by the unaltered behaviour observed in highly contaminated crystals. In addition, the experiments on mixed crystals containing TCB-d_2 indicates that the scattering is unaffected by the presence of chemically induced energy barriers, as well as traps.

The X_1 and X_2 traps which remain in the emission spectrum of purified TCB are presumed to originate from structural defects in the crystal. In particular it was established via EPR experiments that the X_2 traps are associated with slightly rotated TCB molecules (see Section 6.4). On the other hand experiments by Zwemer *et al.* have shown that the Y-trap emission also originates from a perturbed TCB molecule that is influenced by an adjacent chemical impurity [60]. This conclusion is corroborated by the vibrational progression of the Y-trap phosphorescence, which corresponds to that of a TCB molecule. Thus the distinction made between structural and chemical impurities may be illusory. In any event, none of the procedures adopted to improve the structural quality of the crystal seem to be effective on the microscopic level over which scattering takes place. Any structural improvements achieved by annealing and slow cooling are not reflected either in a changed proportion of X-trap emission or an increase of the scattering times.

As the probability of finding the triplet excitons in the lowest k-state increases with decreasing temperature a trapping process dependent on k should show a marked response. If the trap is induced by an impurity substituted directly in a molecular chain, then the trapping rate from a state k is given according to the calculation of Benk and Silbey [54; cf. equation (3.29)] by

$$K_k = Bc(\Delta E)^3 \sin^2 ka \qquad (30)$$

In this expression, c represents the impurity concentration, ΔE denotes the energy difference between the trap and band state k, a is the intermolecular distance, and B is a proportionality constant. The trapping rate observed as a function of temperature,

$$K(T) = \frac{\Sigma_k K_k \exp\{-E_k/k_B T\}}{\Sigma_k \exp\{-E_k/k_B T\}} \qquad (31)$$

is obtained by summing over all possible states of origin, each weighted by its Boltzmann factor. The temperature dependence according to this model is illustrated in Figure 20. Here the relative trapping rates have been scaled so that the theoretical and experimental values are equal at 0.35 K. Since $\sin^2 ka$ attains its maximum value at the band centre and its minimum value at the edges, equation (31) predicts a continuous decrease in trapping rates between 0.80 K and 0.35 K, as most of the population falls into the lowest level, $k \approx \pm \pi/a$. For example, for an 8 cm^{-1} trap, trapping at 0.35 K should be 60 per cent slower than at 0.80 K. The clear temperature independence of the X_1 and X_2 trapping rate proves however that the substitutional model of Benk and Silbey does not adequately describe the situation in TCB. The two X-traps therefore cannot be treated as being consolidated into the exciton band, if the chain is to be regarded as being of infinite length.

Concluding this section we can say that the observed k-independence of trapping to X_1 and X_2 seems consistent with the failure of the substitutional model to account for the scattering process in TCB. However, we should be extremely careful with drawing conclusions about this system. First the presence of even shallower traps, as suggested by the low temperature phosphorescence spectrum, makes it impossible to pinpoint X_1 and X_2 as the only scattering centres. Nevertheless the results of Botter et al. [39] (see Figure 12) indicate that the X_2 traps are involved. Second, it may even be necessary to consider rather complex structures, such as substitutional traps distributed in finite chains or funnel-like traps [61], in order to model exciton scattering and trapping realistically.

6.4. INTRA- OR INTERCHAIN SCATTERING? A STUDY OF THE SPIN-LATTICE RELAXATION IN THE PRESENCE OF A MAGNETIC FIELD

The results discussed so far do not allow us to decide whether the triplet exciton scattering is an intra- or an interchain process, because in zero-magnetic field we can not distinguish between the two inequivalent sites. However, several years ago it was suggested that interchain hopping is responsible for the spin-lattice relaxation (SLR) in triplet exciton systems [11]. Therefore a study of the SLR rate could, in principle, provide a means of ascertaining the interchain hopping rate, and its relative importance in the scattering process.

The first measurements of SLR in a linear chain triplet exciton system were performed by Schmidberger and Wolf [20] on 1,4-dibromonaphthalene, and later by Zieger and Wolf [33] on TCB. In both cases the spin-lattice relaxation rates were derived from saturation curves of the ODMR and EPR lines for only one direction of the external magnetic field. In the case of TCB SLR rates of about 10^6 s^{-1} at 4.2 K were obtained, and it was concluded that the interchain hopping rate had to be $\leq 4 \times 10^5$ s^{-1}. Later, experiments on the same system in very high magnetic fields were reported by Wolfrum et al. [62]. These authors found at 5 Tesla and 2 K a SLR rate of only 2×10^3 s^{-1}; from the B^2—B^3 field dependence they concluded that a one-phonon process is responsible for the SLR rate. Experiments in zero-magnetic field by Botter et al. [39], using ESE techniques, yielded SLR rates of 3×10^3 s^{-1} at 1.15 K, but later Lutz et al. [63] derived a much lower value of only a few s^{-1} at 4.2 K from an analysis of the phosphorescence decay curves. Clearly, there was no agreement on the magnitude of the SLR rates in the TCB triplet exciton system.

Although these experimental results are not conclusive, it is tempting to explain the SLR in the triplet exciton system of TCB as arising from a hopping process between the magnetically inequivalent stacks of type A and B. The idea being that any transfer between stacks A and B leads to a redistribution of the populations of the triplet sublevels, because the spin eigenvectors in the two stacks are spatially inequivalent. In a magnetic field the relative orientation of the spin eigenvectors in A and B depend on the strength and orientation of the applied magnetic field and hence one predicts a characteristic orientational dependence of the SLR rates. A comparison between the observed orientational dependence of the SLR rates and the results of a numerical calculation of the relative orientation of the spin eigenvectors should provide a good test for the reliability of this model. The experiments to be discussed in this section show that indeed a striking resemblance exists between the experimentally observed and theoretically predicted orientational dependence of the SLR rates. From the results we can obtain an estimate of the contribution of interchain hopping to the observed scattering rate in zero field [64].

The SLR rates were measured with ESE techniques in an X-band (9.3 GHz) spectrometer upon laser flash excitation at 375 nm in the 0–0 band of the $T_0 \leftarrow S_0$ absorption. The results of the experiments are shown in Fig. 21 where the open circles represent the values of the average SLR rate $\bar{w}$ as a function of the direction of the magnetic field **B** for the high-field transition in the yz plane of one of the two inequivalent stacks. It is seen that a remarkable orientational dependence exists which, as we believe, is caused by a hopping process between the two inequivalent stacks. Let us first discuss the experimental procedure used to obtain the results presented in Fig. 21.

In the first step we ascertained the directions of the spin axes of the X-traps with conventional ODMR and continuous illumination. This orientational study is relatively simple since at 1.2 K the ODMR signals of these traps, which are about 18 cm^{-1} below the exciton band, are very strong. To our surprise we found that the angles between the corresponding spin axes of the two sites differ from one sample to the other. In three out of five samples we found that the directions of the

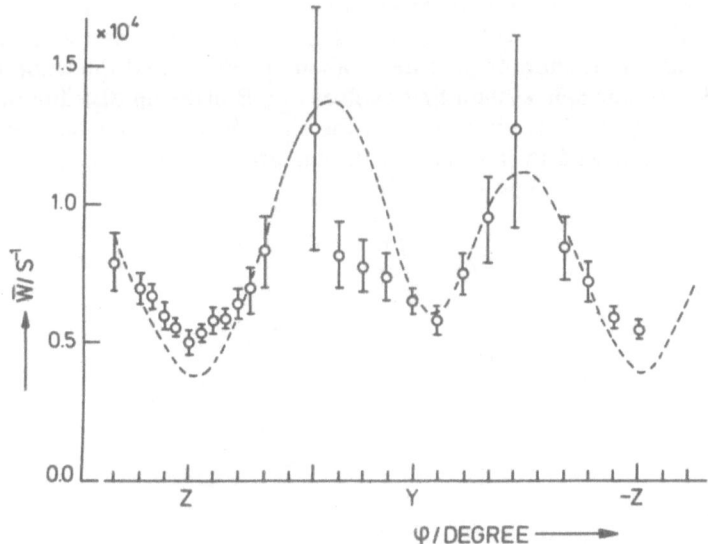

Fig. 21. The open circles represent the angular dependence of the spin-lattice relaxation rates in the triplet excitons of TCB in the yz-plane by measuring the evolution of the ESE signal in the high-field transition upon the laser flash excitation. The microwave frequency is 9.3 GHz and the temperature is 1.2 K. For the direction of the exciton spin axes see Fig. 23.

spin axes corresponded to the high temperature monoclinic crystal structure, while in the other two samples they corresponded to the low temperature triclinic structure. The phase transition in TCB occurs at 188 K [50, 51], and we therefore conclude that it is possible to undercool the high temperature phase.

Once the directions of the principal axes of the two X-trap sites in a particular sample are known, we switch from ODMR detection of the traps to ESE detection of the triplet excitons. In this experiment we generate the ESE signal by two microwave pulses following the 15 ns laser flash tuned to the 0—0 band of the $T_0 \leftarrow S_0$ exciton absorption. Since at temperatures below 2 K the exciton trapping time is in the order of 1—10 ms [29, 30], either the exciton or the trap signal can be observed. This is done simply by varying the delay t_d between the laser flash and the first microwave pulse while keeping the interval τ between the microwave pulses constant. With $t_d \ll 10$ ms only the exciton signals are observed whereas for times longer than 10 ms the trap signals dominate. In this way it can be established that the directions of the spin axes of the X-traps deviate a few degrees from those of the excitons. The fact that a selection can be made between excitons and traps is very useful because the ESR transitions of excitons and traps overlap in almost all orientations.

In Fig. 22 we illustrate the time resolving power of the ESE technique by presenting the ESE detected low-field $\mathbf{B} \| \mathbf{y}$ signal of the triplet excitons with $t_d = 5$ μs and 1 ms. The change in sign is the result of SLR in the triplet exciton system, and can be understood with the aid of the energy level scheme in Fig. 23, in which we have also indicated the relative populating rates of the sublevels. Immediately following the laser flash the $m_s = 0$ sublevel carries a strong overpopulation resulting in the emissive low-field signal at $t_d = 5$ μs. Since at 1.2 K SLR is fast compared with the exciton decay rate of about 1200 s^{-1} and the trapping rate of 100 s^{-1} [29, 30], the spin system first evolves to a Boltzmann distribution over the sublevels. As a result, it is observed that the signal at $t_d = 1$ ms has changed sign and also has decreased in intensity. With increasing delay the ESE signal of the

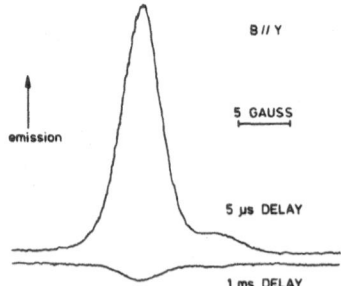

Fig. 22. An illustration of the time evolution of the low-field electron spin echo signals of the triplet excitons in TCB with $\mathbf{B} \| \mathbf{y}$. The strong signal is recorded 5 μs after the laser flash when a large excess population exists in the $m_s = 0$ sublevel. The small signal is taken at a delay of 1 ms when a Boltzmann distribution is established. The small signal at the right of the large signal is due to a small piece of the crystal broken off the sample.

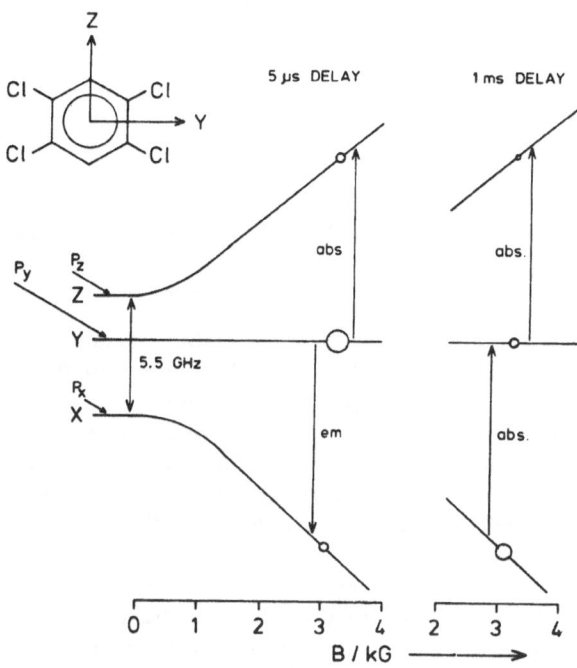

Fig. 23. The energy level scheme of the triplet state of the excitons in TCB with the magnetic field **B** parallel to the *y* axis. We have illustrated the situation in which the SLR is faster than the exciton decay rate. As a result the ESE detected low-field signal changes from emission to absorption as shown in Fig. 22.

exciton decreases further, owing to trapping and decay. Simultaneously we observe the growth of the broader X-trap signal which corresponds to a Boltzmann distribution over its sublevels. The trap signal subsequently decays with a time constant of 35 ms, the average X-trap lifetime. By combining the results for the high- and low-field transitions in the three canonical orientations, we deduce that the relative populating rates P_x, P_y and P_z of the zero-field levels of the exciton upon excitation from S_0 are $P_y > 2P_z \gg P_x$.

In the first few hundred microseconds after the laser excitation, during which we study the SLR, the total population of the triplet exciton is approximately constant. We therefore expect the decay of the ESE signal to contain two time constants, λ_1 and λ_2, from which the average SLR rate $\bar{w} = (\lambda_1 + \lambda_2)/6$ is obtained (see Ch. I). In practice, however, it is difficult to distinguish the two time constants and only a mono-exponential decay can be observed with decay constant λ. In Fig. 21 we have plotted $\bar{w} = \lambda/3$ for a sample with a triclinic structure. For the monoclinic structure the $\bar{w}$ values differ only slightly and the same general picture results.

The most complete orientational dependence of the SLR rate was obtained for the high-field signal in the *yz* plane and is shown in Fig. 21. For orientations of the external magnetic field 40° to 50° away from the canonical orientations, the populating rates of the triplet sublevels are almost equal, and the signals are so

TABLE I

The average spin-lattice relaxation rates at 1.2 K of the triplet excitons in
TCB in an external magnetic field at the various canonical orientations

$\bar{w}$ (s^{-1})	low field	high field
$B \parallel x$	$(6.7 \pm 1.3) \times 10^3$	$(5.6 \pm 0.9) \times 10^3$
$B \parallel y$	$(3.7 \pm 0.2) \times 10^3$	$(6.5 \pm 0.4) \times 10^3$
$B \parallel z$		$(4.9 \pm 0.4) \times 10^3$

weak that it is difficult to obtain reliable data for the relaxation rates. For the same
reason the orientational dependence of the SLR rates in the other two planes
could only be measured in a range of $\pm 20°$ around the principal axes. In Table I
we give the observed values of $\bar{w}$ with **B** parallel to the three principal axes.

In addition we have also measured the average SLR rate in zero-magnetic field.
This experiment is performed in the same way as in magnetic field. Again the
decay is characterized by one time constant λ, which depends somewhat on the
particular zero-field transition studied, and moreover varies somewhat between
different samples. Once more we take for $\bar{w} = \lambda/3$ and obtain $\bar{w} = (3.2 \pm 0.5) \times 10^3 \, \text{s}^{-1}$. It should be mentioned that in the zero-field experiment we have no
means of determining whether we are dealing with a monoclinic or a triclinic
crystal structure.

As we mentioned before, our aim is to test whether the SLR among the spin
levels of the triplet exciton is caused by a hopping process between the two
inequivalent stacks. In Ch. I, it is shown that such a hopping leads to a characteristic orientational dependence of the SLR rates in the presence of an external
magnetic field. These authors studied SLR in the lowest triplet state T_0 of
naphthalene in durene. They proved that this relaxation is caused by thermal
excitation to and decay from a nearby local phonon state T_e where the spin axes
are rotated with respect to those in the lowest triplet state T_0. Since spin angular
momentum is conserved during the excitation, a triplet spin originally residing in
an eigenstate of T_0 is transferred to a non-stationary superposition of the eigenstates of T_e. During the lifetime of T_e the system evolves in time and, on returning
to T_0, the spin does not end up in the spin state from which it departed. The SLR
rate resulting from this exchange process depends on the angle between the
corresponding pairs of spin eigenvectors in T_0 and T_e. By calculating the average
values of these angles for the three pairs of eigenstates, an orientational dependence was found similar to the experimentally observed one.

In our system of TCB, assuming a hopping process between the two inequivalent stacks A and B, we have a situation which is almost identical to the one
described above. An important difference is that the "lifetimes" of both sites A and
B are long (at least in the order of a few microseconds) so that after a transfer
from one stack to the other, the non-stationary terms have completely averaged
out before the next transfer occurs. Moreover, since no measurable energy
difference exists between the two stacks the probability P_{AB} for jumping from stack

A to B may be assumed to the equal to P_{BA}. If the angles $\phi_{u(a)u(B)}$ between corresponding spin eigenvectors $u(A)$ and $u(B)$ $(u = |+\rangle, |0\rangle, |-\rangle)$ of triplet exciton A and B are small (in our case the $\phi_{u(A)u(B)} < 4°$ in a magnetic field of about 0.3 T), we may write for the average SLR rate (see Ch. I),

$$\bar{w} = \frac{1}{2} P_{AB} \times \frac{1}{3} \sum_{\phi_{u(A)u(B)}} \sin^2 2\phi_{u(A)u(B)} \approx 2P_{AB} \langle \phi_{u(A)u(B)}^2 \rangle \qquad (32)$$

At this point it should be mentioned that the expression given by Verbeek *et al.* applies to the SLR rate T_1^{-1} in a two-level system, and therefore differs by a factor two. Equation (32) can be rewritten

$$\bar{w} \approx 2P_{AB} \langle 1 - |P_{u(A)u(B)}|^2 \rangle \qquad (33)$$

Here $|P_{u(A)u(B)}|^2$ stands for the scalar product of two corresponding eigenvectors of the triplet excitons A and B.

The first aim of this study is to check whether the observed angular dependence of the SLR rates in the presence of a magnetic field can be explained by this model. We have therefore made numerical calculations of the factor $\langle 1 - |P_{u(A)u(B)}|^2 \rangle$ as a function of the direction of the magnetic field in the yz plane for the triclinic structure. To our satisfaction this theoretical curve shows maxima and minima at the orientations where we found them experimentally, although the predicted amplitude is too large. The best fit is obtained with $P_{AB} = 2.6 \times 10^5 \text{ s}^{-1}$, and is indicated by the dotted curve in Fig. 21.

The next test is to check whether the value of P_{AB} derived from the magnetic field experiments can account for the observed average SLR rate in zero-magnetic field. Here we have to realize that the angles $\phi_{u(A)u(B)}$ between corresponding eigenvectors in stack A and B are not very small (they range from 4° to 10° for the triclinic, and from 5° to 25° for the monoclinic structure) and that the approximation used for deriving (32) does not strictly apply. However, this appears to be a minor problem in the interpretation of the data, as we shall see when comparing the theoretical and experimental results. On the basis of equation (33) and using $P_{AB} = 2.6 \times 10^5 \text{ s}^{-1}$, the average SLR rate $\bar{w}$ in zero field is predicted to be $\bar{w} = 3 \times 10^4 \text{ s}^{-1}$ for the monoclinic, and $\bar{w} = 0.7 \times 10^4 \text{ s}^{-1}$ for the triclinic crystal structure, whereas the observed SLR rate $\bar{w} = (0.32 \pm 0.05) \times 10^4 \text{ s}^{-1}$. These numbers suggest that in the zero-field experiment a sample with a triclinic structure was studied. This does not seem unreasonable since the cooling of the samples was done very slowly in a period of 12 hours, in contrast to the high-field experiments. In the latter the cooling is done much more quickly, thus increasing the probability that the room temperature phase will be undercooled. Nevertheless, there is a discrepancy of at least a factor of two.

It is interesting to compare the estimate for the interchain hopping rate $P_{AB} = 2.6 \times 10^5 \text{ s}^{-1}$ with the rates for $k - k'$ scattering as derived from our zero-field experiments. At 1.2 K the scattering probability for the $k \approx 0$ excitons $T_S^{-1}(k \approx 0) = 4 \times 10^5 \text{ s}^{-1}$, and for the $k \approx \pm \pi/a$ excitons $T_S^{-1}(k \approx \pi/a) = 1 \times 10^5 \text{ s}^{-1}$. We see that the interchain hopping rate P_{AB} is of the same order of magnitude as the average scattering rate. Hence we can not exclude the possibility that the scattering among the k-states observed in the zero-field experiments is caused to a large extent by interchain processes.

6.5. THE WIDTH Δk OF THE TRIPLET EXCITON WAVE PACKETS

An important problem, not discussed so far is the width Δk of the triplet exciton wave packets. Since the exciton k states correspond with different zero-field splittings we expect that this width Δk gives rise to a well-defined spread in resonance frequency of the zero-field transitions. So from the homogeneous linewidth $(\pi T_M)^{-1}$ of the triplet spins, obtained from a measurement of their phase memory time T_M, we may expect to derive information about the width of the exciton wave packets [68].

The measurement of T_M was performed by varying the interval τ between the two microwave pulses while keeping the delay t_d between the laser flash and the first microwave pulse at a constant value. However, to obtain the maximum signal-to-noise ratio it was necessary to optimize the value of t_d for every microwave frequency. As we know, the $k \approx 0$ excitons are strongly populated immediately following the laser flash whereas the $\pm \pi/a$ excitons reach an appreciable population only after a scattering process through the exciton band. Therefore it is advantageous to use a delay t_d for the $k \approx 0$ excitons as short as possible (about 0.5 μs) whereas for the $\pm \pi/a$ excitons a delay t_d of 10 μs was used.

In Fig. 24a we present the values for the inverse of the phase memory time T_M^{-1}, obtained from the decay of the two-pulse echo signals, as a function of the resonance frequency in the $T_y - T_x$ zero-field transition at 1.15 K. For comparison we show in Fig. 24b the shape of the optically detected $T_y - T_x$ transition at the same temperature under continuous illumination, i.e. in Boltzmann equilibrium. It is observed that T_M^{-1} depends on the microwave frequency, in other words on the value of the quasi-momentum k of the triplet excitons. In addition to the $T_y - T_x$ transition we have also measured the variation of T_M^{-1} over the $T_z - T_y$ zero-field transition. The results are presented in Fig. 25a and for comparison we show in Fig. 25b the optically detected lineshape of the same transition under steady state illumination.

The relation between the width in k space of an exciton wave packet and the corresponding spread in zero-field resonance frequency in the absence of any scattering mechanism is relatively easy thanks to the relation (14) derived by Francis and Harris for the variation of the zero-field splitting frequency with k. From (14) we derive,

$$\Delta \nu = (\pi T_2)^{-1} = -2\mu \Delta(ka) \sin ka. \tag{34}$$

Here $\Delta(ka)$ is the width of the exciton wave packet which for convenience has been expressed in radians, and $\Delta \nu$ is the corresponding spread in zero-field resonance frequency, which we have related to an "intrinsic" dephasing time T_2. For the $T_y - T_x$ transition $\mu = -2.5$ MHz, and for the $T_z - T_y$ transition $\mu =$

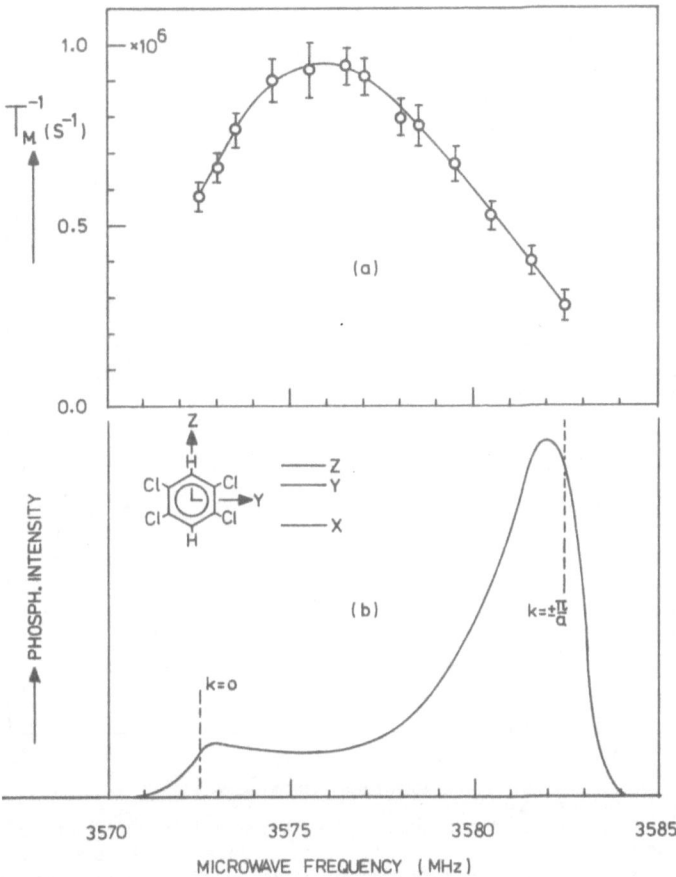

Fig. 24. (a) The inverse phase memory time T_M^{-1} as a function of frequency for the $T_y - T_x$ triplet exciton zero-field transition. $T = 1.15$ K. (b) A schematic indication of the lineshape of the optically detected $T_y - T_x$ triplet exciton zero-field transition observed under steady state illumination. The resonance frequencies of the $k \approx 0$ and $\pm \pi/a$ excitons occur at 3572.5 MHz and 3582.5 MHz respectively, and are indicated by the broken lines. $T = 1.15$ K.

−1.4 MHz. It is seen that in the absence of any further broadening mechanism one predicts a sinusoidal variation of $\Delta\nu$ over the zero-field lineshape.

The phonon induced scattering of the exciton wave packets will cause a further broadening of the spin packets in the zero-field lineshapes. The probability for scattering from the $k \approx 0$ states, $T_S^{-1}(k \approx 0)$, is known from the decay of their excess population following the laser flash excitation; $T_S^{-1}(k \approx 0) = (0.45 \pm 0.04) \times 10^6$ s^{-1}. The rate $T_S^{-1}(k \approx \pm \pi/a)$ has been measured with spin-locking experiments as described in Section 6.3; $T_S^{-1}(k \approx \pm \pi/a) = (0.10 \pm 0.01) \times 10^6$ s^{-1}.

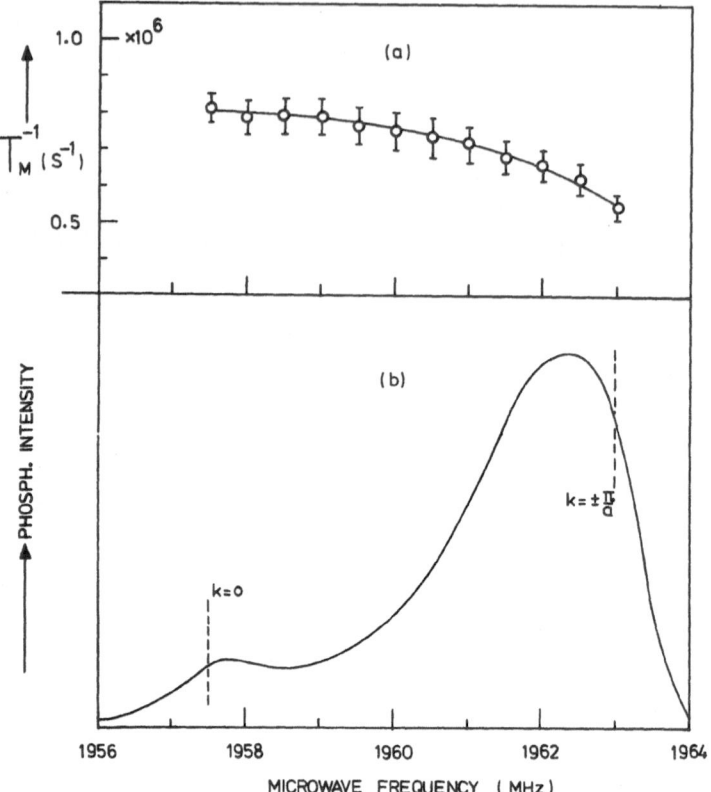

Fig. 25. (a) The inverse phase memory time T_M^{-1} as a function of frequency for the $T_z - T_y$ triplet exciton zero-field transition. $T = 1.15$ K. (b) A schematic indication of the lineshape of the optically detected $T_z - T_y$ triplet exciton zero-field transition under steady state illumination. The resonance frequencies of the $k \approx 0$ and $k \approx \pm \pi/a$ excitons occur at 1957.5 MHz and 1963.0 MHz respectively and are indicated by the broken lines. $T = 1.15$ K.

We now assume that the total homogeneous width of the spin packets is the sum of these two effects i.e.,

$$(\pi T_M)^{-1} = \Delta \nu + (\pi T_S)^{-1} = (\pi T_2)^{-1} + (\pi T_S)^{-1}. \tag{35}$$

To check whether this relation indeed applies we have drawn in Fig. 26a the variation of T_S^{-1} over the $T_y - T_x$ zero-field lineshape. The point at $k \approx 0$ was taken from the measurements but the rest of the curve was calculated using the expression (19) with the same parameters as in the model calculations of Section 6.2. By subtracting T_S^{-1} from T_M^{-1} we obtain the values indicated by the open circles in Fig. 26b and it is seen that a form results which resembles a sinusoidal function as predicted by equation (34). The broken line is a fit to this curve according to,

$$T_2^{-1} = C + 2\pi \mu \Delta(ka) \sin ka, \tag{36}$$

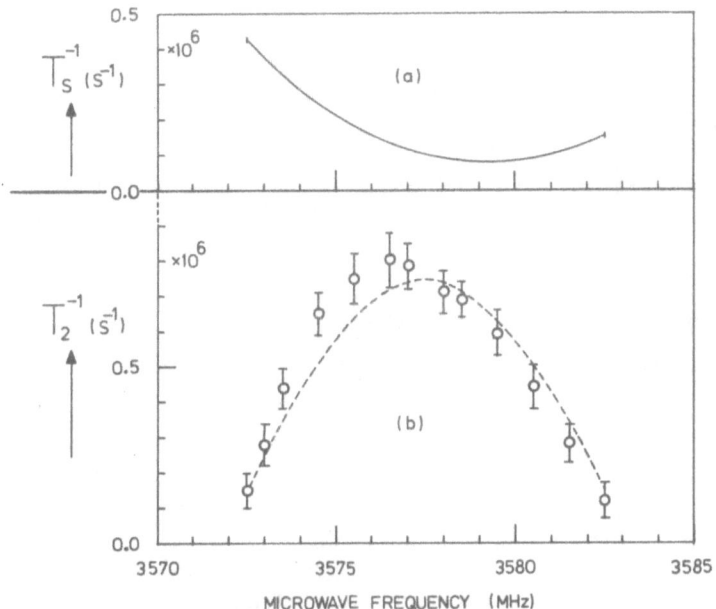

Fig. 26. (a) The variation of the scattering rate T_S^{-1} with the zero-field resonance frequency of the $T_y - T_x$ transition. The value T_S^{-1} for $k \approx 0$ has been obtained from the measurements discussed in Section 6.1; the other values are derived using expression (19) (see Section 6.2). (b) The open circles represent the value of T_2^{-1} obtained by subtracting the value of T_S^{-1} from the observed inverse phase memory time T_M^{-1} in Fig. 24a. The best sine fit of these data is represented by the broken line.

with $C = (0.13 \pm 0.07) \times 10^6 \, \text{s}^{-1}$ and $2\pi|\mu|\Delta(ka) = (0.59 \pm 0.02) \times 10^6 \, \text{s}^{-1}$. Since $|\mu| = 2.5 \times 10^6 \, \text{s}^{-1}$ for the $T_y - T_x$ transition we derive $\Delta(ka) = (0.038 \pm 0.005)$ radians.

We have applied the same method to analyze the results for T_M^{-1} obtained on the $T_z - T_y$ transition. First, we have again substracted the term T_S^{-1} from T_M^{-1}. The result is shown in Fig. 27b by the open circles and can be fitted by,

$$T_2^{-1} = C' + 2\pi\mu\Delta(ka) \sin ka, \qquad (37)$$

with $C' = (0.37 \pm 0.05) \times 10^6 \, \text{s}^{-1}$ and $2\pi|\mu|\Delta(ka) = (0.29 \pm 0.01) \times 10^6 \, \text{s}^{-1}$. For this $T_z - T_y$ transition the amplitude of the sine function is about two times smaller than for the $T_y - T_x$ transition but since the value of μ is also about two times smaller, the resulting value for $\Delta(ka) = (0.032 \pm 0.005)$ radians is extremely close to the one derived for the $T_y - T_x$ transition.

The constants C and C', necessary to fit the curves in Figures 26b and 27b, probably represent contributions to the dephasing of fluctuating dipolar interactions with surrounding nuclear spins just as in the case of isolated triplet molecules [69]. This interaction is known to give rise to dephasing times for isolated TCB molecules of about 4 μs [39, 70] close to the value derived from our analysis.

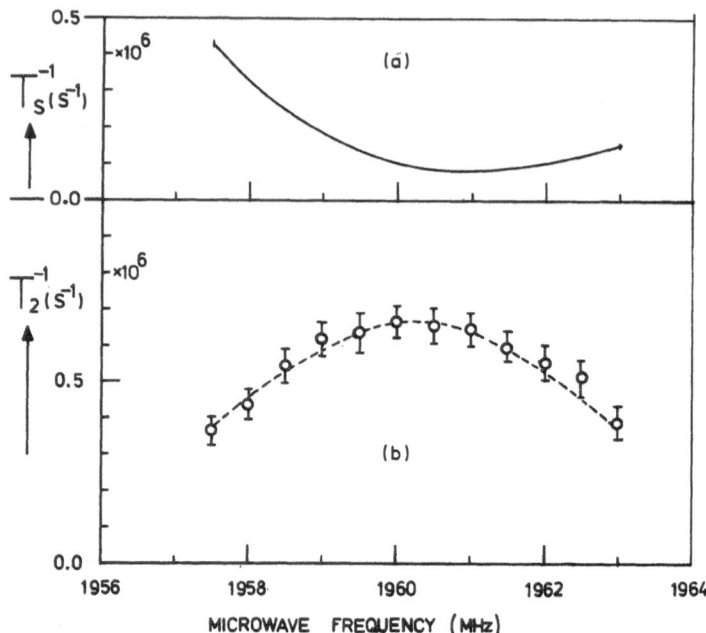

Fig. 27. (a) The variation of the scattering rate T_S^{-1} with the zero-field resonance frequency of the $T_z - T_y$ transition. (b) The open circles represent the value of T_2^{-1} obtained by subtracting T_S^{-1} from the observed inverse phase memory time T_M^{-1} in Fig. 25a. The best sine fit is again indicated by the broken line.

Concluding this section we can say that the simple model outlined above gives a satisfactory description of the observed variation of T_M^{-1} over the $T_y - T_x$ and $T_z - T_y$ zero-field transitions. In particular the observation of a sinusoidal variation of T_2^{-1} makes us believe that it arises from a k independent width Δk of the exciton wave packets. From the analysis we derive that this width amounts to 0.5 per cent of the Brillouin zone. This would mean that the triplet excitation is delocalized over about 200 TCB molecules.

6.6. RELAXATION OF VIBRATIONALLY EXCITED TRIPLET EXCITONS IN TCB

In the last set of experiments on TCB we have measured the evolution of the $T_y - T_x$ zero-field transition after excitation into vibrationally excited states of the triplet exciton. First, we recorded the excitation spectrum by scanning the laser (with a bandwidth of 0.7 cm^{-1}) from the 0—0 absorption line to shorter wavelengths while detecting the intensity of the ESE signal of the $k \approx \pm \pi/a$ excitons at 3582.5 MHz at a long delay after the laser flash. This spectrum, shown in Fig. 28, is essentially the same as the one published by George and Morris [28] obtained by conventional spectroscopy, and from whom we have also adopted the assignments. Then we measured the evolution of the $k \approx 0$ and $k \approx \pm \pi/a$ ESE signals with the laser set at the various peaks in the excitation spectrum. In Fig. 29 we show the evolution of the ESE signals of the $k \approx 0$ and $k \approx \pm \pi/a$ excitons

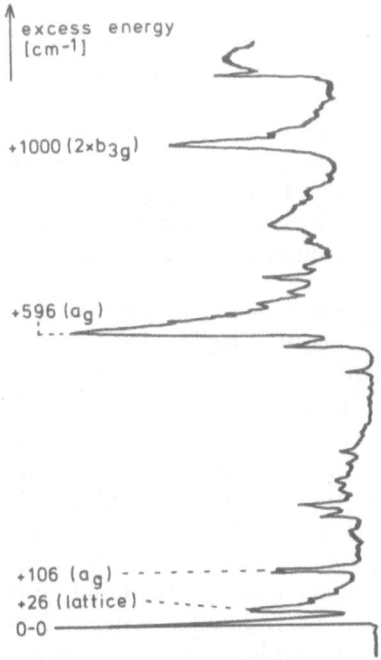

Fig. 28. The $T_0 \leftarrow S_0$ absorption spectrum of TCB obtained by scanning the pulsed dye laser from the 0—0 transition to shorter wavelengths and by detecting the ESE signal of the $k \approx \pm \pi/a$ excitons at a delay of 10 μs after the laser flash.

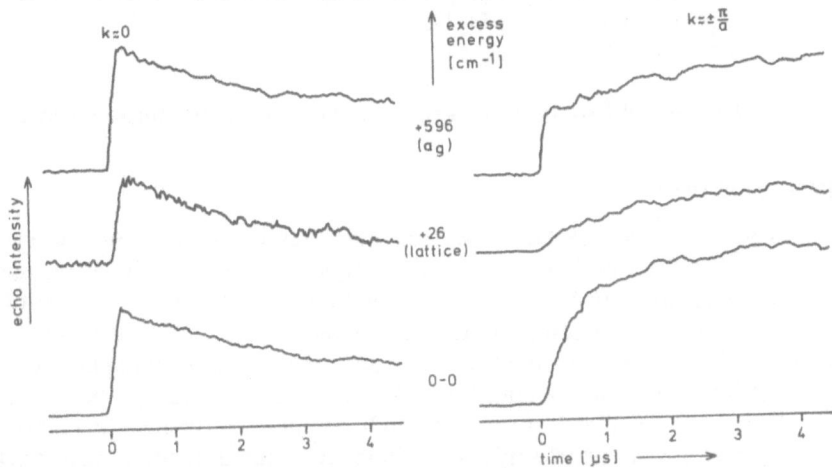

Fig. 29. The time-dependent behaviour of the zero-field transition of the $k \approx 0$ excitons at 3572.5 MHz and the $k \approx \pm \pi/a$ excitons at 3582.5 MHz after excitation into the electronic origin and two higher lying states, corresponding to a lattice vibration of 26 cm^{-1} and an a_g molecular vibration of 596 cm^{-1} respectively. $T = 1.15$ K.

after excitation in the 0—0 band and two higher lying bands corresponding to a lattice vibration of 26 cm^{-1} and a molecular vibration of 596 cm^{-1} respectively. It is seen that excitation into the lattice vibration leads to the same behaviour as into the 0—0 band, i.e. the $k \approx 0$ excitons exhibit an overpopulation upon the laser flash excitation, whereas the signal corresponding to the $k \approx \pm \pi/a$ excitons grow in slowly. After excitation into the molecular vibrational state however, the $k \approx \pm \pi/a$ exciton population first builds up with the same fast time constant as the $k \approx 0$ excitons followed by a slow evolution of a few microseconds. The same behaviour was found for all molecular vibrational states accessible with the dye laser.

The most remarkable result of this study is the difference between the behaviour of the molecular vibrations and the crystal vibration. We think that the explanation is found in the difference of the vibrational energies in the excited triplet state and the singlet ground state. For instance, the a_g vibration of 596 cm^{-1} has an energy of 685 cm^{-1} in the singlet ground state [28]. So the vibrationally excited state can only separate into a ground state vibron and a vibrationless triplet exciton by generating a multi-phonon process in the crystal. In contrast, the decay from the crystal vibrational state can occur without the creation of a multi-phonon process since the vibrational energy in the ground state is the same as in T_0.

Further it is interesting to note that the fast build-up of the $k \approx \pm \pi/a$ excitons after excitation in a molecular vibrational state, is followed by a slower decay. Clearly the vibrational relaxation leads to a uniform population distribution over the k states. Since this is not the thermal equilibrium situation we observe an evolution to the Boltzmann population distribution over the exciton band with a characteristic scattering time of a few microseconds. These observations explain why in the early experiments of Botter *et al.*, discussed in Section 6.1, no preferential populating process of the $k \approx 0$ excitons was present after exciting high above the absorption origin with the N$_2$ laser.

7. The Dynamical Properties of Triplet Excitons in p-Dichlorobenzene.

7.1. INTRODUCTION

In the preceding section we have focussed our attention entirely on the system TCB. The reason for this is clear; the TCB crystal is an example of an almost perfect one-dimensional triplet exciton system with the inherent advantage of the simplicity of its band structure. Another advantage is that at liquid helium temperatures the lifetime of the triplet excitons and their related spin-spin and spin-lattice relaxation times are so long that EPR and ESE signals can be observed. Perhaps the most attractive feature is the SOC induced k dependent microwave dispersion, which allows us to study the behaviour of the individual wave packets in k space.

It would be of great value to test the validity of the models used for describing the behaviour of the triplet excitons in TCB on related one-dimensional systems

like 1,4-dibromonaphthalene (DBN), p-dichlorobenzene (p-DCB), p-dibromoben-
zene (p-DBB) and 9,10-dichloroanthracene (DCA). In this section we present
results of a study on p-DCB [71, 72]. Unfortunately it proved to be impossible to
observe ESE signals of the triplet excitons in this system. Nevertheless from an
optical study and an ESE study of the dominant X-traps it is possible to conclude
that probably the bandwidth of the triplet excitons is smaller than that in TCB.
Further that a very fast SLR exists among the triplet sublevels in combination with
a very fast trapping process even at temperatures as low as 1.2 K. A possible
explanation of these effects based on the difference of the crystal structures of
TCB and p-DCB will be discussed.

7.2. EXPERIMENTAL RESULTS

A complicating factor in the study of p-DCB is that three different phases can
exist. A triclinic β phase above 30.8 °C, a monoclinic α phase between 30.8 °C
and 0 °C and finally the monoclinic γ phase below 0 °C [73—75]. By slowly
cooling the sample from its melting temperature through the phase transition at
30.8 °C, large α phase neat crystals are obtained. By rapid cooling from 10 °C
through the α—γ phase transition at 0 °C the α phase is preserved as will be
shown by the ODMR experiments at liquid helium temperatures.

In Figs. 30a and 30b we show the steady-state phosphorescence spectra around
the 0—0 region at 1.2 K and 4.2 K respectively. The peak at 27888 cm^{-1} coincides,
within the experimental accuracy of 1 cm^{-1}, with the value of the 0—0 absorption
and hence originates from the excitons. The emission becomes visible at temperatures above 2.8 K. The intense peak 24.6 cm^{-1} below the exciton 0—0 emission is
attributed to a structural defect trap (labeled X). At 4.2 K (Fig. 30b) another
transition appears 50 cm^{-1} below the exciton 0—0 band, which is attributed to a
second defect trap. The 0—1 (b_{2g}) vibronic transitions, about 308 cm^{-1} to the red,
at 4.2 K exhibit a remarkable FWHM of 5.5 cm^{-1} compared with 3 cm^{-1} for the
0—0 emission lines. In Fig. 30c we present the 0—1 region at 4.2 K. It is observed
that the exciton 0—1 line is asymmetric, as shown in greater detail in Fig. 30d.

The ratio of the exciton and X-trap emission intensity (I_{exc}/I_{trap}) was measured
between 2.8 and 4.2 K and showed an exponential temperature dependence. If
there is a fast communication between the trap and exciton states we expect this
intensity ratio to be described by

$$I_{exc}/I_{trap} = N \exp(-\Delta E/kT).$$

where N denotes the number of exciton states per trap and ΔE is the mean
band-trap energy separation. From the results we derive that $\Delta E = (23 \pm 2)$
cm^{-1}, in good agreement with the spectroscopically determined exciton-trap
separation of 24.6 cm^{-1}. Further it is found that $N = (2.0 \pm 0.5) \times 10^3$.

To obtain information about exciton trapping rates, experiments were performed
in which the time evolution of the exciton and trap emission was monitored
following laser flash excitation in the 0—0 band of the $T_0 \leftarrow S_0$ absorption. At
1.2 K the exciton emission is extremely weak but nevertheless it was possible to

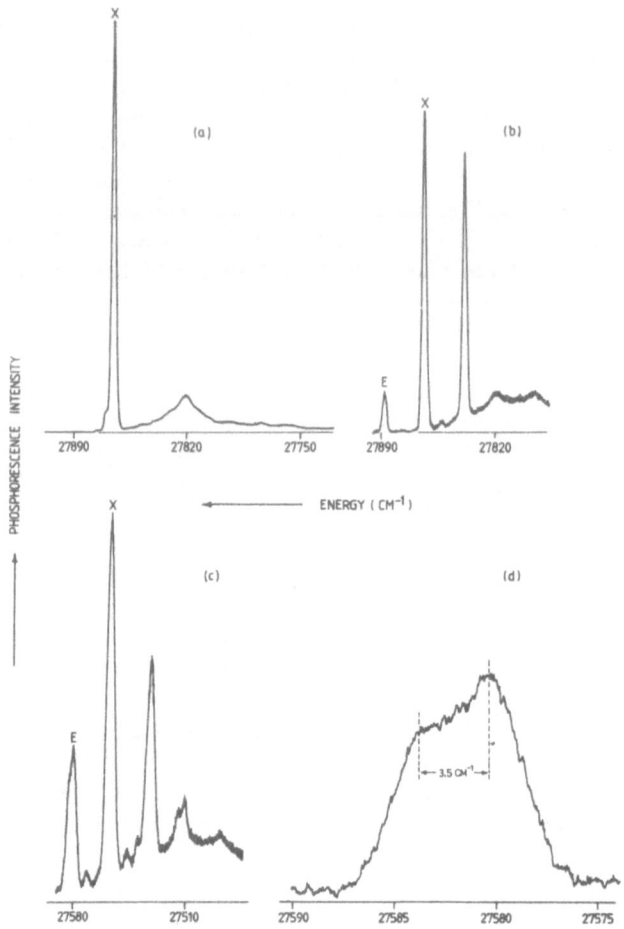

Fig. 30. Steady state phosphorescence spectra of p-DCB; (a) and (b) represent the 0—0 region at 1.2 K and 4.2 K respectively, (c) the 0—1 region at 4.2 K and (d) the 0—1 excitonic emission at 4.2 K. The bandwidth of the monochromator is 1 cm^{-1}.

establish that the decay time is about 3 μs. The X-trap emission however is much stronger and it was found that the ingrowth of its intensity proceeds with a time constant of about 5 μs followed by a much slower decay with a characteristic time of (17 ± 1) ms [71, 72]. At 4.2 K the exciton and X-trap decay with the same time constant of (600 ± 100) μs, indicating a rapid communication between exciton and traps.

To establish whether the crystal at 1.2 K was in its (undercooled) α phase or in its γ phase X-band ODMR experiments were performed to ascertain the directions of the spin axes. Moreover, since p-DCB diluted in a para-dibromobenzene (p-DBB) host crystal exhibits an appreciable distortion in its triplet state [76, 77], it was interesting to check the extent to which this distortion would occur in neat crystals of p-DCB. The ODMR signals in p-DCB at 1.2 K originate from the X-traps. This is evidenced by the hyperfine structure which is observed with the magnetic field parallel to, for instance, the spin axes y_s. From the observation that the out-of-plane spin axes of the two sites make an angle of 125° we conclude that the samples are in the α phase. Apparently the rapid cooling from 10 °C to liquid N_2 temperatures indeed causes an undercooling of the α phase.

In Fig. 31 we present, in a stereographic projection, the directions of the spin axes x_s, y_s and z_s with respect to the crystal a', b and c axes and the molecular axes x_m, y_m and z_m. From the observation that x_{1s} and x_{2s} make an angle of 125°, it is concluded that $x_{1s} \| x_{1m}$ and $x_{2s} \| x_{2m}$. For the relation between the in-plane spin axes and the molecular axes, there are two possibilities depending on the choice whether one views along the positive or the negative direction of b. It was decided that the configuration in Fig. 31 with an angle of $(15 \pm 2)°$ between in-plane spin axes and the molecular axes is the most likely one.

In zero-field ODMR signals of the same X-traps were found, but it was impossible to observe signals of the excitons even when the temperature was lowered to 0.4 K [72]. The X-trap zero-field transitions have a linewidth of about 3.5 MHz and occur at 1758 MHz, 3604 MHz and 5362 MHz. These values correspond very well with the zero-field splittings reported by Buckley *et al.* [78].

When exciting the $T_0 \leftarrow S_0$ 0—0 absorption with a strong laser flash at 1.2 K, strong ESE signals are observed at the zero-field frequencies of the X-trap. An extensive search for ESE signals of the triplet excitons, in a region of $\pm$ 200 MHz around the X-trap zero-field frequencies at delay times as short as 0.5 μs, was

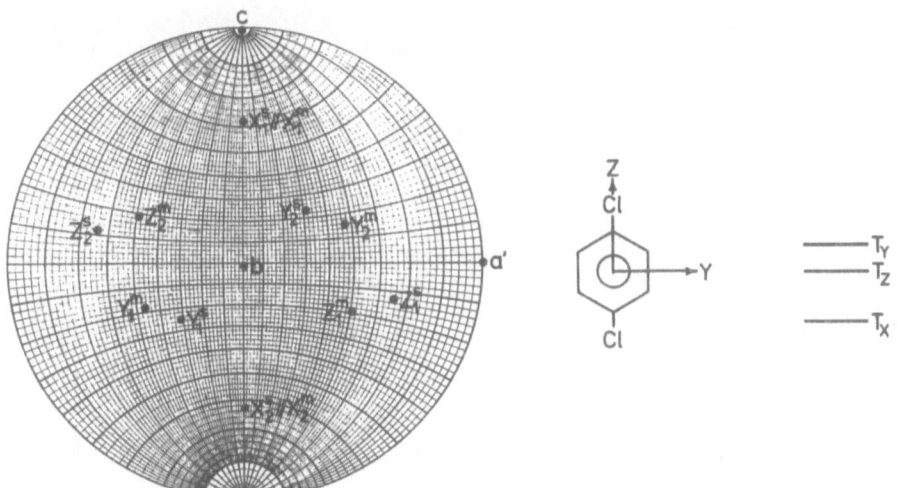

Fig. 31. The directions of the spin axes x_s, y_s and z_s of the p-DCB X-traps with respect to the a', b and c crystal axes as represented in a stereographic projection. The directions of the molecular axes x_m, y_m and z_m represent the most probable of two alternative choices. All points are indicated on the front half of the sphere. Also depicted is the ordering of the triplet sublevels.

unsuccessful. The ESE technique was then applied to study the exciton trapping by monitoring the evolution of the X-trap population as a function of the delay time t_d.

In Fig. 32 the time evolution of the ESE signal of the $T_y - T_x$ transition is shown. The fast ingrowth of an emissive signal with a time constant of (0.40 ± 0.05) μs (Fig. 32b) is followed by a decay with a time constant of (6.5 ± 0.5) μs

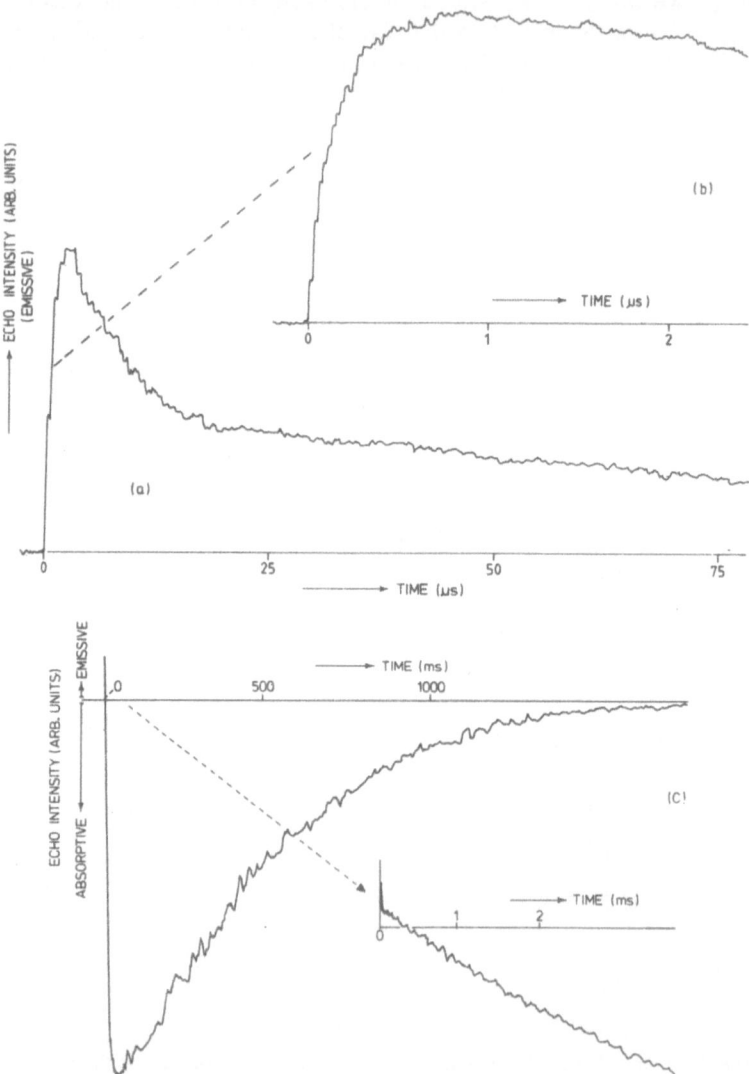

Fig. 32. (a) The short-time behavior of the ESE signal of the X-trap $T_y - T_x$ transition at 5362 MHz. The same evolution is shown in the inset (b) at an expanded time scale. (c) The change from emission to absorption of this signal on a millisecond time scale. $T = 1.2$ K.

to a smaller emissive signal (Fig. 32a). On a millisecond time scale (Fig. 32c) the echo signal changes from emission to absorption and finally decays to zero in (300 ± 50) ms. Similar experiments on the $T_y - T_z$ transition shows almost the same behaviour except for the fact that the final decay constant is (2.6 ± 0.3) ms. When tuning the spectrometer to the $T_z - T_x$ zero-field transition an ingrowth in (1.2 ± 0.2) µs to an emissive signal is observed without an overshoot and a subsequent long-time behaviour identical to that observed for the $T_y - T_x$ transition. For experimental reasons it is not possible to detect the evolution of the ESE signal in the first 150 ns following the laser flash.

7.3. DISCUSSION

Since the intensity ratio of the exciton and X-trap emission between 2.8 and 4.2 K shows an exponential behaviour with an activation energy very close to 24.6 cm^{-1}, it is concluded that in this temperature region a rapid communication exists between exciton and X-trap. The number of exciton states per trap $N = (2.0 ± 0.5) \times 10^3$ is of the same order as that found for TCB [30, 39]. As the emission of the deep trap at 1.2 K is unobservable it may be concluded that the capture probability of this trap is much smaller than that of the X-trap.

When comparing the lineshapes of the 0—0 with the 0—1 emission to a b_{2g} vibrational state of S_0 [28, 75, 76] (Figs. 30b, 30c) a remarkable difference in linewidth is observed. Moreover the 0—1 excitonic line exhibits two peaks separated by 3.5 cm^{-1}, in contrast to the trap 0—1 emissions which have a Gaussian lineshape. These effects can be understood by assuming that we observe the emission to a b_{2g} vibrational exciton also with a strong one-dimensional structure and with a bandwidth larger than that of the one-dimensional triplet excitons. Since a $\Delta k = 0$ selection rule applies for a transition between these two bands, the two peaks in Fig. 30d then are interpreted as the $k \approx 0$ and $k \approx \pm \pi/a$ regions of the vibrational exciton band, where the density of states is a maximum. The bandwidth of the triplet excitons can be obtained from the temperature dependence of the intensities of the two peaks, which must be given by a Boltzmann relation. Although in the paper by van Kooten and Schmidt [71] a value of 0.9 cm^{-1} was given as the triplet exciton bandwidth, later measurements by Jongenelis et al. down to 0.4 K showed that the bandwidth must be smaller than 0.1 cm^{-1} [72]. The width of the vibrational exciton band is estimated to be $\approx$ 4.5 cm^{-1}. The width and the shape of the 0—1 emission lines of the traps may be explained by realizing that the emission is from a localized excitation to a superposition of all k states of the vibrational excitons [79]. It is worth mentioning that this interpretation differs from the model by Bellows et al. [80] who, on the basis of a Raman study of the vibrational states in p-DCB, conclude that the double peak structure represents a Davydov splitting, i.e. that it is caused by interaction between inequivalent molecules.*

---

* *Note added in proof.*

Recent calculations [81] have shown that the model of Bellows et al. [80] is correct and that the b_{2g} vibrational exciton is not one-dimensional.

The time-resolved optical experiments at 1.2 K reveal that the lifetime of the excitons is only 3 μs i.e. more than three orders of magnitude shorter than in TCB at the same temperature. The fact that the build-up time of the X-trap is slightly longer (5 μs) is an indication of the presence of an intermediate, shallower trap as was also found in TCB.

More information about the dynamic properties of the triplet excitons is provided by the ESE experiments. From the long-time behaviour it follows that the SLR rates in the X-trap at 1.2 K are in the order of or slower than the decay rates of the spin levels. From the work by Buckley et $al.$ [78] it is known that the lifetimes of T_y, T_z and T_x are 13 ms, 46 ms and 600 ms respectively. Thus the change from emission to absorption of the $T_y - T_x$ signal arises mainly from the decay of T_y and the subsequent decay to zero from the decay of the non-radiative level T_x.

The evolution of the ESE signals within microseconds after the laser flash must reflect the combined effect of SLR and trapping of the triplet excitons. This short-time evolution can be simulated by assuming that the laser flash predominantly populates the T_y sublevel of the triplet excitons (this is reasonable since T_y is the sublevel with the highest transition probability in the $T_0 \leftarrow S_0$ absorption [78]) and that a SLR time of a few hundred nanoseconds is present among the sublevels of the exciton, together with a trapping time of about 3—5 μs. The idea that the SLR in the triplet exciton band is fast compared with the trapping is supported by the observation that the maximum in the absorptive part of the ESE signal of the $T_y - T_x$ transition is about 10 times larger than the emissive part. This proves that upon trapping T_y and T_x receive about the same population, i.e. the spin alignment in the triplet exciton band is almost completely lost before trapping takes place.

It is remarkable that in p-DCB the SLR rates and the trapping rates are so much faster than in TCB. In zero-field it was found that the SLR rates in TCB are about $(150 \text{ μs})^{-1}$ compared with $(0.2 - 0.5 \text{ μs})^{-1}$ for p-DCB, whereas the trapping rates were found to be about $(10 \text{ ms})^{-1}$ compared with $(3-5 \text{ μs})^{-1}$ for p-DCB. In analogy with the discussion in Section 6.4 we suggest that the SLR in p-DCB is caused by a fast inter-stack hopping. The average SLR rate being given by expression (32). Since we know the angles between pairs of corresponding spin axes in the two inequivalent stacks we estimate the interstack hopping rate $P_{AB} \approx 10^7 \text{ s}^{-1}$. This should be compared with an inter-stack hopping rate $P_{AB} \approx 2 \times 10^5 \text{ s}^{-1}$ in TCB at the same temperature of 1.2 K. The very fast inter-stack hopping rate in p-DCB probably is caused by a larger exchange interaction between inequivalent molecules. In TCB it is known that $\beta_\perp / \beta_\parallel < 3 \times 10^{-4}$ [45]. When comparing the two values for P_{AB} we would estimate the ratio $\beta_\perp / \beta_\parallel$ in p-DCB to be $10^{-2}—10^{-3}$. It is not clear whether this larger value of $\beta_\perp$ also is an explanation for the faster trapping in p-DCB. The fact that the ratio of the SLR rates in the two systems is close to the ratio of their trapping rates suggests that the two effects might be related.

7.4. CONCLUSION

The optical and ESE experiments in p-DCB reveal that the triplet exciton bandwidth in this system $4\beta < 0.1$ cm^{-1} compared with $4\beta = 1.34$ cm^{-1} in TCB. The SLR rates and trapping rates at 1.2 K are three orders of magnitude faster than in TCB. The difference in SLR rates is explained by assuming that in p-DCB a faster inter-stack hopping takes place which is the result of a larger value of the inter-stack interaction $\beta_\perp$. We suggest that the faster trapping rate in p-DCB is related to the same process. A possible explanation for the larger value of $\beta_\perp$ is found when considering the mutual orientation of the two inequivalent molecules. In the case of p-DCB a much larger overlap of the π orbitals is predicted than in the case of TCB.

8. Summary

One of the most important results of the study of TCB is that in this system the $k - k'$ scattering of the triplet excitons at very low temperatures is a process induced by impurities that break the translational symmetry of the exciton-phonon coupling. To our surprise we find that chemical impurities do not play an important role and we conclude that structural defects most probably are responsible for the observed scattering. Further it appears that one-phonon processes dominate the scattering at temperatures where $k_B T$ is smaller than the width of the exciton band and that at higher temperatures two-phonon Raman type processes take over. It is not clear whether in the latter process impurities also play a role.

The results of the SLR measurements on TCB indicate that the rate of inter-stack hopping is of the same order of magnitude as the average scattering rate. A striking difference in SLR rate is found in the related one-dimensional system p-DCB. Here the SLR rates are three orders of magnitude faster at the same low temperature of 1.2 K. The much faster inter-stack hopping in p-DCB responsible for this fast SLR is thought to be caused by a larger exchange interaction between inequivalent molecules. This seems reasonable when comparing the mutual orientation of inequivalent molecules which favours a larger overlap of the π-orbitals in p-DCB than in TCB.

Although the results presented here have solved a number of problems related to the dynamical properties of one-dimensional triplet excitons several questions still remain to be solved. For instance it is not clear whether the scattering is restricted to one stack or that it is caused by an inter-stack process. Nor is it clear whether the collective excitation moves through the crystal along the chain. Further it remains obscure what causes the trapping in particular the enormous difference in trapping rates between the two related systems TCB and p-DCB.

Hopefully all these questions will get an answer in the coming years.

9. Acknowledgement

This work was performed as part of the research program of the "Stichting voor Fundamenteel Onderzoek der Materie (FOM)" with financial support from the "Nederlandse Organisatie voor Wetenschappelijk Onderzoek (NWO)".

References

1. J. Frenkel, *Phys. Rev.* **37** (1931), 17; 1276.
2. A. S. Davydov, *Theory of Molecular Excitons*, McGraw-Hill, New York (1962).
3. A. S. Davydov, *Phys. Stat. Sol.* **20** (1967), 143.
4. A. S. Davydov, *Theory of Molecular Excitons*, Plenum Press, New York, London (1971).
5. D. P. Craig and S. H. Walmsley, *Excitons in Molecular Crystals*, Benjamin, New York (1968).
6. R. S. Knox, 'Theory of Excitons', in *Solid State Physics Supplement 5*, eds. H. Ehrenreich, R. Seitz and B. Turnbull, Academic Press, New York (1963).
7. D. M. Burland and A. H. Zewail, *Advances in Chemical Physics*, Vol. XL, eds. I. Prigogine and S. A. Rice, John Wiley and Sons, New York (1979).
8. V. L. Broude, E. I. Rashba and E. F. Sheka, *Spectroscopy of Molecular Excitons*, Springer Verlag, Berlin (1983).
9. J. Schmidt and J. H. van der Waals, *Time Domain Electron Spin Resonance*, eds. L. Kevan and R. N. Schwartz, John Wiley and Sons, New York (1979).
10. D. Haarer, D. Schmid, and H. C. Wolf, *Phys. Stat. Sol.* **23** (1967), 633.
11. D. Haarer and H. C. Wolf, *Phys. Stat. Sol.* **33** (1969), K117; *Mol. Cryst. Liq. Cryst.* **10** (1970), 359.
12. L. Yarmus, J. Rosenthal and M. Chopp, *Chem. Phys. Lett.* **16** (1972), 477.
13. W. Bizarro, L. Yarmus, J. Rosenthal and N. F. Berk, *Chem. Phys. Lett.* **55** (1978), 49.
14. P. Reineker, *Phys. Stat. Sol.* **B70** (1975), 189 and 471.
15. H. Haken and G. Strobl, *The Triplet State*, ed. A. Zahlan, Cambridge University Press (1967); *Z. Physik* **262** (1973), 135.
16. J. Rosenthal, L. Yarmus, N. F. Berk, and W. Bizarro, *Chem. Phys. Lett.* **56** (1978), 214.
17. H. Port and D. Rund, *Chem. Phys. Lett.* **54** (1978), 474.
18. H. Port, K. Mistelberger and D. Rund, *Mol. Cryst. Liq. Cryst.* **50** (1979), 11.
19. H. Port and D. Rund, *Chem. Phys. Lett.* **69** (1980), 406.
20. R. Schmidberger and H. C. Wolf, *Chem. Phys. Lett.* **16** (1972), 402; **25** (1974), 185; **32** (1975), 18 and 21.
21. A. H. Francis and C. B. Harris, *Chem. Phys. Lett.* **9** (1971), 181 and 188.
22. R. M. Hochstrasser and J. D. Whiteman, *J. Chem. Phys.* **56** (1972), 5945.
23. D. M. Burland, *J. Chem. Phys.* **59** (1973), 4283.
24. D. M. Burland, U. Konzelman, and R. M. Macfarlane, *J. Chem. Phys.* **67** (1977), 1926.
25. J. Klafter and J. Jortner, *J. Chem. Phys.* **68** (1958), 1513.
26. H. Port, private communication.
27. D. D. Dlott and M. D. Fayer, *Chem. Phys. Lett.* **41** (1976), 305.
28. G. A. George and G. C. Morris, *Mol. Cryst. Liq. Cryst.* **11** (1970), 61.
29. R. M. Shelby, A. H. Zewail, and C. B. Harris, *J. Chem. Phys.* **64** (1976), 3192.
30. W. Güttler, J. U. von Schütz, and H. C. Wolf, *Chem. Phys.* **24** (1977), 159.
31. C. B. Harris and M. D. Fayer, *Phys. Rev.* **B10** (1974), 1784.
32. D. M. Burland, D. E. Cooper, M. D. Fayer, and C. R. Gochanour, *Chem. Phys. Lett.* **52** (1977), 279.
33. J. Zieger and H. C. Wolf, *Chem. Phys.* **29** (1978), 209.
34. R. D. Wieting and M. D. Fayer, *J. Chem. Phys.* **73** (1980), 744.
35. W. G. Breiland and M. C. Saylor, *J. Chem. Phys.* **72** (1980), 6485.
36. D. D. Dlott, M. D. Fayer, and R. D. Wieting, *J. Chem. Phys.* **67** (1977), 3808.
37. R. D. Wieting, M. D. Fayer, and D. D. Dlott, *J. Chem. Phys.* **69** (1978), 1996.
38. D. A. Zwemer and C. B. Harris, *J. Chem. Phys.* **68** (1978), 2184.
39. B. J. Botter, A. I. M. Dicker, and J. Schmidt, *Mol. Phys.* **36** (1978), 129.
40. A. J. van Strien, J. F. C. van Kooten, and J. Schmidt, *Chem. Phys. Lett.* **76** (1980), 7.
41. G. H. Wannier, *Phys. Rev.* **52** (1937), 191.
42. A. H. Francis and C. B. Harris, *J. Chem. Phys.* **57** (1972), 1050.
43. J. Schmidt and J. H. van der Waals, *Chem. Phys. Lett.* **3** (1969), 546.

44. M. J. Buckley and C. B. Harris, *Chem. Phys. Lett.* **5** (1970), 205; *J. Chem. Phys.* **56** (1972), 137.
45. S. J. Sheng and D. M. Hanson, *Chem. Phys. Lett.* **33** (1975), 451.
46. C. A. van 't Hof, Thesis, University of Leiden (1976).
47. L. E. Erickson, *Phys. Rev.* **143** (1966), 295.
48. M. Mehring and F. Freysoldt, *J. Phys. E.; Sci. Instrum.* **13** (1982), 894.
49. J. F. C. van Kooten, Thesis, University of Leiden (1984).
50. C. Dean, M. Pollak, B. M. Craven, and G. A. Jeffrey, *Acta Cryst.* **11** (1958), 710.
51. F. H. Herbstein, *Acta Cryst.* **18** (1965), 997.
52. P. W. Anderson, *J. Phys. Soc. Japan.* **9** (1954), 316.
53. A. J. van Strien, J. Schmidt, and R. Silbey, *Mol. Phys.* **49** (1982), 151.
54. H. Benk and R. Silbey, *J. Chem. Phys.* **79** (1983), 3487.
55. D. R. Yarkony and R. Silbey, *J. Chem. Phys.* **67** (1977), 5818.
56. V. M. Kenkre and R. S. Knox, *Phys. Rev.* **B9** (1974), 5279.
57. B. Jackson and R. Silbey, *J. Chem. Phys.* **77** (1982), 2763.
58. J. F. C. van Kooten, M. G. Munowitz, and J. Schmidt, *Mol. Phys.* **52** (1984), 1397.
59. J. C. Brock, *Chem. Phys. Lett.* **55** (1978), 267.
60. D. A. Zwemer, C. B. Harris, and H. C. Brenner, *Chem. Phys. Lett.* **57** (1978), 505.
61. H. Benk, H. Haken, and H. Sixl, *J. Chem. Phys.* **77** (1982), 5730.
62. H. Wolfrum, K. F. Renk, and H. Sixl, *Chem. Phys. Lett.* **68** (1979), 90.
63. D. R. Lutz, K. A. Nelson, R. W. Olsen, and M. D. Fayer, *J. Chem. Phys.* **69** (1978), 4319.
64. A. J. van Strien and J. Schmidt, *Chem. Phys. Lett.* **86** (1982), 203.
65. D. Antheunis, Thesis, University of Leiden (1974).
66. P. J. F. Verbeek, H. J. den Blanken, and J. Schmidt, *Chem. Phys. Lett.* **60** (1979), 358.
67. P. J. F. Verbeek and J. Schmidt, *Chem. Phys. Lett.* **63** (1979), 384.
68. J. F. C. van Kooten, A. J. van Strien, and J. Schmidt, *Chem. Phys. Lett.* **90** (1982), 337.
69. C. A. van 't Hof and J. Schmidt, *Mol. Phys.* **38** (1979), 309.
70. H. B. Levinsky and H. C. Brenner, *Chem. Phys. Lett.* **78** (1981), 177.
71. J. F. C. van Kooten and J. Schmidt, *Chem. Phys. Lett.* **117** (1985), 77.
72. A. P. J. M. Jongenelis, E. H. Abramson, and J. Schmidt, to be published.
73. P. A. Reynolds, J. K. Kjems, and J. W. White, *J. Chem. Phys.* **56** (1972), 2928.
74. M. Ghelfenstein and H. Zware, *Mol. Crystl. Liq. Cryst.* **14** (1971), 283.
75. B. W. Gash, D. B. Hellmann, and S. D. Colson, *Chem. Phys.* **1** (1972), 191.
76. G. Castro and R. M. Hochstrasser, *J. Chem. Phys.* **46** (1967), 3617.
77. C. von Borczyskowski, *Chem. Phys. Lett.* **85** (1982), 293.
78. M. J. Buckley, C. B. Harris, and R. M. Panos, *J. Am. Chem. Soc.* **94** (1972), 3692.
79. C. Aslangul and Ph. Kottis, *Phys. Rev.* **B13** (1976), 439.
80. J. C. Bellows, P. N. Prasad, E. M. Monberg, and R. Kopelman, *Chem. Phys. Lett.* **54** (1978), 439.
81. A. P. J. M. Jongenelis, T. H. M. van den Berg, J. Schmidt and A. van der Avoird, *J. Chem. Phys.* submitted.

SPECTRAL HOLE-BURNING IN CRYSTALLINE AND AMORPHOUS ORGANIC SOLIDS. OPTICAL RELAXATION PROCESSES AT LOW TEMPERATURE

SILVIA VÖLKER

Center for the Study of the Excited States of Molecules, Huygens and Gorlaeus Laboratories, University of Leiden, Leiden, The Netherlands.

1. Introduction to Hole-Burning

With the development of tunable lasers in the early 1970s, the interest in the study of optical relaxation processes of impurity ions and molecules in solids by means of coherent optical techniques has grown rapidly. Two types of information are sought in these studies, first the relaxation or dephasing mechanism of an individual guest molecule or ion due to its interaction with the host, and second the propagation of optical excitation within the solid due to mutual interaction or energy transfer between different guests. Whereas both processes lead to broadening and shift of the optical linewidth, the second process is revealed by an additional time evolution of the spectral profile.

The need for coherent optical techniques arises from the fact that spectral lineshapes in solids are seldom determined by dynamical interactions, but mainly by strain or structural disorder. This spread in local interactions between the guest molecules and their environment causes a distribution of frequencies of the electronic transition which gives rise to inhomogeneous broadening of the spectral band. In *crystalline hosts* at low temperature the *inhomogeneous* width, Γ_{inh}, usually is of the order of $0.1-5$ cm^{-1}, whereas in *glasses*, $\Gamma_{inh} \approx 100-500$ cm^{-1}. Two examples of inhomogeneously broadened spectra are given in Fig. 1 [1]. The guest is an organic molecule (free-base porphin, H_2P) embedded in an n-alkane crystal (a), and in an organic glass (b) at 4.2 K. Notice the much narrower linewidth (3 cm^{-1}) in the crystalline host as compared to the glassy host (150 cm^{-1}). Great improvement in narrowing of spectral lines, mainly of molecules incorporated in glasses at low temperature, can be achieved by site-selection spectroscopy [2—4]. The excitation energy of a given transition is selected by means of a laser, and the fluorescence or phosphorescence signal is then detected in the $0-0$ transition, or a vibrational band, with a monochromator of high resolution. In a site-selected excitation spectrum, the emission wavelength is fixed and the laser scanned; conversely, in a site-selected fluorescence spectrum (a technique called fluorescence line-narrowing [3]) the laser excites the sample at a fixed frequency, while the monochromator is scanned (for a review, see ref. [4]).

In order to get information on the intrinsic (*homogeneous*) optical lineshape, various coherent laser techniques have been developed in both the time domain and the frequency domain. To the first category belong photon echoes [5—9] and

113

J. Fünfschilling (ed.), Relaxation Processes in Molecular Excited States, 113—242.
© 1989 *by Kluwer Academic Publishers.*

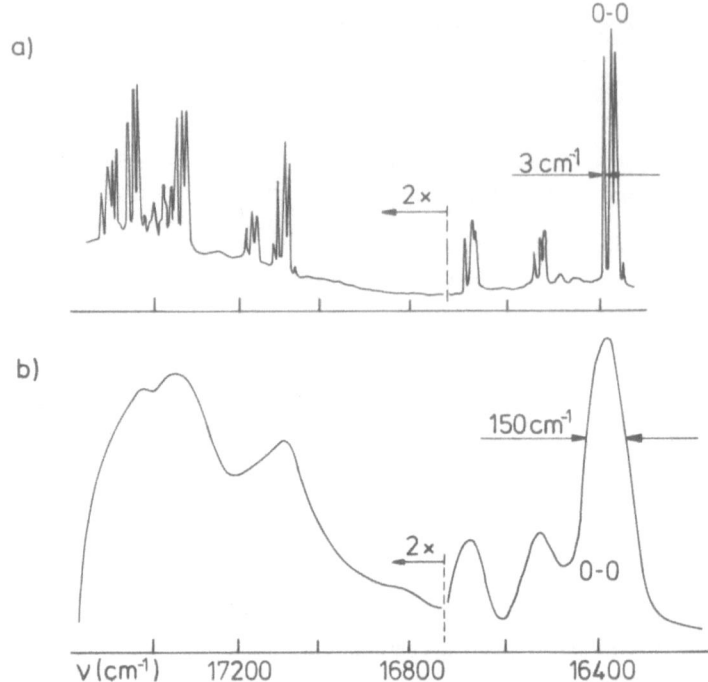

Fig. 1. Excitation spectra of free-base porphin (H_2P) at $T = 4.2$ K with broad-band detection of the fluorescence signal: (a) in a crystal of n-heptane; (b) in 2-methyltetrahydrofuran (MTHF) glass. Notice the difference of almost two orders of magnitude in the inhomogeneous linewidth between a crystal and a glass [1].

optical free induction decay [10—13], to the second, fluorescence line-narrowing [2, 3, 14] and spectral hole-burning [15—20]. The homogeneous linewidth, Γ_{hom}, and the effective dephasing time, T_2, of an optical transition are related by the expression $\Gamma_{hom} = (\pi T_2)^{-1}$, and as will be shown below time- and frequency-domain techniques are complementary, each having certain advantages and disadvantages.

The intention of this chapter is to present a review of the literature on specific aspects of spectral hole-burning (HB), in particular of organic solids. The systems to be discussed are organic molecules diluted as guests in *crystalline* as well as in *glassy* host matrices at low temperature. In Section 2, a comparison of the various techniques used for hole-burning will be given, and special emphasis will be placed on the need to work at very low burning fluences if one wants to determine the true homogeneous linewidth. In Section 3, a large number of organic and a few inorganic systems, classified according to their hole-burning mechanism, will be discussed. Because many HB-experiments have been performed on porphyrins and related molecules of biological relevance, we have devoted special attention to them (see 3.1). Section 4 deals with some selected hole-burning applications.

Optical relaxation and dephasing processes of electronically excited states of crystalline and amorphous systems, at temperatures between ~ 0.3 and ~ 20 K, will be discussed in more detail, and a comparison will be made between hole-burning and photon echo results. We have further included some recent hole-burning experiments on materials like organic molecules incorporated in semi-crystalline polymers and adsorbed on the surface of porous silica glass, and rare earth ions (Eu^{3+}) embedded as guests in organic glasses. The purpose of these experiments was to test models for optical dephasing in amorphous solids. Other hole-burning applications to be presented are the effects of external fields on holes burnt in zero field in crystalline and glassy systems. It will be shown that Zeeman and Stark effects can be accurately measured with MHz-resolution in ordered as well as in disordered materials. Finally, some technical applications of hole-burning, such as frequency domain optical storage, attractive from an industrial point of view, will be briefly discussed (see 4.1.5). The reader interested in this subject and other practical HB-applications is referred to recent reviews [21, 22].

It should be mentioned that parts of the contents of this chapter have been published in refs. [23, 24]. A number of other review articles complement different experimental and theoretical aspects of hole-burning. For organic systems, the reader may consult refs. [21, 23—30], for inorganic systems, refs. [21, 30—32], for optical spectroscopy of glasses, refs. [21, 23—34], for external field effects, refs. [23, 35], for hole-burning in the IR-regime and technological applications, refs. [21, 22], for optical dephasing in crystals, refs. [23, 25] and [36] for experiments and theory, respectively.

1.1. BASIC CONCEPTS

The influence of the host dynamics on an isolated probe molecule is reflected in the width of the homogeneous line, Γ_{hom}, of an optical transition. Γ_{hom} can be obtained by solving the time evolution of the density matrix of a two level system (with ground state S_0, and electronically excited state S_e) coupled to a coherent electric field, and to a fluctuating heat bath with which the optical system can exchange energy. This represents the optical analogue of the Bloch equations for a precessing spin (see also chapters by J. Schmidt, and R. Silbey in this book). The effective optical dephasing time, T_2, is then related to the homogeneous linewidth, Γ_{hom} (in Hz), by

$$\Gamma_{hom} = (\pi T_2)^{-1} = (2\pi T_1)^{-1} + (\pi T_2^*)^{-1}, \qquad (1)$$

where T_1 is the lifetime of the excited state S_e (also called longitudinal relaxation time in magnetic resonance), and T_2^* is the phase coherence or "pure dephasing" time determined by fluctuations of the optical transition frequency (usually due to phonon scattering, and electron and nuclear spin fluctuations, if present). On extrapolation to zero temperature, $2T_1$ generally provides a limiting value for T_2 (or Γ_{hom}).

As mentioned above, information on the homogeneous linewidth which is hidden under the inhomogeneous spectral band can be obtained, among others, with the hole-burning (HB) technique. The essence of HB is that the transition energy of the guest molecule embedded in a solid matrix (crystalline or glassy) changes after absorption of monochromatic light. By irradiating the inhomogeneously broadened spectral band with a narrow band laser of width Γ_ℓ and frequency ν_1, those molecules that are resonant with the laser may undergo a phototransformation such that the product absorbs at a different frequency. This creates a hole or dip in the original absorption band at ν_1 (see Fig. 2). The photoproduct can either be stable at low temperature and a permanent hole is formed, or it can be a metastable state that acts as a population storage level, by which a *transient* hole is created. The hole is subsequently probed in a second step by means of a tunable laser, which is scanned over the frequency region of the hole. The intensity of the latter should be low enough to avoid further burning of the absorption band. The hole represents, after minor corrections, a negative replica of the homogeneous spectral transition and the holewidth, if carefully measured (see Section 2.2), yields the unknown quantity Γ_{hom}. Thus, with HB optical resolutions can be obtained that are 10^3–10^5 times higher than those reached with conventional techniques [1, 19, 23–34]. In this way, for example, it is possible to study guest-host interactions at

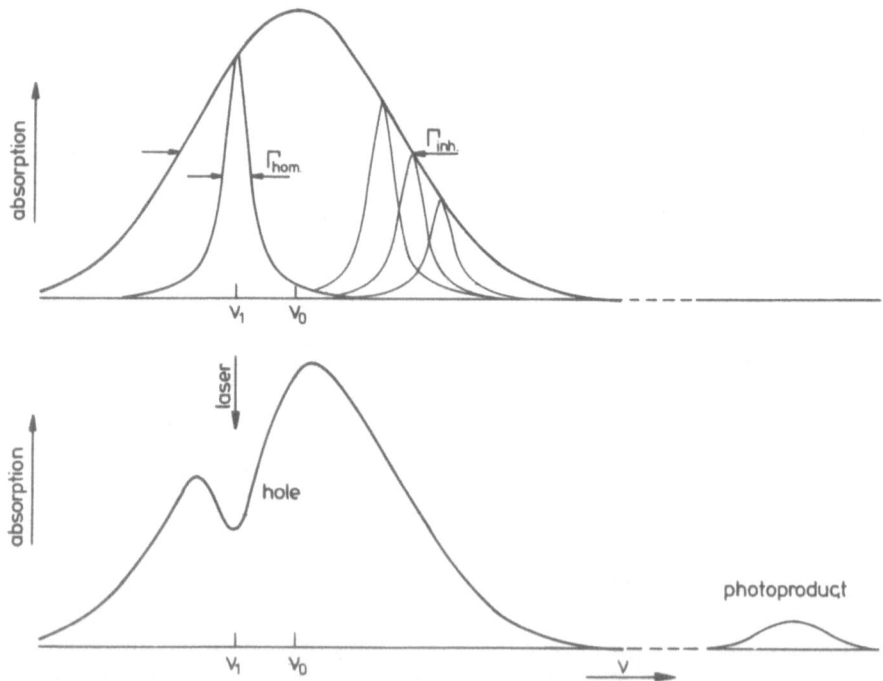

Fig. 2. Top: Diagram of an inhomogeneously broadened absorption band of width Γ_{inh}. It is the envelope of individual electronic transitions of homogeneous width Γ_{hom}. Bottom: Laser-induced "hole" burnt at frequency ν_1 at low temperature. The photoproduct absorbs at a different frequency, here assumed to be outside the inhomogeneous band.

very low temperatures, measure relaxation times of vibronically excited states, observe the shift of a hole under pressure, or perform Zeeman and Stark spectroscopy in the MHz-regime.

Because the optical transition energy depends on the environment in a complicated way, the laser is not selecting molecules in a specific environment, but a set of molecules in different environments, all absorbing at the same frequency ν_1. Furthermore, since the correlation between transition energy and environmental parameters is, in general, different for the photoproduct and the original molecule, it is expected that the photoproduct band or "anti-hole" will be broader than the hole.

1.2. MECHANISMS

Three mechanisms are known to be responsible for hole burning: (a) *photochemical* hole-burning (PHB) [1, 15, 17—19], (b) *non-photochemical* hole burning (NPHB) [16, 20, 26], and (c) *transient* hole-burning THB [37—39].

(a) *Photochemical hole-burning.* The absorption of the photoproduct of molecules that undergo *PHB* is usually well separated from the original absorption band. Within *PHB* one can distinguish between *intramolecular* photochemical reactions, which take place inside the guest molecule [1, 15, 17—19, 40—46] or *intermolecular* reactions, that occur between the guest and the host [27, 47—49]. Within the first group, photochemistry may be reversible, in the sense that irradiation into the photoproduct fills in the hole and reconstitutes the original spectrum (e.g. photoautomerism in free-base porphyrins [1, 18, 19, 23—25, 28, 41—46], see Section 3.1), or it may be irreversible (e.g. photofragmentation or dissociation of dimethyl-s-tetrazine [17] and s-tetrazine [40], see Section 3.2.2). Further, the photochemical process can be the result of the absorption of *one photon* as observed for porphyrins and quinizarin [27, 47, 48], or *two-photons* as is the case for dimethyl-s-tetrazine [50] and carbazole in boric acid glass [51]. Photochemical holes are usually persistent for hours or days, as long as the samples are kept at sufficiently low temperature.

(b) *Non-photochemical hole-burning (NPHB)* is characteristic for amorphous systems. After excitation with a narrow band laser, a slight structural rearrangement of the local environment of the guest molecule seems to take place by which the guest-host interaction changes [16, 20, 26, 52—56]. A tunneling mechanism has been proposed by Small and coworkers to explain this phenomenon, which is depicted in Fig. 3 [20, 26, 52]. It is assumed that the guest molecule is coupled to a two-level system (TLS) of the glass (TLS are very low frequency excitations in amorphous solids, see Section 4.2) and, after narrow-band laser excitation, tunneling takes place in the electronically excited state which is followed by relaxation to a different ground state configuration. As a consequence, the optical transition is shifted, and a *non-photochemical* hole is burnt in the absorption band. In this model the absorption frequency of the "photoproduct" was assumed to be very

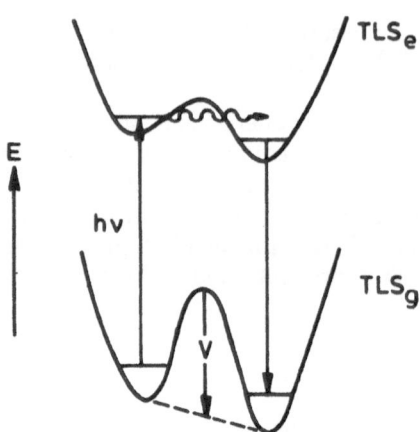

Fig. 3. Diagram of the non-photochemical (NPHB) mechanism according to refs. [20, 26, 52]. TLS_g and TLS_e represent potential energy curves of a two-level system (TLS) of a glass interacting with the ground and electronically excited states of a guest molecule. After laser irradiation, NPHB is thought to occur through tunneling in the excited state.

close (< 2 cm^{-1}) to that of the original molecule [20], thus, within the inhomogeneously broadened absorption band. Holes which are the result of NPHB are permanent at temperatures that are much lower than that corresponding to the ground state barrier V (Fig. 3). Examples of photostable organic molecules known to undergo NPHB in glassy matrices are perylene [16], tetracene [20, 26, 52], pentacene [55, 56], and tetracene-doped anthracene glasses [54]. A similar type of hole-burning, in the sense that it is photophysical and not photochemical in nature, has also been observed in inorganic glasses doped with rare earth ions [53].

It should be remarked that the time scales on which PHB- and NPHB-experiments are performed (usually seconds to minutes) are long compared to those of photon echo experiments (picoseconds to milliseconds). One of the reasons for it is that hole-burning quantum efficiencies are often low, and consequently a hole burnt in a short time would not be detectable.

(c) *Transient hole-burning (THB)* [37—39, 57] is a mechanism by which population is transferred from the ground state to a metastable state (e.g. a long-living triplet state or hyperfine level). The lifetime of the hole (usually microseconds to seconds) is determined by the decay time of the metastable state. Examples that show transient hole-burning behaviour will be given in Section 3.1.4.

2. Experimental

2.1. METHODS AND TECHNIQUES

Persistent holes (*photochemical* and *non-photochemical*) are usually burnt with narrow-band dye lasers at a given temperature. Subsequently, the hole is probed either at the same temperature [1, 15—20, 23, 24, 38, 39, 43—45, 49, 52, 53, 55, 56, 58—64, 67], as is generally the case, or at a different temperature [48, 54, 65—68]. If one wants to get information on the homogeneous linewidth, Γ_{hom}, an additional stringent condition is that $\Gamma_{laser} \ll \Gamma_{hom}$. Various probing schemes have been used, among which transmission spectroscopy [16, 20, 26, 47, 48, 52, 55, 59, 62, 64—71] and excitation spectroscopy [1, 15, 17—19, 24, 38—46, 49, 53, 54, 56—58, 60, 61, 63] are the most common ones. In the latter case, the spectrum is generally scanned with the same laser used for burning, but at 10^{-3}—10^{-1} of its original intensity. Hole-burning experiments probed in transmission with a narrow-band laser have been reported [62, 64, 67, 68, 70, 71], but in many experiments a lamp in combination with a monochromator have been used in the probing step, which introduces significant errors in the determination of the holewidth (see Section 2.2). When the sample emits sufficient light, the excitation method is to be preferred over the transmission method, because samples of low optical quality and low optical density can be measured with high sensitivity using very low laser burning powers (for a critical comparison of the two methods, see refs. [24, 56]). The only pre-requisite of the excitation technique is that part of the absorbed energy in the sample is emitted as fluorescence or phosphorescence.

Fig. 4 shows a schematic diagram of the experimental arrangement for *PHB* and *NPHB* measurements as used in refs. [1, 24, 43, 45, 49, 56, 72]. The sample,

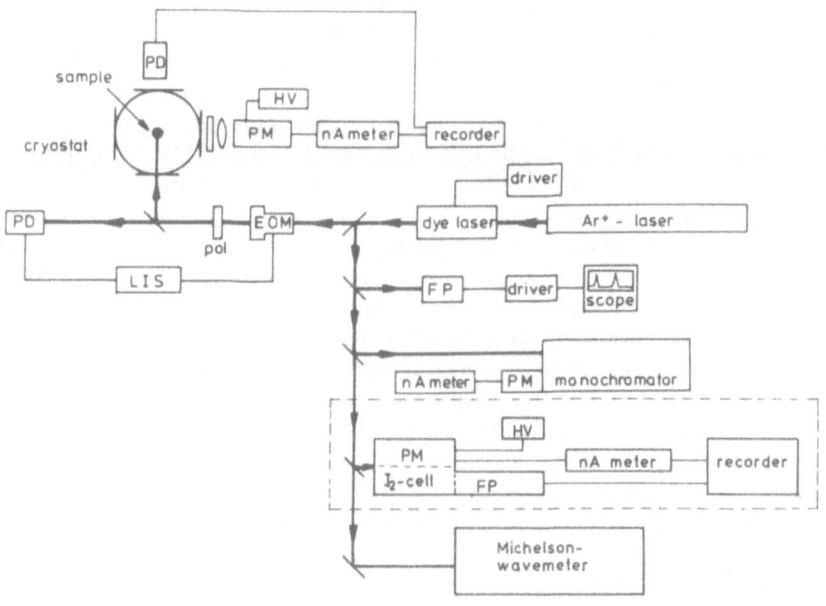

Fig. 4. Experimental set-up for PHB and NPHB, for details see text and ref. [72]. FP: Fabry-Pérot interferometer; PM: photomultiplier; EOM: electro-optic modulator; LIS: light intensity stabilizer; PD: photodiode. The part of the scheme enclosed by a dashed rectangle represents a home-built optical wavemeter (see text). The frequency of the cw-dye laser was either checked with the latter or with a Michelson-type interferometric wavemeter (accuracy ~ 50 MHz).

immersed in liquid helium, is irradiated with a single-frequency cw-dye laser (Coherent Radiation, model 599—21), pumped by an Ar^+-ion laser. The frequency jitter of the dye laser is ≤ 2 MHz, and its frequency scanning range covers 30 GHz. In order to have a rough frequency reference for the cw laser, part of the laser beam is directed through a 0.75 m monochromator (Jarell-Ash, bandwidth ~ 1 cm^{-1}) tuned to the frequency of the absorption band of interest. For a wavelength accuracy of about 50 MHz, another part of the beam is directed either through a Michelson-type interferometric wavemeter [73], or through a home-built optical wavemeter consisting of an iodine cell in series with a high finesse (F = 400) Fabry-Pérot interferometer [74]. The mode structure of the laser and its scan length can be analyzed and calibrated by means of a spectrum analyzer (FSR = 1.5 GHz) connected with a scanning interferometer driver to an oscilloscope. The amplitude of the laser output is stabilized to $< 0.5\%$ by an electro-optic modulator (Coherent, model 28) driven by home-built electronics. The laser beam may be focused or defocused on the sample, depending on the power density needed for burning.

The holes are subsequently probed by excitation spectroscopy. The fluorescence signal from the sample is detected in a direction perpendicular to the excitation beam by a cooled photomultiplier (EMI, type 9658 R). In order to separate the fluorescence from the excitation light, the emission from the sample is passed through a cut-off filter before it reaches the photomultiplier. In special cases, the fluorescence signal is selectively detected by means of a 0.85 m-double monochromator (Spex 1402) set at a vibrational band (resolution ~ 1 cm^{-1}). The signal from the photomultiplier is measured by an electrometer (Keithley, model 610 CR) and fed into a x-t recorder. For some amorphous systems, the hole shapes were probed simultaneously in *fluorescence* (see above), and in *transmission* through the sample with a photodiode (EG&G, Model HUV 4000) [56].

In cases where it is impossible to probe holes in fluorescence excitation spectroscopy because of absence of emission from the sample, a special type of low temperature photoacoustic spectroscopy may be used. This photoacoustic technique is based on resonant detection of second sound in superfluid ^{4}He by means of a resistance thermometer in a Helmholtz resonator [75]. This reasonator consists of two cavities connected to each other by a narrow tube (total volume ~ 3 cm^3, Q $\simeq$ 120) placed in a bath with liquid helium below its λ-point (2.17 K). One of the cavities contains the sample, the other the thermometer. The excitation source is a cw-dye laser modulated by an acousto-optic device at the resonance frequency of the cell of $\nu \simeq 120$ Hz. For hole-burning experiments, the cw-single frequency dye laser mentioned above was used, whereas for site selected spectra, a cw-broad band dye laser (Spectra Physics, model 375, bandwith $\simeq 1$ cm^{-1}) tunable over the whole dye output was taken. The absorption of periodic light pulses by the sample leads to the generation of second sound waves (undamped temperature oscillations) picked up by the thermometer. For a detailed description of the apparatus, and examples of high resolution and hole-burning spectra of crystalline and amorphous samples at 1.65 K, see ref. [75].

Other hole probing techniques that have the advantage of yielding signals against zero background (for a review, see ref. [76]), are laser frequency modulation (FM) spectroscopy [77—79], ultrasonic modulation of holes at kHz [80a] and MHz rates [80b], polarization spectroscopy [81, 82], and holographic detection of holes in which two crossed laser beams induce spatial and temporal periodic changes of the spectral features [83, 84].

For *transient* hole-burning, either one or two single-frequency cw-dye lasers are used, depending on the lifetime of the metastable level [37—39]. In the latter case, the first laser is used at a fixed frequency to burn the hole, the second one, at a lower intensity, to scan the excitation spectrum of the hole. The rest of the experimental set-up may be similar to that described above for PHB experiments.

Absorption spectra at low temperature are usually taken before starting hole-burning experiments, in order to determine the frequency positions of the electronic transitions of the guest molecules in their respective hosts. These spectra are often recorded in the form of fluorescence excitation spectra. The experimental arrangement is then similar to that used for hole-burning, but with a different excitation source, for example, a tunable dye laser pumped by a pulsed nitrogen laser (Molectron, DL-200, bandwidth ~ 0.7 cm^{-1} in refs. [1, 24, 43, 45, 49, 56, 72]). The emission signal is detected either broad-band as explained above, or by site-selection using a fixed wavelength set by the 0.85 m double monochromator (bandwidth $1-2$ cm^{-1}).

A gas flow cryostat was used for temperatures above 4.2 K in refs. [1, 85, 86]. Between 4.2 and 1.2 K, the measurements were carried out in a conventional four-window ^{4}He bath cryostat in which the temperature, controlled via the vapour pressure, was measured by a calibrated carbon resistor with an accuracy of ± 0.01 K [1, 24, 39, 43, 45, 49, 56, 60, 72, 75, 86—88]. For temperatures below 1.2 K a home-built ^{3}He-cryostat, placed inside the ^{4}He-cryostat, was used [24, 49, 72, 87, 88]. For technical details of the ^{3}He-system, see refs. [72, 89].

In order to get an accurate determination of Γ_{hom}, a series of holes should be burnt as a function of laser power and burning time. Whenever the relative depth of the hole is shallow (a few %), and $\Gamma_{hom} \ll \Gamma_{inh}$, the value of Γ_{hom} can be estimated from the relation $\Gamma_{hom} \approx 0.5\Gamma_{hole} - \Gamma_{\ell}$. Here Γ_{hole} is the measured total holewidth at half height of an approximately Lorentzian hole, and Γ_{ℓ} is the laser jitter (~ 2 MHz). Since the hole is a convolution of a burning step and a probing step, the relation $\Gamma_{hole} = 2\Gamma_{hom}$ is valid for holes burnt and probed at the same temperature (see appendix of ref. [19], [27, 90]).

2.2. PITFALLS IN THE DETERMINATION OF THE HOMOGENEOUS LINEWIDTH

Hole-burning is a rather straightforward technique, but high laser powers and long burning times may cause serious problems like local heating, specially in the study of amorphous systems which are known to have poor thermal conductivity [91]. PHB and NPHB experiments, in which the influence of laser power and burning time on the holewidths has been studied [43, 49, 56, 86], suggest that many of the holewidths reported in the literature [20, 47, 48, 52, 54, 55, 59, 65, 66, 69, 92—94] may suffer from saturation and/or heating effects. In order to determine the value of the homogeneous linewidth, Γ_{hom}, it is not sufficient to verify that the hole shape is Lorentzian, but one should measure the holewidth as a function of laser power and burning time in a series of experiments. This is illustrated in Fig. 5, where three holes burnt for 20 seconds in the $S_1 \leftarrow S_0$ 0—0 transition of free-base porphin (H_2P) in polyethylene (PE) at various burning powers, at $T = 1.2$ K, are represented [43]. Notice that the holes broaden by a factor of almost 20, for a power increase of about 10^5. The hole shape, however, remains Lorentzian, independent of burning power and time.

Two other examples of holewidths as a function of laser power and burning time are shown in Fig. 6. The holewidth of the $S_1 \leftarrow S_0$ 0—0 transition of H_2P in ethanol at 4.2 K seems to increase in an approximately linear fashion at very low burning powers, whereas it becomes saturated at high powers (Fig. 6a) [43]. Fig. 6b shows results of dimethyl-s-tetrazine (DMST) in PE at 2.3 K. The three curves, traced through the data by means of a least square fit, extrapolate to just one value of the holewidth for zero burning time [43]. The dependence on burning power and time has been studied in a very systematic way for the NPHB-system pentacene in polymethylmethacrylate (PMMA) [56]. A plot of the holewidth, $\frac{1}{2}\Gamma_{hole}$, as a function of burning time t is shown in Fig. 7a for two different burning powers at 1.2 K. Notice that the values of $\frac{1}{2}\Gamma_{hole}$ on the lowest curve extrapolate smoothly to $\Gamma_{hom} = 0.5$ GHz when $t \to 0$, whereas the two upper curves reach the same value of Γ_{hom} but with a much steeper slope. In the latter case, much shorter burning times are needed in order to get the right extrapolation value. Further, observe that the holewidths become saturated at longer burning times t and/or higher burning powers P (the values of P and t depend on the system investigated) in a similar way as in Fig. 6a for a PHB-system. Fig. 7b is a log-log plot of $\frac{1}{2}\Gamma_{hole}$ versus burning fluence, Pt, for the same NPHB system as in Fig. 7a. The data

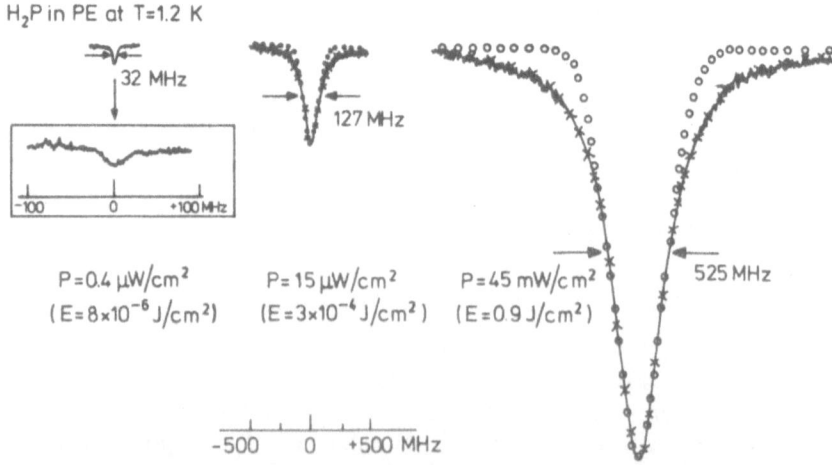

Fig. 5. Holes burnt in the $S_1 \leftarrow S_0$ 0—0 transition of H_2P in polyethylene (PE) at $T = 1.2$ K, for 20 s at various burning powers. Lorentzian (crosses) and Gaussian (open circles) fits are shown. Notice that an increase in power of about 10^5 produces a hole broadening of a factor ≈ 20 [43].

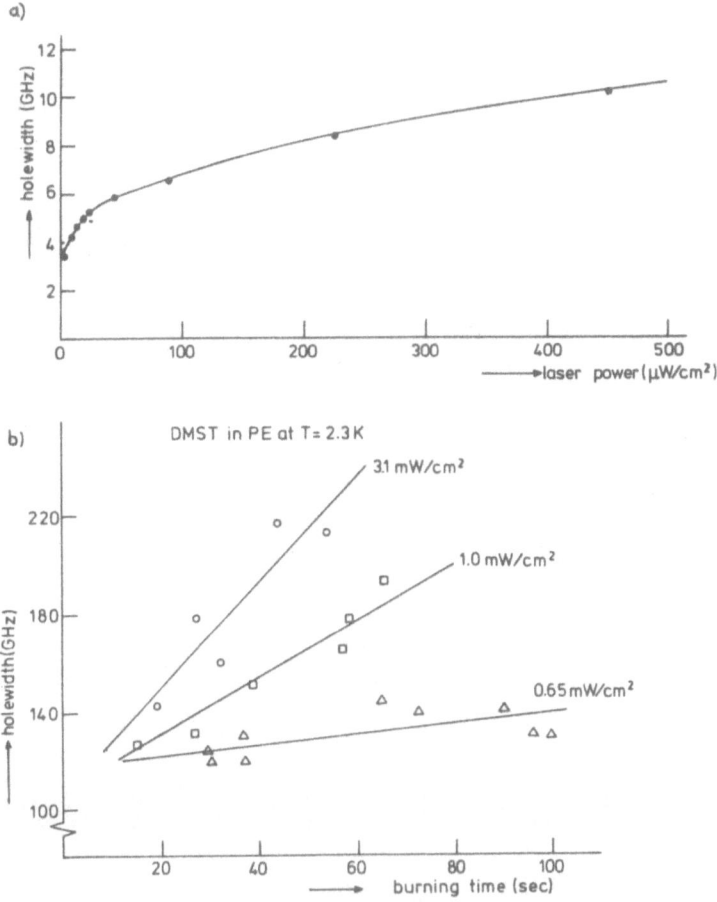

Fig. 6. (a) Holewidth as a function of laser burning power for H$_2$P in ethanol at $T = 4.2$ K, burnt for 20 s [43]. Notice the saturation effect on the hole width at higher powers. (b) Holewidth versus burning time plotted for three different burning powers for dimethyl-s-tetrazine (DMST) in PE at $T = 2.3$ K [43]. The extrapolation to $t = 0$ yields the true value of Γ_{hom}.

follow a relation $\frac{1}{2}\Gamma_{hole} = c(Pt)^{0.23}$ over at least 3 decades of Pt-values, and extrapolate linearly (for $Pt < 0.07$ J cm^{-2}) to $\frac{1}{2}\Gamma_{hole} \simeq \Gamma_{hom} = 0.5$ GHz when $Pt \rightarrow 0$ [56]. This type of behaviour has by now been observed in many HB-systems [86]. From the results shown in Figs. 6 and 7 it follows that the "true value of Γ_{hom}", at a given temperature, is only obtained at the lowest possible laser fluences Pt. A discussion about the "true value of Γ_{hom}" is postponed to Sections 4.2.3 and 4.2.4, where we will treat slow relaxation processes in glasses ("spectral diffusion").

Another critical point in this type of experiments is the hole detection technique. Holes probed in transmission through the sample, by using a monochromator in combination with a lamp, can only be observed if very high burning

a)

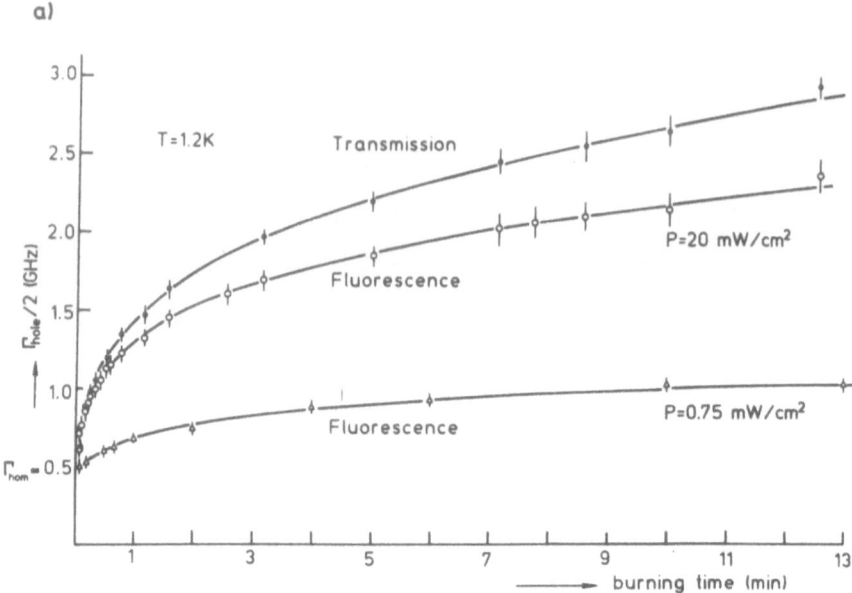

b)

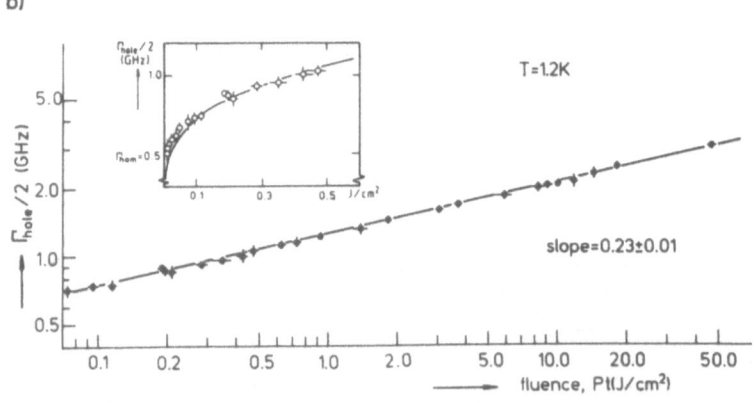

Fig. 7. (a) $\frac{1}{2}\Gamma_{hole}$ as a function of burning time for pentacene in polymethylmethacrylate (PMMA), at $T = 1.2$ K for two laser burning powers. At $P = 20$ mW/cm^2, holes were detected simultaneously via the fluorescence excitation signal and the transmission through the sample. At $P = 0.75$ mW/cm^2, holes were observed in fluorescence only. For $t \to 0$, $\frac{1}{2}\Gamma_{hole} \simeq \Gamma_{hom} = 0.50$ GHz. (b) Log-log plot of $\frac{1}{2}\Gamma_{hole}$ versus Pt for pentacene in PMMA at $T = 1.2$ K. Holes were probed in fluorescence excitation. The data follow a relation $\frac{1}{2}\Gamma_{hole} = c(Pt)^{0.23}$ for values of $Pt \geq 0.07$ J/cm^2 (at least up to 50 J cm^{-2}). Insert: $\frac{1}{2}\Gamma_{hole}$ versus burning fluence for $Pt \leq 50$ J cm^{-2}, at $T = 1.2$ K. Notice that $\frac{1}{2}\Gamma_{hole} \approx \Gamma_{hom} = 0.50$ GHz when $Pt \to 0$. For $Pt < 0.07$ J cm^{-2}, the linewidths follow a linear dependence of Pt, with a slope $a = 7.6 \pm 0.1$ when Γ_{hole} and Γ_{hom} are given in GHz [56].

powers have been used [20, 47, 48, 52, 55, 59, 65, 66, 69, 94]. These powers are orders of magnitude larger than those needed when one probes the hole by means of the fluorescence excitation method [24, 43, 49, 56], or in transmission by scanning a narrow band laser through the spectral region of the hole [56, 62, 70, 71]. Furthermore, the resolution is much higher with the latter technique, because it is given by the bandwidth of the laser (of a few MHz), instead of being limited by the slit of the monochromator (about 3—20 GHz). An example that illustrates this point has been taken from the literature, and is given in Fig. 8 [48, 67]. The sample investigated was quinizarine (1,4-dihydroxyanthraquinone) in ethanol-methanol glass [47, 48, 65, 67]. The left part of Fig. 8 represents the holewidth versus temperature for holes probed in transmission with a lamp and a mono-chromator of resolution 0.15—0.7 cm^{-1} [48]. As a comparison, the right part of Fig. 8 shows the holewidth probed in transmission with a pressure scanned nitrogen-pumped dye laser of resolution 0.02 cm^{-1} [67]. Notice that at $T \approx 1$ K the holewidths in the two figures differ by almost an order of magnitude, and that they extrapolate to about $\Gamma_0 \approx 1$ cm^{-1} in the left plot of Fig. 8 [48], and to $\Gamma_0 \approx 0-10^{-3}$ cm^{-1} in the right plot of Fig. 8 [67], when $T \to 0$. Further, the tempera-

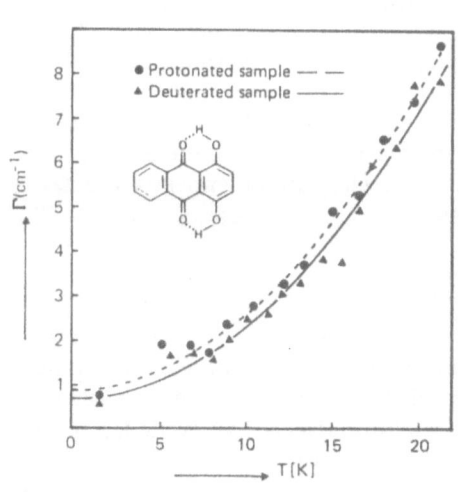
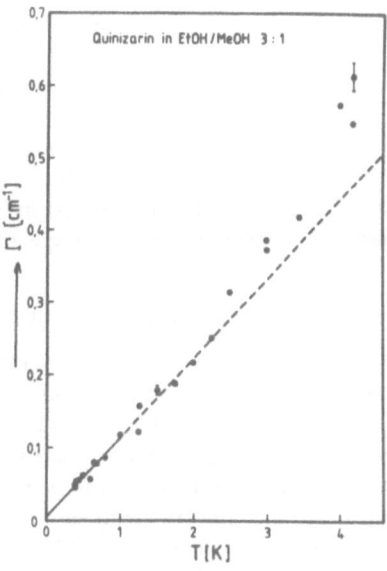

Fig. 8. Holewidth versus temperature for quinizarin (1,4-dihydroxyanthraquinone) in ethanol-methanol glass. Left: the holes were probed in transmission with a lamp and a monochromator of resolution ~0.15 − 0.7 cm^{-1} [48]. Right: the holewidths were probed in transmission with a pressure scanned nitrogen pumped dye laser of resolution $\approx$ 0.02 cm^{-1} [67]. Notice that the data extrapolate to $\Gamma_0 \approx 1$ cm^{-1} in the left plot, and to $\Gamma_0 \approx 0 - 10^{-3}$ in the right plot, when $T \to 0$ (see text). This example shows that the true homogeneous linewidth is only obtained if the instrumental bandwidth is much smaller than the width to be determined.

ture dependence in the two cases is very different, T^2 on the left [48] versus $T^{1.3}$ to $T^{1.0}$ on the right [67]. This is a clear example of a hole-burning experiment done on the same sample by the same group, which shows that probing the holewidths with a lamp and a monochromator does not yield the true value of Γ_{hom}. In order to obtain a reliable value of Γ_{hom}, a laser is needed to burn and probe the holes, with a bandwidth much smaller than the homogeneous linewidth to be determined, otherwise inaccuracies are introduced by deconvolution [95].

Another reason for artifacts in the evaluation of holewidths is local heating of the sample. This plays a crucial role in amorphous materials, because glasses have poor thermal conductivity [91]. In fact, many hole-burning experiments have been performed at rather high burning powers in flow cryostats, where the sample was only in contact with helium gas [20, 47, 48, 52, 55, 59, 64—66, 96] and the hole-detection was done in transmission through the sample. It is not improbable that in such cases the temperature in the bulk of the sample was higher than assumed, and therefore the holes were too broad.

3. Organic Systems Studied

Most of the systems to be discussed here are organic molecules incorporated either in crystalline or in glassy hosts. Special emphasis has been given to porphyrins and related compounds, because of their relevance in biological processes (see Section 3.1). Section 3.2 is divided in three groups of materials according to their hole-burning-mechanisms: intermolecular PHB, two-photon intramolecular HB and NPHB. A few inorganic materials that undergo "photon-gated" spectral HB, are treated in Section 3.2.2. They are potential candidates for frequency-domain optical storage applications.

3.1. PORPHYRINS AND BIOLOGICAL SYSTEMS

It is well known that porphyrins, like chlorophylls, and the heme group in hemoglobin and in cytochrome, play an important role in electron transfer processes in nature. The optical properties of these compounds in the visible and near UV are largely determined by the porphyrin nucleus. The cyclic tetrapyrrole of a peripherally unsubstituted porphyrin is called *porphin* [97]. Such a molecule can easily be incorporated into *n*-alkane crystalline hosts (of comparable dimensions) and in organic glassy hosts. As will be shown in the next section, free-base porphyrins, in which the central metal has been replaced by two protons, are prototype molecules for *photochemical hole-burning*. By contrast, metal porphins usually do not undergo photochemistry, but are able to undergo *transient hole-burning* (Section 3.1.4). Many complex biologically active molecules, like chloro-phyllide proteins, antenna pigments and some photosynthetic bacterial and green plant reaction centers prove to undergo permanent and/or transient hole-burning. These processes will be discussed in Sections 3.1.2 and 3.1.3.

3.1.1. *Free-Base Porphin and Derivatives: Examples of Intramolecular PHB*

The first hole-burning experiments, reported more than a decade ago, were carried out on free-base porphin derivatives: H_2-phthalocyanine in an *n*-octane matrix [15], free-base porphin (H_2P) in *n*-octane [18, 19] and H_2-tetra-4-tert-butylphthalocyanine in solid tetradecane [98] and nonane [99]. Other *free-base porphyrins* that undergo a similar PHB process, as will be shown below, are free-base chlorin [41, 44], pheophytin [100], octylpyropheophorbide [23, 101] and octaethylporphin [102].

The mechanism that induces photochemical hole-burning in these compounds is an *intramolecular* tautomerism, by which the two inner-hydrogens switch from one pair of opposite nitrogens to the other at 90°. This proton motion, known from NMR measurements in liquid solutions at room temperature [103, 104], ceases at low temperature, but the effect can then still be induced by light. This was first observed by studying the fluorescence of tetrabenzoporphyrin in an *n*-octane matrix at 77 K, and interpreted as a displacement of the inner protons being equivalent to an in-plane rotation of the entire molecule over 90° [105]. When unsubstituted *porphin* molecules (H_2P) (Fig. 9) are slowly frozen into an *n*-alkane matrix at low concentrations, well oriented single crystals can be grown [106, 107]. From ESR experiments, it is known that H_2P is incorporated in such a crystal in two distinct orientations, which correspond to those with the N—H · · · H—N axes of the molecule at right angles to each other [108]. This leads to highly resolved fluorescence and excitation spectra at liquid helium temperature. The spectra consist of two series of prominent bands mutually displaced by approximately $\sim 60-100$ cm^{-1} [18, 19, 108]. Such a fluorescence spectrum is shown in Fig. 10 for the so-called A-site of H_2P in a crystal of *n*-octane (n-C_8) at 4.2 K. The two 0—0 fluorescence lines at the left side correspond to the two H_2P tautomers. They can be reversibly transformed into each other by selective laser excitation [18, 19]. The widths of these lines, although only of a few cm^{-1}, are still inhomogeneously broadened, as will be discussed below.

The mechanism of the photoisomerization of H_2P and its deuterated D_2P was studied in a systematic way in ref. [18]. Single 0—0 transitions and vibronic levels of H_2P incorporated in a crystal of *n*-octane at low temperature (4.2 and 77 K)

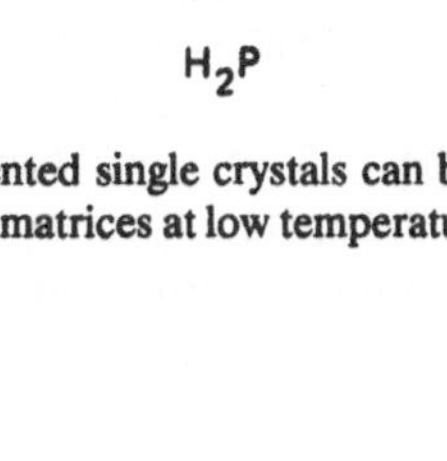

H_2P

Fig. 9. Free base porphin (H_2P). Oriented single crystals can be grown when H_2P is incorporated in *n*-alkane matrices at low temperature [106].

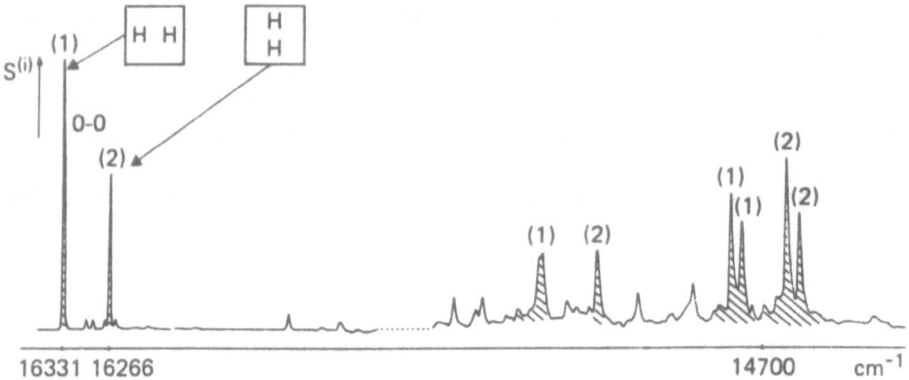

Fig. 10. Fluorescence spectrum of the $S_1 \rightarrow S_0$ transition of H_2P in *n*-octane (A-site) at 4.2 K. The two 0—0 lines labeled (1) and (2) on the left side correspond to two tautomers of H_2P, which can be reversibly transformed into each other by selective laser irradiation [18, 19].

were selectively excited with a tunable dye laser. During the course of these experiments, unexpected holes were burnt accidentally into the 0—0 bands at 4.2 K. The width of these holes was limited by the ~0.7 cm^{-1} bandwidth of the pulsed dye laser used (see Fig. 11, top right). Subsequent high resolution experiments with a narrow-band cw-dye laser ($\Gamma_\ell \simeq 20$ MHz) revealed that the holewidth yields the homogeneous linewidth, Γ_{hom}, of the $S_1 \leftarrow S_0$ 0—0 transition of H_2P in a specific site (A-site) of *n*-octane at 4.2 K (see Fig. 11, bottom) [19]. In the last decade, many high (MHz)-resolution hole-burning studies have been performed on this molecule as guest in crystalline [19, 23, 41, 42, 58, 60, 85] and glassy hosts [1, 24, 28, 43, 45, 49, 72, 86]. The results will be discussed in the course of this chapter.

The photochemical quantum yield for H_2P is of the order of $\leq 1\%$, whereas it is only $\simeq 2\%$ for D_2P [18]. Thus, in order to burn a detectable hole, the molecules must be optically pumped many times [19]. The effect of deuterium substitution of the inner protons on the photochemical rate is further evidence that the HB process is due to a NH-tautomerism. A mechanism for such a photoisomerization was proposed in ref. [18]. It is illustrated in Fig. 12, where potential energy curves for the two lowest triplet (T_0, T_1) and singlet (S_0, S_1) states of H_2P are shown as a function of the nuclear coordinate Q. The latter describes the rotation of the two inner protons. It is known that crossing from one potential well to the other neither takes place in S_1 [18] nor in the lowest triplet state T_0 [109], but must occur during intersystem crossing from $S_1 \rightarrow T_0$, as a consequence of vibronic mixing of the T_1 and T_0 potential curves. This allows to pass from $-Q_0$ to Q_0 with a small but finite probability [18]. The barrier for proton rotation in the ground state is known from NMR to be ~3500 cm^{-1} for tetraphenylporphyrin [104]. A comparable barrier would, therefore, be expected for H_2P. This implies that each tautomer should be stable at liquid helium temperature in the absence of light, which is in fact observed.

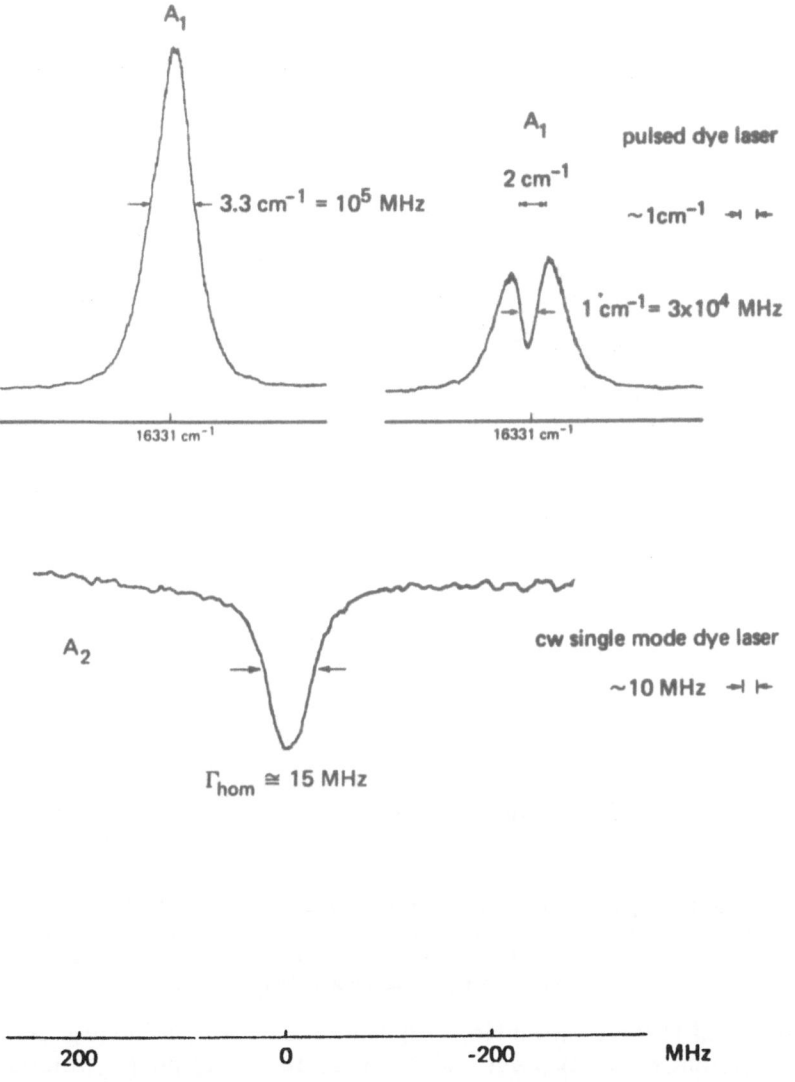

Fig. 11. Top left: Excitation spectrum of the $S_1 \leftarrow S_0$ 0—0 transition of H_2P in crystalline n-octane (A_1-site) at 4.2 K, before burning. Top right: same spectrum as on the left side, after burning a hole with a pulsed dye laser of width $\simeq$ 1 cm^{-1} [18]. Bottom: A hole burnt in line A_1 at 4.2 K with a cw laser of width $\approx$ 20 MHz [19].

Different n-alkane hosts induce different crystal field interactions with the guest molecules and give rise to a variety of sites (called A, B, C, ...) [46, 110]. The fluorescence and excitation spectra of H_2P in polycrystalline n-alkanes are, in general, more complicated than the double spectrum shown in Fig. 10. For even n-alkanes, for example, the number of sites strongly depends on the cooling rate of the sample. This effect is depicted in Fig. 13, where fluorescence spectra of the

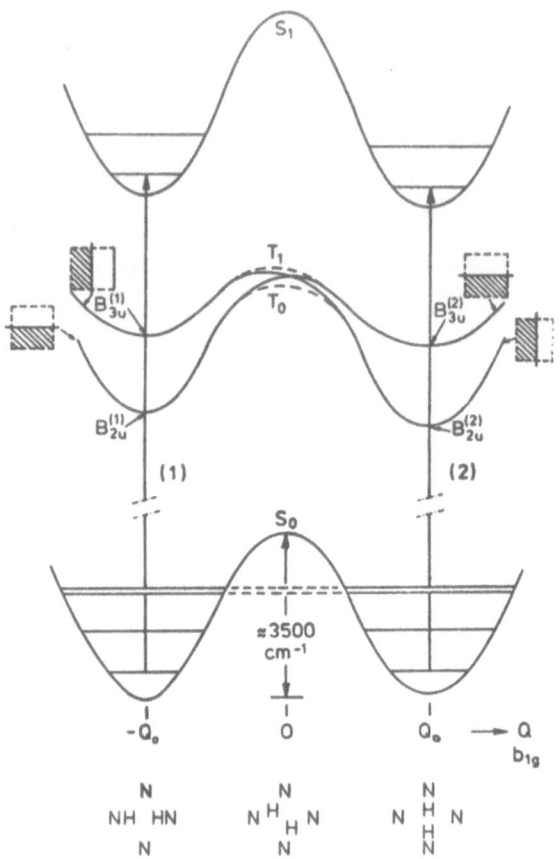

Fig. 12. Schematic diagram of the potential energies of the two lowest singlet (S_0, S_1) and triplet (T_0, T_1) states of H_2P in an n-alkane crystal as a function of the nuclear symmetry coordinate Q, involving a rotation of the two inner hydrogens. At $Q = -Q_0$, the situation is depicted for a molecule in site (1), while $Q = Q_0$ represents a molecule in site (2) [18].

0—0 region of H_2P are shown [46, 110]. Notice that for each site (e.g., A, B, . . .), pairs of tautomers (1, 2) occur which can be reversibly phototransformed into each other [46, 110]. The energy splitting between these tautomers of ~2 to 100 cm^{-1} depends on site and host: the tighter the fit of the molecule in the host lattice, the larger the splitting [46]. The spectral lines in these crystals are only a few cm^{-1} wide. This is due to a rather weak interaction between the H_2P guest and the n-alkane host. Similar narrow line spectra have been observed for metal porphins in the same hosts [106, 107, 110]. The results indicate that the guest may be considered as an almost isolated and well oriented molecule in the crystal lattice of the n-alkanes [111, 112]. As mentioned above, these spectral linewidths are inhomogeneously broadened. A way to get information on the homogeneous linewidth, and thereby on the dynamical processes of the excited states of porphyrins, is high-resolution hole-burning (see examples in Section 4).

Fluorescence of H_2P in even n-alkanes

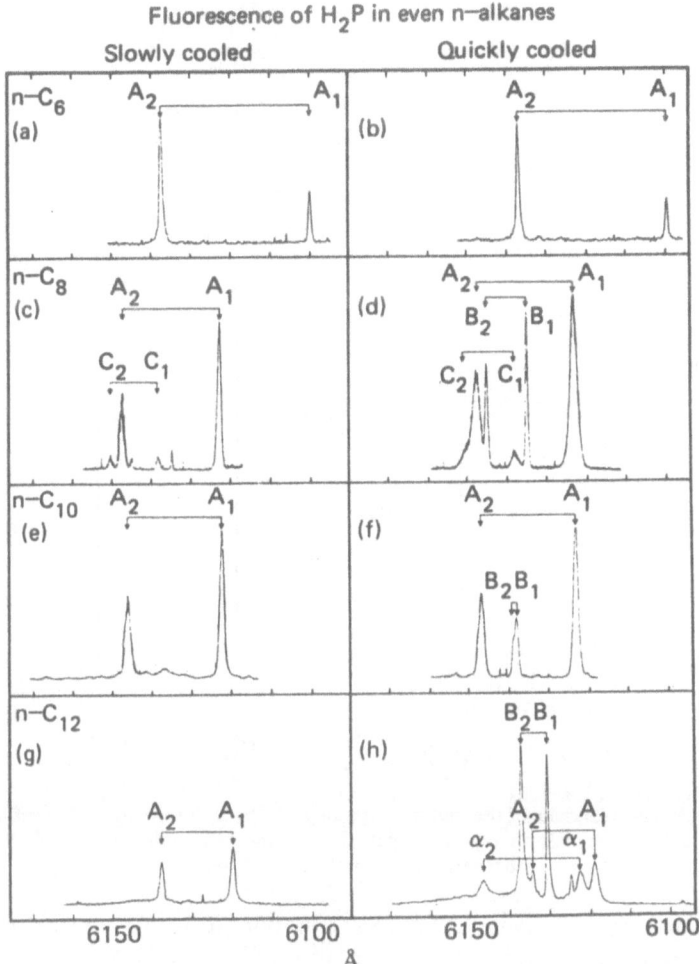

Fig. 13. Fluorescence spectra of the $S_1 \rightarrow S_0$ transition of H_2P in n-alkane matrices with an even number of carbon atoms. The arrows indicate pair of lines originating from the two tautomeric forms of H_2P in a given site. In slowly cooled samples a single site is dominantly occupied (A-site), whereas rapid cooling leads to the presence of multiple sites [46].

Free-base chlorin (H_2Ch) is a derivative of H_2P in which a double bond in one of the pyrrole rings is reduced. This hydrogenation introduces an intrinsic asymmetry in the molecule. As a consequence, the two tautomers of H_2Ch become chemically inequivalent and absorb at very different frequencies. A large tautomer splitting of $\Delta_2 - \Delta_1 \approx 1600$ cm^{-1} is observed (see Fig. 14, right) [41, 44], in contrast to H_2P, where $\Delta_2 - \Delta_1 \leq 100$ cm^{-1}, as just seen above (see Fig. 14, left). The stable tautomeric form of H_2Ch has the NH $\cdots$ HN axis parallel to the reduced bond [41, 44, 103, 113]. A spectrum of the $S_1 \leftarrow S_0$ 0—0 region of the stable chlorin

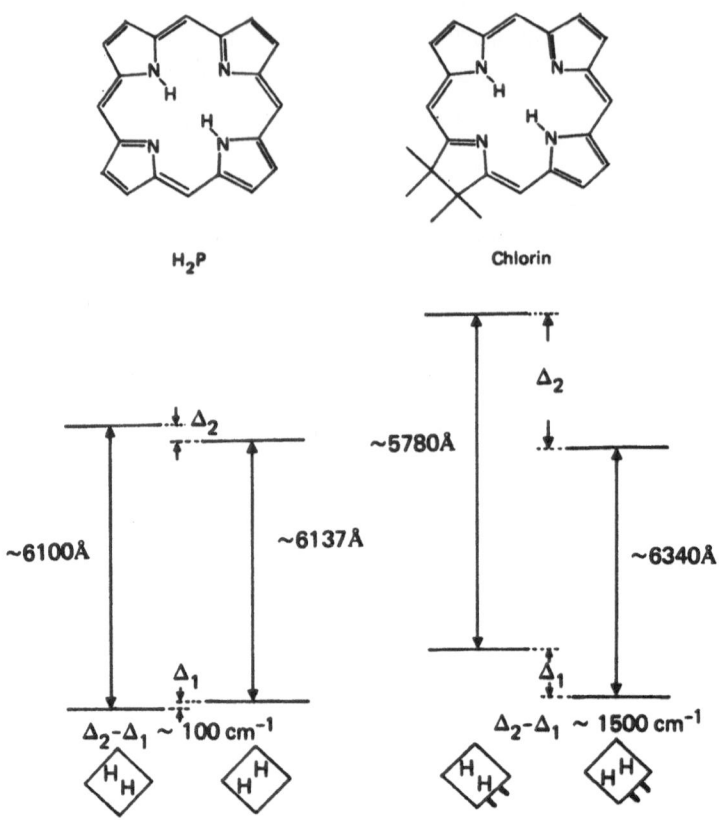

Fig. 14. Energy level diagrams of the tautomer splittings $\Delta_2 - \Delta_1$ for the $S_1 \leftarrow S_0$ 0—0 transition of H_2P (left) and free-base chlorin (H_2Ch) (right). Notice the large difference in $\Delta_2 - \Delta_1$ values between the two molecules [41, 44].

molecule in n-hexane is given in Fig. 15a. The four electronic transitions observed around 634 nm correspond to four orientations (1, 2, 3, 4) of H_2Ch in the n-C_6 host [44]. The less stable tautomer of chlorin, H_2Ch^*, can be produced by photoexcitation. Each of the four lines of Fig. 15a then transforms into those of H_2Ch^*, which absorb around 578 nm. This is shown for line 2 in Fig. 15a and b [41, 44]. Irradiation of the photoproduct (line 2') restores the original stable tautomer (line 2). The photochemical hole-burning rate for the back reaction from the unstable tautomer, H_2Ch^*, to the stable one, H_2Ch, is three orders of magnitude higher than that for the forward reaction, the rate of which is comparable to that of H_2P [44]. This large difference in rates is most probably due to an additional path, $S_1(H_2Ch^*) \rightarrow S_1(H_2Ch)$ which accounts for $\sim 6\%$, and does not take place in H_2P. Since evidence was found for the existence of only one photoproduct, it was assumed that the latter has its two inner protons opposite to each other and perpendicular to the outer single bond [44, 113]. Narrow holes of about ~ 40 MHz could be burnt in the inhomogeneous lines of both tautomers at

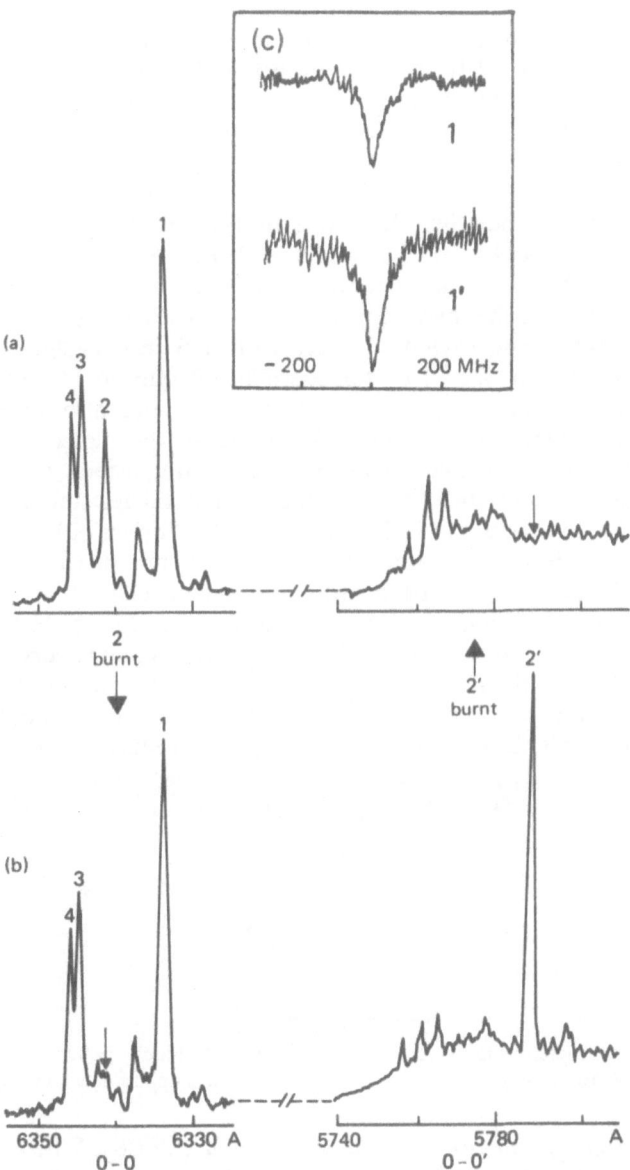

Fig. 15. Phototautomerism of free-base chlorin (H_2Ch) in crystalline n-hexane (n-C_6). (a) The $S_1 \leftarrow S_0$ 0—0 lines of the low energy tautomer at 4.2 K are seen on the left-hand side, whereas the spectral region of the photoproduct 0—0 lines is seen on the right-hand side. (b) After burning line 2 (left), a photoproduct line 2′ appears at $\simeq$ 1500 cm⁻¹ higher energy (right). The process is reversible. Similar effects have been observed for the other lines 1—4. (c) Holes burnt in lines 1 and 1′ at 1.6 K with a narrow-band ($\sim$ 2 MHz) cw-dye laser [44].

1.6 K (see Fig. 15, top). The width of these holes is consistent with the fluor-escence lifetime of H_2Ch, $T_1 = 8$ ns [44]. From temperature dependence measure-ments between 1.5 and 4.2 K [44], and from the application of electric [114] and magnetic fields [115, 116] to holes burnt in zero field in the 0—0 transitions, it was concluded that chlorin occupies A-like sites in the n-alkane crystalline hosts studied (n-C_6, n-C_8, n-C_{10}).

Bacteriochlorin (H_2B), another derivative of H_2P, is the parent compound of bacteriochlorophyll-a and b and bacteriopheophytins [117]. In this molecule, two peripheral double bonds of the macrocyclic porphyrin skeleton are removed by hydrogenation at opposite pyrrole rings (Fig. 16). From high resolution site-selected fluorescence and excitation spectra of H_2B in crystalline n-hexane, n-octane and n-decane at 4.2 K, it was possible not only to identify the internal molecular vibrations of H_2B, previously unknown, but also to observe low-frequency modes related to torsional oscillations of the molecule in the lattice [23, 118]. A schematic diagram of the vibrational frequencies and their relative intensities in the ground state, S_0, and the excited singlet states, S_1 and S_2, is shown in Fig. 17. In contrast to H_2P [42], the Q_y 0—0 origin of H_2B is very strong [118]. From its bandwidth, if assumed homogeneously broadened, a lower limit for the $S_2 \rightarrow S_1$ lifetime, $\tau \approx 0.7$ ps, could be estimated [118]. By irradiation of the Q_x 0—0 band at ~725 nm, no evidence of a photoproduct of H_2B in n-alkanes was found. This result suggests that either the photoproduct is very unstable, with a lifetime shorter than the time scale of the experiment (a few minutes to burn and probe a hole), or that an isomerisation of the two inner protons would cause such a loss of conjugation in the macrocycle, that the photoproduct does not exist [118].

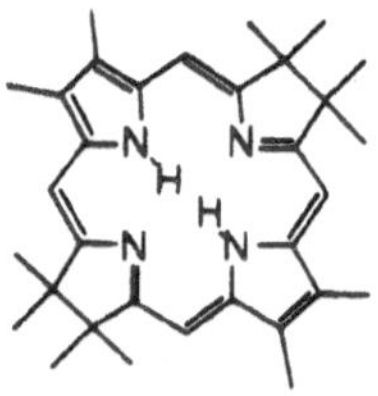

Bacteriochlorin

Fig. 16. Free-base bacteriochlorin (H_2B). Two peripheral double bonds of the macrocyclic porphyrin skeleton are removed by hydrogenation at opposite pyrrole rings.

Pheophytin-a (pheo) and octylpyropheophorbide-a (OPPB). When the inner Mg of chlorophyll-a is replaced by two protons, one obtains pheophytin-a (Fig. 18a). In order to have a model compound for pheophytin-a that would fit in an n-alkane host, octylpyropheophorbide-a (OPPB) (Fig. 18b) was prepared [119]. In this molecule, an n-octyl chain replaces the long phytyl tail present in pheophytin-a, and the carbomethoxy-group on ring V of pheophytin-a is removed. The purpose of these chemical changes was to check whether the long phytyl chain was responsible for the broad structureless background observed in the spectra of

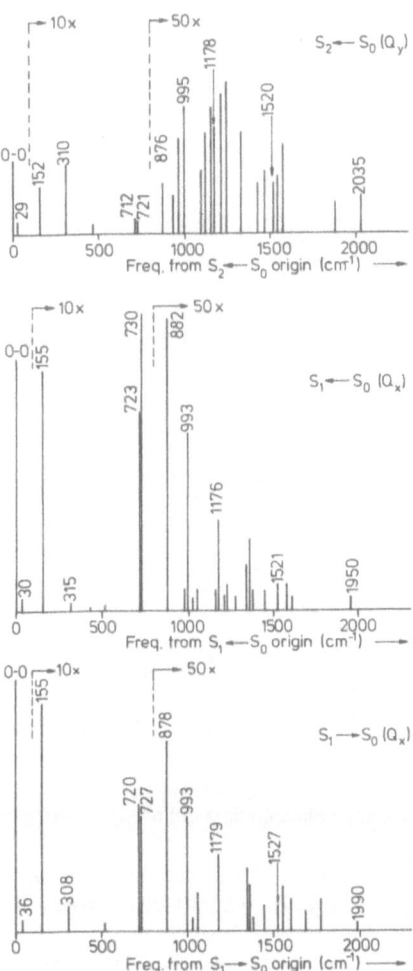

Fig. 17. "Stick-diagram" representing the positions and intensities of the molecular vibrational frequencies of H_2B in site 1 of n-C_6 at 4.2 K, for the transitions $S_1 \rightarrow S_0$ (Q_x), $S_1 \leftarrow S_0$ (Q_x) and $S_2 \leftarrow S_0$ (Q_y). The intensities have been corrected for the dye output response of the laser versus wavelength [118].

pheophytin-a [120]. Fig. 19 (top) shows a fluorescence spectrum, broad-band excited at ~ 400 nm, of the $S_1 \rightarrow S_0$ 0—0 region of OPPB in n-C_{10} at 4.2 K. Only five large sites, a few smaller sites, and very little background are present. A site-selected fluorescence spectrum (site 1) of the same region is also given in Fig. 19 (bottom) [101]. Another nice example of the high resolution possible with this compound is the site-selected excitation spectrum (site 5) of Fig. 20, which shows the region from the origin $S_1 \leftarrow S_0$ transition up to ~ 1600 cm^{-1}. From these type

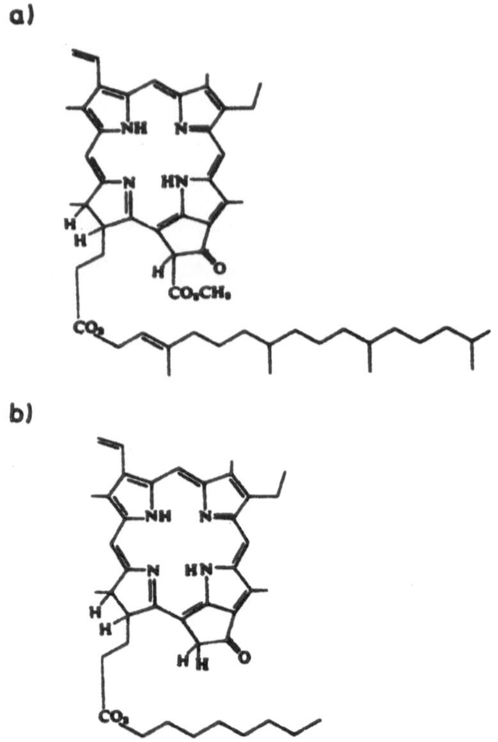

a)

b)

Fig. 18. Structure of (a) pheophytin-a, (b) octylpyropheophorbide-a (OPPB).

of spectra, vibrational frequencies of the ground state S_0 up to ~ 1600 cm^{-1}, and of the excited singlet states up to ~ 5500 cm^{-1} from the origin were identified [101].

In contrast to statements in the literature [120], well defined single-site spectra with very small background can also be obtained for *pheophytin-a* in *n*-decane [101]. The condition required is that very low concentrations ($\sim 10^3$ times lower than for OPPB) are used. From the similarity of the spectra of pheophytin and OPPB, it was concluded that the latter appears to be a good model compound for pheophytin [101]. It has the advantage that it can be incorporated in oriented single crystals of *n*-alkanes and, thus, its electronic states might be accurately studied in magnetic and electric fields.

A reversible, selective photochemical reaction, and hole-burning was observed in OPPB at 4.2 K [101]. The photoproduct absorption bands are very narrow and its 0—0 transitions (at ~ 640—650 nm) lie ~ 650 cm^{-1} higher in frequency than the 0—0 transitions of the chemically stable form of OPPB (at ~ 670—680 nm). Fig. 21 illustrates photochemical burning of two sites (1 and 5) of the 0—0 transition and the appearance of a photoproduct. This photoreaction has been

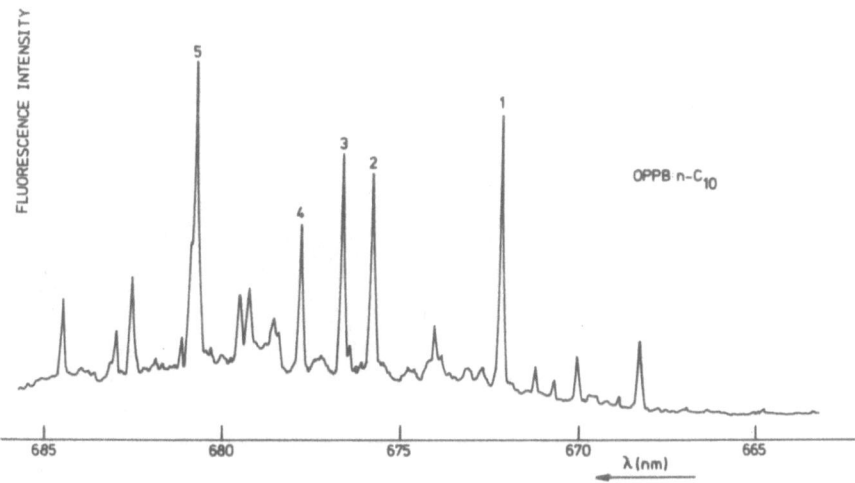

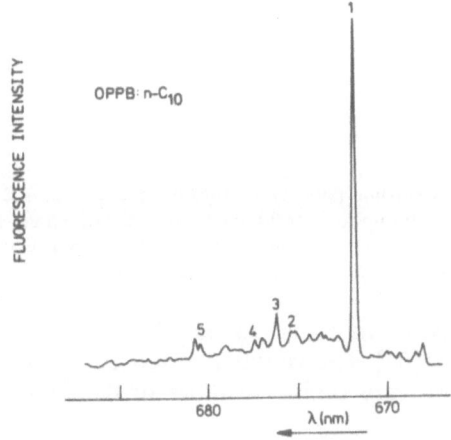

Fig. 19. Fluorescence spectrum of the $S_1 \rightarrow S_0$ 0—0 region of OPPB in n-decane (n-C_{10}) at 4.2 K. Top: Broad-band excited in the Soret band at ~ 400 nm. Bottom: Site-selected excitation of the 775 cm^{-1} vibronic band of site 1 at 638.9 nm [101].

attributed to tautomerism of the inner protons from one pair of opposite nitrogens to the other [101], similarly as observed in H_2P [18] and H_2Ch [44]. The backward reaction in OPPB was found to be faster than the forward reaction, but only by a factor of about $1 : 50$ [101], as compared to $\sim 1 : 10^3$ in chlorin [44]. This is probably related to the asymmetry introduced by ring V in OPPB and chlorophyll derivatives. This asymmetry also tends to enlarge the crystal field splitting, which

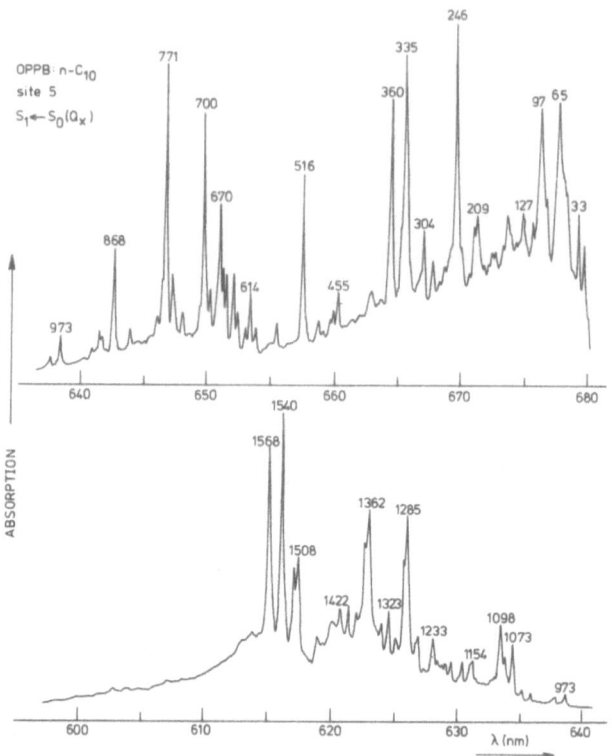

Fig. 20. Single-site excitation spectrum (site 5) of OPPB in n-C_{10} at 4.2 K. The vibrational frequencies of the $S_1 \leftarrow S_0$ transition up to ~ 1600 cm^{-1} above the origin are given. The intensities of the upper and lower spectra are neither matched to each other nor corrected for the dye output response of the laser versus wavelength [101].

in the 0—0 region of H_2Ch was found to be 33 cm^{-1} [41, 44], whereas it is ~ 200 cm^{-1} and in OPPB [101]. The vibrational frequencies of the S_1 state of the photoproduct of OPPB are very similar to those of the stable molecule (within ~ 15 cm^{-1}) [101].

The photochemistry that takes place in the system *pheophytin-a* in an n-decane crystal is like that of OPPB. The stable form absorbs at ~ 670 nm, and the various narrow-line sites of the photoproduct lie between 480 and 580 cm^{-1} higher in energy [101]. Spectra of pheophytin-a in ether glass at 4.2 K have been reported in the literature [100]. They show a very broad and structureless band at ~ 500 cm^{-1} above the original 0—0 band after burning in the latter. Also holes of ≤ 0.15 cm^{-1} width were observed in this photoproduct [121].

When comparing the photochemistry of H_2P [18] with that of H_2Ch [44] and OPPB [101] (see Fig. 22), there seems to be a correlation between the frequency difference of the 0—0 transitions of the original molecule and its photoproduct, and the relative photochemical rates of forward and backward reaction: a larger

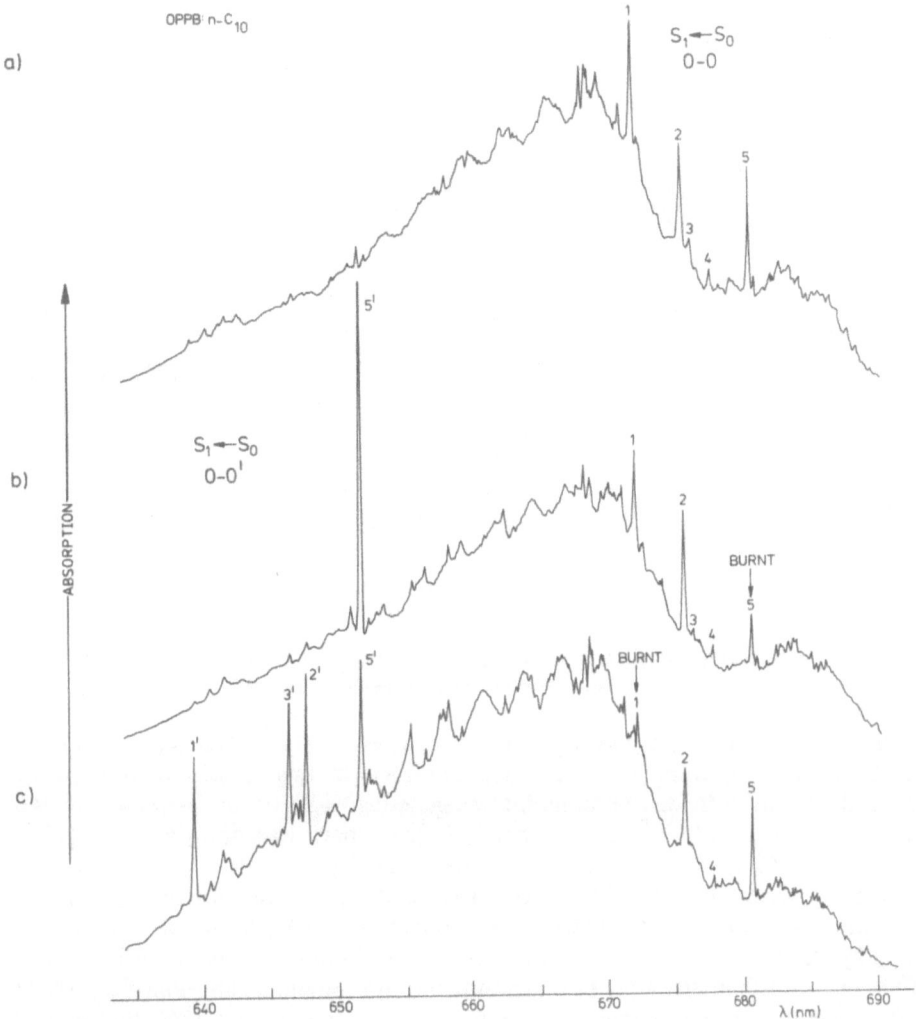

Fig. 21. Broad-band detected excitation spectra of OPPB in n-C_{10} and its photoproduct at 4.2 K. (a) Before burning. (b) After burning into the 0—0 line of site 5 at 680.0 nm. (c) After burning into the 0—0 line of site 1 at 672.2 nm [23, 101].

distance between tautomers induces a faster backward reaction. For H_2Ch and OPPB, the back reaction seems to take place not only through the triplet state, as in H_2P (with a yield $\leq$ 1%), but also in the excited singlet state. As mentioned above, it was found that ~ 6% of the photoproduct molecules of H_2Ch come back to the stable form in this way [44].

Another free-base porphin derivative studied by hole-burning is phthalocyanine

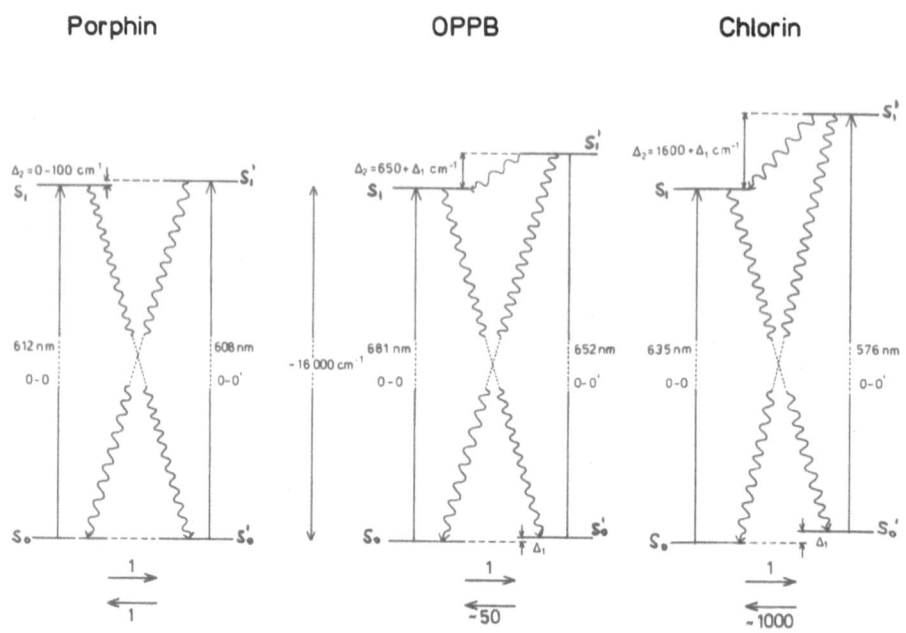

Fig. 22. Schematic energy level diagrams of NH-phototautomerism for free-base porphin, OPPB and chlorin in *n*-alkane crystals at liquid helium temperature.

(H$_2$Pc), a tetrabenz-tetraaza-porphyrin. Its $S_1 \leftarrow S_0$ 0—0 transition occurs at ~690 nm. Hole-burning has first been observed in this transition in *n*-octane [15], and later in PMMA where the inhomogeneous width of the 0—0 band is 340 cm^{-1} and a series of holes of ~0.7 cm^{-1}, 2 cm^{-1} apart, was burnt at 4.2 K [122]. This system was thought to be an interesting candidate for frequency storage of information, since theoretically ~10^3 holes could be burnt into the 0—0 band. From the temperature dependence of the holes burnt in H$_2$Pc in polystyrene (PS) and polyethylene (PE), it was concluded that the homogeneous linewidth, Γ_{hom}, differs by more than an order of magnitude in both polymers, and that $\Gamma_{\text{hom}} \propto T^2$ in PS, whereas an anomalous behaviour was reported for PE [66]. Subsequent hole-burning studies on H$_2$Pc in Ps, PE and PMMA as a function of burning time and laser power yielded values for the Debye-Waller factor of ~0.6—0.9 (see Section 3.2.1), and a photochemical quantum yield of ~1.1 × 10^{-3} [123a]. The latter is similar to the value found for H$_2$Pc in PE by another group [123b], and is almost an order of magnitude smaller than for H$_2$P [18, 109]. Recent thermal cycling experiments between 4 and 25 K on the same polymer systems suggested that very little spectral diffusion takes place at these temperatures [68] (see Section 4.2.4).

Free-base phthalocyanine was also studied in its protonated form in concentrated sulfuric acid glass at 830 nm between 1.2 and 10 K. Homogeneous linewidths of about 1.0 GHz at 1.2 K were reported, which were substantially larger than the fluorescence lifetime-limited value. The holewidths apparently followed a $T^{1.8 \pm 0.1}$ temperature power law [64] (see Section 4.2.2).

Other porphyrins that are characterized by reversible light induced NH-tautomerism are asymmetric *cyclopentanporphyrins* and their covalently linked dimers in isotropic solutions at low (77 K) and higher temperatures (500 K) [122]. The five ring isocycle in these compounds is the same as that in chlorophylls, and participates in the formation of pigment aggregates. These porphyrins have been synthesized as model systems for the study of natural chlorophyll aggregated forms, like the "special pair" in photosynthetic reaction centers. Although photo-induced tautomer transformations have been observed [122], no photochemical hole-burning has been reported so far.

3.1.2. *Chlorophyll-Like Molecules*

Chlorophylls (Chl) (Fig. 23), which play an important role in the primary process of photosynthesis, have thoroughly been studied by means of optical techniques. For example, selective laser excitation and fluorescence spectroscopy [4] as well as hole-burning [125a] at low temperature have been used to investigate the electronically excited states of these molecules and some of their derivatives. The first site-selected spectra of *Chl-a* and *Chl-b* in glassy matrices were reported in 1975 [125b, c, 126], followed by those of *protochlorophyll* (*PChl*) and *pheophytin* (*Pheo*) [120, 127—129]. All these spectra show an intense structureless background due to broad overlapping phonon wings [120, 129]. Some excited state and a few ground state vibrational frequencies up to $\sim 1500 \ cm^{-1}$ from the origin at $T \approx 4$ K could be obtained from these spectra. Qualitatively, these frequencies are similar to those found for porphyrins [44, 130, 131] and chlorin [44], because they are principally determined by the π-electron system localized on the tetrapyrrole ring. Due to their low symmetry, however, no strict selection rules for vibrations in chlorophylls exist [125a].

The first approach to study the homogeneous linewidth in chlorophyll-like molecules was reported by Avarmaa *et al.* in 1981 [132]. This group carried out a

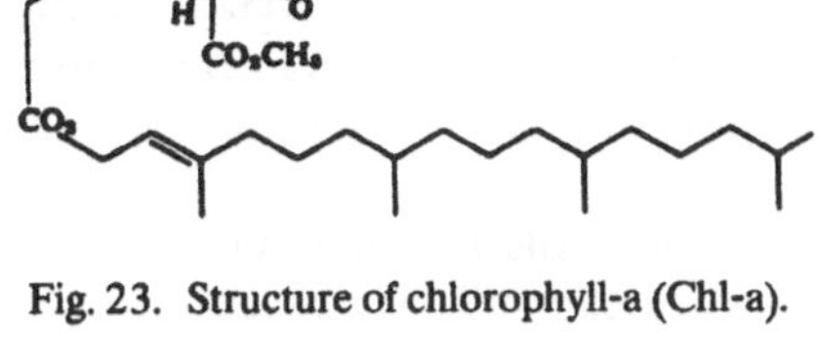

Fig. 23. Structure of chlorophyll-a (Chl-a).

hole-burning experiment on *protochlorophyll* in ether, ether-butanol and *n*-heptane hosts at 1.8 K. A fixed He-Ne laser was used to burn the holes, which were then probed by means of a fluorescence spectrum with an instrumental resolution of ~ 0.1 cm^{-1}. Holewidths of 0.3—0.5 cm^{-1} were found, depending on the host. Experiments performed later by the same group with a single-frequency cw-dye laser, used for both burning and probing the holes, yielded holewidths of 0.01 cm^{-1} for chlorophyll-a in ether-butanol at 1.8 K [121]. This width, although smaller than before, is still an order of magnitude larger than the lifetime-limited value of 35 MHz ($T_1 \simeq 8$ ns) of chlorophyll-a. Reasons for the large difference were not clear at that time.

Two types of mechanisms have been attributed to act simultaneously during the hole-burning process in chlorophylls [132]. One is *transient* hole-burning via the triplet state reservoir population [38, 39] (see Section 3.1.4), the other is *non-photochemical hole-burning* [16, 20, 26] or "site interconversion", as called in ref. [132]. The holes burnt by the first process had a lifetime of about 5—10 milli-seconds, which coincides with the decay time of the triplet state in chlorophylls [133]. The second process leads to hole-filling times of about 10s for PChl in *n*-heptane, and to permanent hole-burning for PChl, Chl and Pheo in glassy hosts [132]. Since the large chlorophyll molecules do not fit in an ordered configuration in any matrix, they probably change their degree of interaction with the surround-ing solvent molecules after excitation into the S_1 singlet state. The result of this NPHB-process should be a slight shift of the 0—0 transition of a few cm^{-1} [20] within the broad inhomogeneous band ($\Gamma_{inhom} \simeq 250$ cm^{-1} [132]).

Hole-burning experiments on chlorophyll-a and b have also been reported by Carter and Small [134]. The holes burnt at 665 nm had a Lorentzian line shape with a width of 0.17 cm^{-1} (and lowest burning power of 130 µW/cm^2) at 1.9 K. They broadened with temperature as $T^{1.3\pm0.2}$ between 1.9 and 8.8 K. Such a $T^{1.3}$ dependence of holewidths seems to be typical for amorphous organic systems, as will be discussed in section 4.2. Thermal cycling experiments up to 8.8 K revealed a strong broadening of the holes, which was interpreted in ref. [134] as ground state "spectral diffusion" caused by relaxation of the glass [48, 65, 68].

It is surprising that the holewidths in this experiment [134] are almost a factor 20 larger than those reported for the same molecule in ref. [121]. The reason for this discrepancy, as discussed in section 2.2, is most probably the hole probing technique: the holes were probed in transmission with a lamp and a mono-chromator of resolution ~ 0.08 cm^{-1}. This resolution is obviously not sufficient for linewidths of ≤ 0.01 cm^{-1}, as measured in ref. [121]. It should be reminded that the first holes reported for chlorophyll-a had a width of 0.3 cm^{-1} obtained with an instrumental resolution of ~ 0.1 cm^{-1} [132]; the holewidths only became 0.01 cm^{-1} when probed with a laser of a few MHz resolution [121]. It is difficult to imagine that the influence of the host, being amorphous in both cases, would introduce a difference of a factor of ~ 20 in the holewidth, as speculated in ref. [134].

In addition to the possible hole-burning mechanisms mentioned above, an *intermolecular* photochemistry can still take place through a relative reorientation of the chlorophyll molecule with respect to the axial ligands attached to the central Mg atom [135]. In fact, reversible photochemical holes have been observed in a Mg-porphin-ethanol complex in n-octane [136], and in a Mg-porphin-pyridine complex in n-octane [39], which were attributed to this mechanism.

3.1.3. Complex Biological Systems

Attempts to obtain highly resolved optical spectra of "in vivo" or native biological systems by selective laser excitation remained unsuccessful until a few years ago. The reason was attributed to strong pigment-protein interactions and energy transfer, leading to "spectral diffusion" [125b]. Narrow-line fluorescence spectra, however, have been obtained in recent years for a few systems, like solubilized iron-free cytochrome c [137], Zn-substituted cytochrome c [138], and *proto-chlorophyllide* (PChl) and *chlorophyll(ide)* (*pigments*) *in etiolated barley leaves* at liquid helium temperature [139]. In the latter two systems it was even possible to burn stable holes at 650 nm [140]. The holewidths were estimated to be 0.3 cm^{-1} at 5 K, but no photoproduct could be identified. Since the chromophore in question (PChl$_{650}$) contains pigments in an aggregated form, it was concluded that pigment-pigment interactions might, in principle, be studied with this high-resolution technique.

Hole-burning has also been observed in other protein complexes, like *chlorophyllide in apomyoglobin* [141]. From experiments carried out between 1.35 and 2.5 K, holewidths of the order of $\sim 1-2$ GHz were obtained which broadened with temperature as $T^{1.3}$. Since similar results had been reported for chlorophyll-a and b in polystyrene [134] (see previous Section 3.1.2), it seems that the hole-burning dynamics in protein matrices is like that in polymers and glassy hosts [24, 43, 45, 49, 56, 72, 86, 87]. In contrast to a glass, however, proteins have a well-defined average structure known from x-ray crystallography [142] and neutron scattering [143]. Probably, glass-like guest-host interactions arise from very small variations in the positions of individual residues of the protein, which are able to produce a broad distribution of low frequency modes [141]. A picture of a large number of nearly degenerate configurations is supported by recent specific heat experiments on DNA in the range 0.5–5 K [144], and other biomolecules such as metmyoglobin [145] and melanin [146]. From the experimental results, a very small density of low energy excitations has been calculated, which is similar to that observed in non-biological glasses [144].

Phycoerythrin (Fig. 24, left) and *phycocyanin* (Fig. 24, right) belong to a class of *antenna pigments* called biliproteins. The role of these pigments in photosynthetic organisms is to act as light absorber and energy transmitter. These systems contain bile-pigment chromophores which are covalently bound to their protein environment. Hole-burning experiments on isolated antenna pigments were first reported in 1981 [147, 148]. In contrast to holes burnt in barley leaves [140], the holes burnt in the lowest energy band of c-phycocyanin (at $\sim 630-650$ nm) and c-phycoerythrin ($\sim 550-570$ nm) in buffer solutions at $\sim 2-5$ K are not stable, but do fill in when a new hole is burnt at higher energy. This has been interpreted as a

Phycoerythrobilin   Phycocyanobilin

Fig. 24. Structure of phycoerythrobilin (left) and phycocyanobilin (right) (from refs. [147, 148]).

photo-reversible proton transfer reaction, whereby it was assumed that the photo-product absorbs at shorter wavelengths than the original molecule [147, 148]. The various absorption bands in the low temperature spectrum of c-phycocyanin (595, 627, and 650 nm) have been considered as independent electronic origins, that correspond to chromophores with different photochemical behaviour [147]. Although the holewidths reported for phycoerythrin and phycocyanin were similar ($\simeq 1.5-2$ cm^{-1} at 1.8 K), a hole burnt in the high energy band (550 nm) of phycoerythrin creates a series of satellite holes at $\sim 600-700$ cm^{-1} lower energy [148], which are not observed in phycocyanin. This result has been attributed to slow energy transfer between the various chromophores of the protein [148]. A more trivial interpretation, however, would be an overlap of vibrations at the excitation frequency, which are reflected as holes in the 0—0 region. From the hole-burning results, it was concluded that the chromophores are rather rigidly attached to the surrounding protein, the latter assumed to be highly structured with a well defined geometry [147, 148]. This interpretation seems in contradiction with recently reported fluorescence experiments on c-phycocyanin, where it has been claimed that either the protein itself is disordered in solution, or that a random interaction takes place between the chromophore and the glass, which is much stronger than the chromophore-protein interaction [149].

One of the most debated problems in photosynthesis of both *bacteria and plant systems* is the mechanism of the primary charge separation, that occurs after light excitation and before the electron transfer process takes place [150]. The pigment complex involved in this first event is a strongly coupled pair of chlorophylls (Chl) in plants, or bacteriochlorophylls (BChl) in bacteria. This dimer, (BChl)$_2$, is the primary electron donor, also called "special pair", which is embedded in the reaction center protein. *Reaction centers (RC) of purple bacteria* contain 4 BChl, 2 bacteriopheophytins (Bph), 2 quinones and one iron atom [150]. Two of the BChl's form the "special pair". From recent crystallization and X-ray studies of the *RC of Rhodopseudomonas (Rps) viridis* [151], it became clear that these pigments are organized within a two-fold symmetry around the C_2 axis, with runs from the "special pair", (BChl)$_2$, to the iron atom Fe (see Fig. 25). The charge separation that produces the oxidation of (BChl)$_2$ and the reduction of Bph takes place in a few picoseconds [152]. It is believed, however, that even earlier intermediate acceptors, like one of the BChl monomers, play a role in this process, although this hypothesis is not corroborated yet [153].

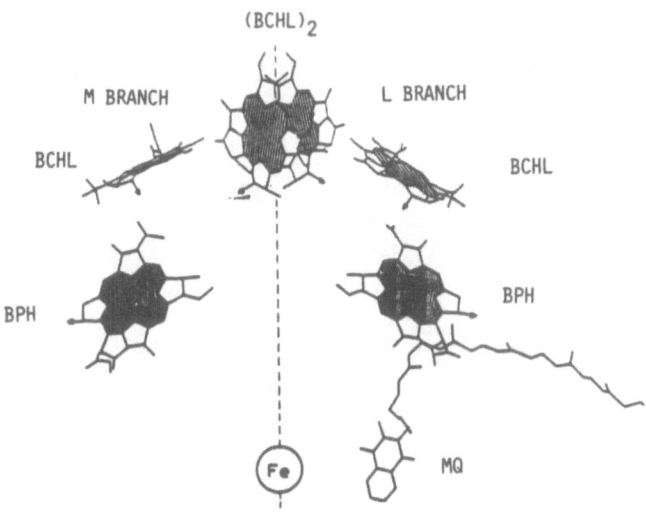

Fig. 25. Structural arrangement of the chromophores in the baeterial reaction center of Rps. viridis [151] (from "Antennas and Reaction Centers of Photosynthetic Bacteria", ed. M. E. Michel-Beyerle, Springer Verlag, Berlin, 1985, p. 352).

Various coherent spectroscopic techniques are being used to solve this puzzle. Hole-burning and accumulated sub-picosecond photon echoes studies of the lowest excited singlet state of the primary electron donor, the $(BChl)_2$ dimer of the reaction center of *Rps. sphaeroides* (P-870) and *Rps. viridis* (P-960) at $T \simeq 2$ K, have been reported [154, 155]. From the photon echo results it was concluded that the dephasing time, T_2, was shorter than the instrumental resolution of ~ 200 fs [154]. In addition, extremely broad holes of ~ 400—500 cm^{-1} burnt at ~ 880 nm (Rps. sphaeroides) and ~ 1000 nm (Rps. viridis) were found by both groups [154, 155]. In these experiments, laser bandwidths of ~ 3—4 cm^{-1} were used for the burning process, and a lamp and a monochromator for probing the holes in transmission through the sample. The absorption intensity recovered completely in less than 0.1 s, and spectra were taken in the form of light-minus-dark difference absorption spectra. The holewidths were found to be independent of burning power, which was varied by a factor of ~ 40, and temperature, which was changed between 1.5 and 2.1 K [155] (see Fig. 26).

Various interpretations of these extremely broad holes have been given [154, 155]. The simplest is that the holewidth is homogeneously broadened and reflects an ultra-fast decay of about 15—25 fs of the excited "special pair", $(BChl)_2^*$, into a charge-transfer state involving charge separation within the BChl dimer [154]. This extremely fast relaxation time could be interpreted as a measure of vibrational relaxation within the excited dimer [154]. The latter, however, is in contrast with vibrational decay times usually found for aromatic molecules ($\tau > 0.1$ ps) [156]. Another possibility would be pure dephasing, but no T-dependence of the hole-widths was observed [155]. Boxer *et al.* put forward an alternative interpretation: if

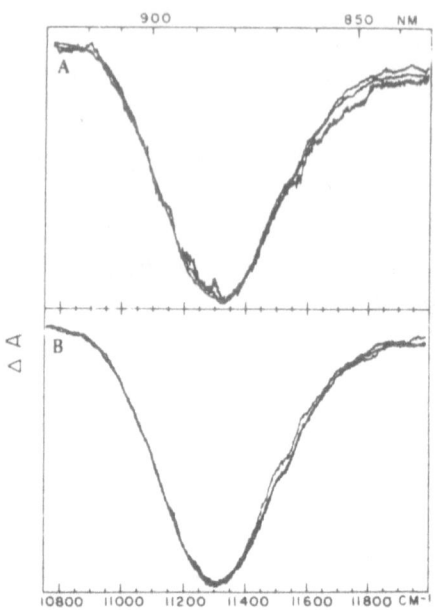

Fig. 26. Light-minus-dark spectra of holes burnt at ~885 nm in the reaction center of Rps. sphaeroides (P-870). (a) Holes burnt with 1.5, 0.9 and 0.3 J/cm^2 at 1.5 K. (b) Holes burnt at 1.5, 1.8 and 2.1 K. No difference in holewidths and areas is observed (from ref. [155]).

the hole is made up of an inhomogeneous distribution of phonon bands with complete suppression of the zero-phonon line, a very large displacement of the potential energy surface of the excited state relative to the ground state would occur [155]. In such a case, the excited state, $(BChl)_2^*$, would have a large dipole moment, i.e. a substantial degree of charge-transfer (C-T) character, which in fact has been confirmed by Stark-effect measurements [157a]. In this model, however, a charge transfer state is not populated within tens of fs following photo-excitation, an hypothesis which is supported by very recent fluorescence experiments in an applied electric field [157b].

These hole-burning results, although very exciting, have not cleared up the role of the BChl monomer as an intermediate in the electron transfer process yet. Furthermore, the role of the surrounding protein remains unknown, and more work is needed to answer these puzzles.

As a consequence of these experiments, theoretical lineshape calculations have been performed by Hayes and Small [158], in which the holes were assumed to be inhomogeneously broadened due to the presence of strong linear electron-phonon coupling of the P-band. They suggested that the unusually large hole-widths observed in refs. [154, 155] arise from a multiphonon broadening process, in which protein-pigment contacts are involved, and are not a manifestation of ultra-fast charge separation within the excited state $(BChl)_2^*$, as claimed in ref. [154].

In contrast to the extremely broad transient holes observed in bacterial reaction centers, narrow permanent holes have been reported for a *plant reaction center* [159]. The experiments were performed on the absorption band of P-700, the primary electron donor state of *photosystem I*, of enriched reaction center particles from spinach chloroplast at 1.6 K. The holewidths ($\Gamma_{hole} \approx 0.12$ cm^{-1}, after deconvolution of the instrumental bandwidth of 0.2 cm^{-1}) were about 3 orders of magnitude smaller than those found in the bacterial photosystems P-870 and P-960 [154, 155]. Similar holes, though somewhat broader (~ 1—2 cm^{-1}), had previously been observed in green algae [160]. One of the obvious questions which arises is whether the electronic structure of the RC of the primary donor state of plant photosystem I is different from that of the bacterial RC's. The experiments performed by Small and coworkers tried to give an answer to this question [159]. The sharp hole observed at 710 nm was superimposed on a very broad hole of width ~ 300 cm^{-1} (see Fig. 27). From the width of the narrow hole, a lifetime of ≈ 90 ps for P-700 was deduced, since no temperature dependence of the hole-width was observed between 1.6 and 3.5 K. The 90 ps were attributed to a very long electron transfer time, which is in contradiction with sub-picosecond transient absorption results, that yield only 1—4 ps for this process [161]. From a fit of the parameters of Hayes and Small's model [158] to the hole spectrum [159], it was concluded that P-700 has a dimer structure with significant C-T character, that it is dominated by strongly coupled low frequency phonons of ~ 30 cm^{-1} [159], as predicted for P-870 and P-960 [158].

It is striking that very similar holewidths have been observed in P-700 [159] and in the Chl-a antenna complex (C-670) of photosystem I at ~ 670 nm by the same group [162a, b], which is probably not a coincidence. As was pointed out in the literature [160, 163], great care has to be taken to select a sample for which one can be certain that the absorption at 700 nm is essentially caused by P-700 and

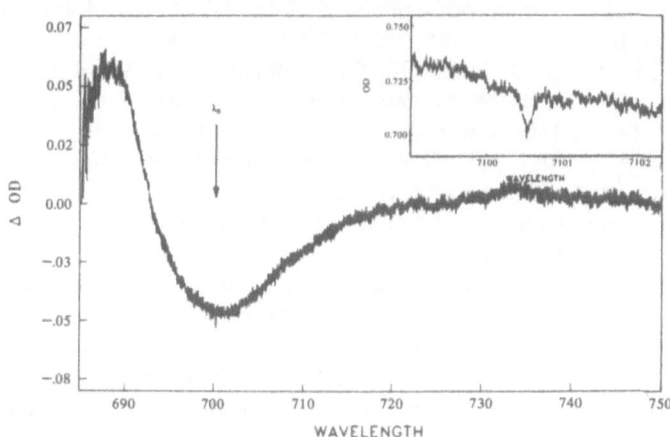

Fig. 27. Difference absorption spectrum of enriched photosystem I particles before and after laser excitation at 700 nm for 10 min with 100 μW/cm², at 1.6 K. The insert shows a narrow hole burnt at ~ 710 nm for 10 min with 1.0 μW/cm², at 1.6 K (from ref. [159]).

not partly by the antenna Chl-a. In fact, contradictory results obtained from recent hole-burning and photon echo experiments on the P-band of the RC of photo-system II of green plants (spinach) at 682 nm have been interpreted along this line [164]. A very large holewidth of 115 cm^{-1} at 1.5 K was attributed to the homogeneous width of the P-band in the RC, whereas a long echo decay time of 500 ps, measured at the same wavelength, was interpreted as due to dephasing of Chl unconnected to the RC. As a consequence of these results, it was suggested that the narrow hole reported in refs. [159, 162a, b] is not part of the hole-burning spectrum of the RC, but is also due to unrelated Chl [164].

3.1.4. *Metal Porphins: Examples of Transient HB*

Transient holes [37] are caused by population transfer from the ground state to a metastable energy level. The hole-burning mechanism can either be a two-level saturation process [37, 165a, b], or involve a triplet state [38, 39, 81, 121, 132, 154, 155, 164, 166, 167] or some nuclear hyperfine levels [168—172]. In such an experiment, the lifetime of the hole is the same as that of the metastable state, which for organic molecules usually is of the order of μs to ms [38, 39, 121, 132, 167]. When the triplet state or nuclear spin-lattice relaxation are involved, the holes may last seconds [81, 166, 168, 170, 171] to hours [81, 169, 172]. In order to observe a transient hole, a pump laser at a fixed frequency has to create a large enough depletion of the ground-state population (saturation), such that by means of a probe laser (which may, in principle, be the same laser) the difference in absorption between pumped and unpumped sites may be detected. The hole can be probed either by fluorescence excitation or transmission spectroscopy, as described above for PHB (see Section 2.1).

Transient hole-burning in inorganic materials was first studied in ruby by Szabo in 1975 [37], and later in colour centers of F_3^+ in NaF [81, 166], bound excitons in GaP : N [165a, b] and various rare earth ions incorporated as impurities in crystal-line [168—175] and glassy hosts [30, 53]. In this section we will discuss transient hole-burning experiments on organic systems, in particular of metal porphyrins: zinc porphin (ZnP) [38] and magnesium porphin (MgP) [39] in *n*-octane, and protochlorophyll (PChl) [132] in *n*-heptane crystals at liquid helium temperature.

It is known that some *metal porphins* have long-lived triplet states and high intersystem crossing yields [176], which makes them well suited for transient hole-burning [38, 39]. On the other hand, they are not generally liable to photochemical reactions, and thus PHB will not be applicable to them. Permanent holes, however, can still be induced in these molecules by means of NPHB if they are embedded in glassy hosts [121, 132]. When a metal porphin is incorporated in a crystalline host, e.g. an *n*-alkane, its lowest degenerate excited singlet state S_1 will split into two orbital components, a lower one, S_{1x}, and a higher one, S_{1y}, separated by a distance δ, the crystal-field splitting which ranges from a few cm^{-1} to a few hundred cm^{-1} [131] (see also Section 4.1.4).

In the transient hole-burning study of *ZnP* in *n*-octane (*n*-C$_8$) (see Fig. 28 for the energy level diagram), the $S_{1x} \leftarrow S_0$ 0—0 transition was excited with a narrow-

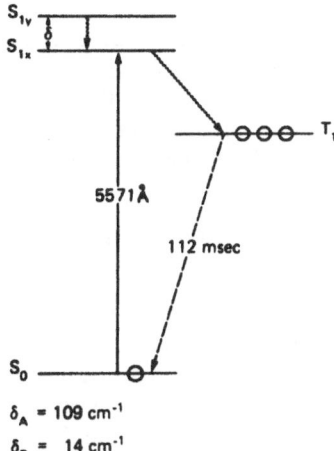

Fig. 28. Energy level diagram of zinc porphin (ZnP) in n-C$_8$ · S_{1x} and S_{1y} represent the two orbital components of the first excited singlet state separated by the crystal field splitting δ · T_1 is the first excited triplet state. Lifetimes and energies were taken from ref. [38].

band cw-dye laser at 557 nm [38]. Because the triplet state of ZnP lives ~ 110 ms, the same laser was subsequently used to probe the hole by scanning it very rapidly over the spectral region of interest. From the temperature dependence of the holewidths, which will be discussed in Section 4.1.1, conclusions were drawn about the optical dephasing mechanism that takes place in this system.

Transient hole-burning experiments on *MgP* in n-C$_8$ [39] were performed with the purpose to solve a dilemma on optical dephasing that arouse between the results of ZnP [38] and those of H$_2$P [58, 60] (see Section 4.1.1). A magnesium-pyridine complex, rather than "pure" MgP, was used [39] because the Mg atom in the porphin molecule, if uncomplexed, scavenges impurities as axial ligands [135, 136, 177, 178]. Transient holes in four sites of the $S_{1x} \leftarrow S_0$ band of MgP in n-C$_8$ were burnt and probed simultaneously with two single-frequency cw-dye lasers.

Not only *transient* holes, but also *photochemical* holes were observed in one of the five largest sites of MgP-pyridine in n-C$_8$ [39]. The photochemistry responsible for this effect is not well understood, but a similar reaction has been observed for a MgP-ethanol complex in n-C$_8$, which was attributed to a reorientation of the ethanol ligand relative to the MgP molecule [136]. Hole-burning experiments performed on a ruthenium phthalocyanine complex with carbonyl-pyridine in glassy matrices at 5 K also suggest that axial ligands play a role in the HB-mechanism through interaction with polar solvents [179].

In addition to the results of optical dephasing in MgP in n-C$_8$, Section 4.1.1 will show that the relaxation time from S_y to S_x can also be determined by hole-burning.

It was mentioned above (see Section 3.1.2) that hole-burning via the triplet state had also been observed for *protochlorophyll (PChl)* in n-heptane [132]. The hole-recovery kinetics in this system was studied by monitoring the fluorescence intensity of the hole during the off-time of the burning laser. The transient holes were filled up in about 5—10 ms at 5 K, in agreement with the triplet state mean lifetime of PChl [133]. Simultaneously with the fast hole recovery, a slow hole recovery of about 10 s duration was observed, which was tentatively attributed to some photochemical and/or NPHB mechanism [132].

3.2. INTERMOLECULAR AND TWO-PHOTON PHB SYSTEMS

Within PHB one can distinguish between *intra-* and *intermolecular* photochemical reactions, i.e. those that occur in the guest molecule and those that take place between the guest and the host. We have already discussed examples of the former, like free-base porphyrins and their derivatives, which undergo a reversible *one-photon* reaction (see Section 3.1.1). In this section, two examples will be given of *intermolecular* photochemical reactions: quinizarin in alcoholic glasses, and ionic dyes in polar amorphous hosts. In addition, various systems will be presented which undergo either *two-photon* or *photon-gated* hole-burning processes.

3.2.1. *Examples of Intermolecular PHB Systems*

a. *Quinizarin.* The photochemistry and many low temperature hole-burning properties of this molecule, also called *1,4-dihydroxyanthraquinone*, have mainly been studied by the group of Friedrich and Haarer [27, 47, 48, 65, 90, 94, 180—183]. The PHB-mechanism is assumed to be an *intermolecular* reaction, in which an intramolecular hydrogen bond of quinizarin is broken and an intermolecular hydrogen bond is formed between the guest molecule and the matrix, as illustrated in Fig. (29) [47, 183]. This mechanism is supported by hole-burning experiments performed on hydrogen-bonded matrices, like ethanol and polyvinylalcohol (PVA) doped with quinizarin. Efficient PHB was observed in these systems at 1.8 K, in contrast to quinizarin embedded in *n*-alkanes and other non-hydrogen bonded matrices which did not exhibit measurable proton transfer [183]. The photo-

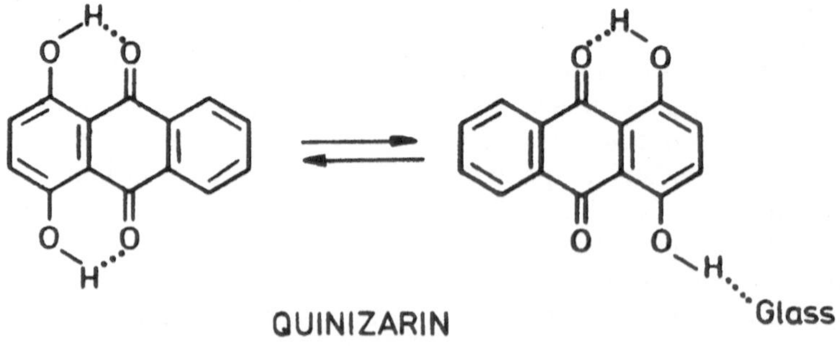

QUINIZARIN   Glass

Fig. 29. PHB-mechanism in quinizarin (1,4-dihydroxyanthraquinone): an intramolecular hydrogen bond is broken by irradiation with light, and an intermolecular hydrogen bond with the alcoholic glassy host is formed (from refs. [47, 183]).

product seems to have a broad absorption band at about 2000 cm^{-1} higher energy, and the photochemistry is thermally reversible at 60—80 K [47]. Recent experiments by Tani and coworkers [184a] have shown that, by substitution of one hydroxy group in the anthraquinone molecule, the hole-burning efficiency decreases by two orders of magnitude, as compared to that in quinizarin. These results suggest that the photoproduct of quinizarin, contrary to what had been assumed in ref. [47] (see Fig. 29), is hydrogen bonded to the matrix in both functional OH-groups [184a]. Further PHB experiments by the same group have shown that the holewidths in quinizarin become larger, the stronger the polarity of the polymer matrices used [184b]. The results were explained by dipole-dipole interaction between the electric dipole of the dye molecule and the number density of electric dipoles localized on the acid unit of the polymer host.

The absorption spectra of quinizarin in alcoholic hosts at 2 K, after PHB at ~ 515 nm, show two types of holes, a zero-phonon hole at the laser frequency and a very broad phonon-side hole at lower energy. The latter is the result of a photochemical reaction initiated in the phonon side-bands of molecules with electronic origins at lower energy [20, 180]. This effect has subsequently been observed in many HB-systems. The ratio between the integrated intensity of the resonant hole, I_0, and that of the total hole plus side-hole, I_{tot}, yields the Debye-Waller factor α, which is related to the electron-phonon coupling strength, S, by $\exp(-S) = I_0/I_{tot} = \alpha$. S is the most probable number of phonons involved in the transition. For quinizarin, $S \simeq 1—2$, which is indicative of an intermediate coupling. The value of S depends strongly on the guest-matrix interaction [180]. This has recently been verified for derivatives of anthraquinone in various acrylic polymers. Hole-burning experiments on these systems have shown that a bulky substituent on the guest molecule induces a stronger interaction with phonons of the host, and therefore, a larger value of S [184c].

The first holes burnt in the system quinizarin in methanol-ethanol glass were very broad, about 2.3 cm^{-1} at 1.3 K, and assumed to be homogeneously broadened because of their Lorentzian line shape [47]. Such large widths were interpreted as corresponding to surprisingly short T_2-relaxation times in glasses of ~ 5—10 ps. Similar broad holes had been reported for tetracene in the same matrix [20]. From the temperature dependence of the holewidth, which was reported to be T^2, also large residual linewidths of $\Gamma_0 \sim 0.6 - 0.9$ cm^{-1} for $T \rightarrow 0$ were derived [48].

Subsequent experiments performed with higher resolution (0.02 cm^{-1}), however, demonstrated that the holewidth could be reduced to at least ~ 0.1 cm^{-1} at $T \simeq 1$ K, and that it extrapolated to a value of $\Gamma_{hole} < 10^{-3}$ cm^{-1} for $T \rightarrow 0$ (after deconvolution, see also Fig. 8) [67]. The temperature dependence of the holewidth in these experiments was found to be $T^{1.3}$ above 2 K, and linear below 2 K. A $T^{1.3}$ law has been found for many organic amorphous systems at low temperatures [24], as will be shown in Section 4.2.2.

From the experiments on quinizarin in alcohol glass, it was concluded that the holewidth consists of two contributions, a reversible one due to the homogeneous linewidth or "pure" dephasing process, and an irreversible contribution which arises from inhomogeneous broadening due to ground state "spectral diffusion" processes [48, 65, 94] (see also Section 4.2.4).

b. *Ionic dyes: resorufin and cresyl violet*. These are charged dye molecules (Fig. 30), which exhibit very broad inhomogeneous spectral bands, $\Gamma_{inh} \sim 500{-}1000$ cm^{-1}, when incorporated in alcoholic glasses and polymers, even at liquid helium temperature [49, 59, 92, 185]. An absorption spectrum of resorufin in ethanol glass at 1.2 K is given in Fig. 31 [186]. Since the inhomogeneous widths are much broader than the frequency differences between vibrational bands, it is difficult to obtain information on the vibronic structure and dynamics of these ionic systems. Such intricate spectra, however, can be unravelled by means of site-selection spectroscopy in combination with hole-burning.

Site-selected spectra of ionic dyes in amorphous hosts between 1.2 and 4.2 K exhibit line profiles different from those of neutral molecules in the same glasses. This is illustrated in Fig. 32, which shows site selected spectra of the $S_1 \leftarrow S_0$ transition of resorufin (upper part) and free-base prophin (lower part) in organic glassy hosts at 4.2 K. Notice that the narrow lines in the upper spectrum are accompanied by rather strong phonon side-bands,which is not the case for H_2P [1]. Similar type of spectra have been reported for resorufin in PMMA [92]. From

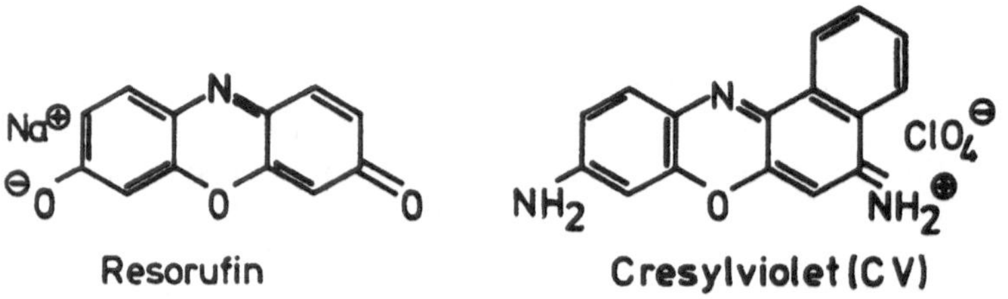

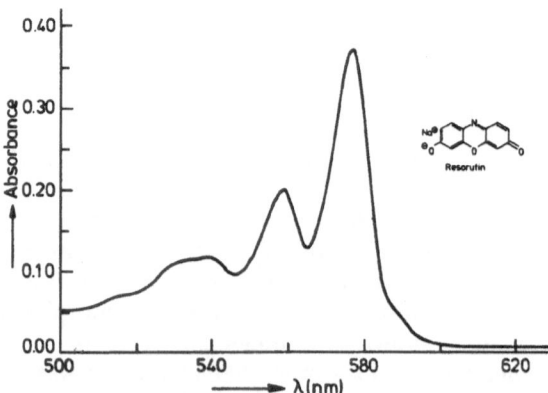

Fig. 30. Structure of the ionic dye molecules resorufin and cresyl violet. The counter ions are Na^+ and ClO_4^-, respectively.

Fig. 31. Absorption spectrum of resorufin in ethanol glass at 1.2 K. Notice the large inhomogeneously broadened bands [186].

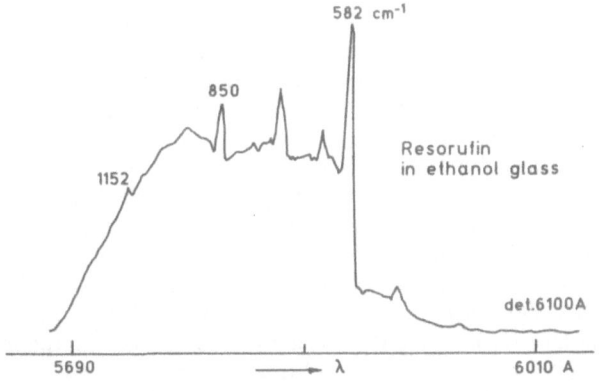

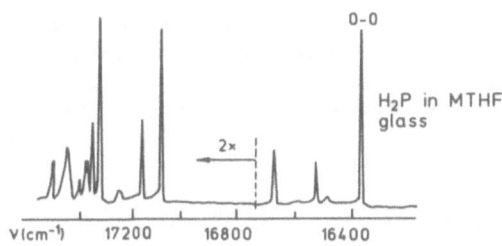

Fig. 32. Site-selected excitation spectra of the $S_1 \leftarrow S_0$ transition of resorufin in ethanol glass (upper part), and free-base porphin (H_2P) in MTHF glass (lower part) at 4.2 K. Notice the strong phonon side-bands in the spectrum of the ionic dye, whereas only zero-phonon lines are observed in the spectrum of H_2P.

such a site-selected spectrum, however, it is still difficult to estimate the electron-phonon coupling strength S, because of the presence of a large background.

Quantitative information on the parameter S can be obtained by an alternative hole-burning method, namely from the ratio of the burnt hole area to the total area of hole plus phonon-side hole, as mentioned above for quinizarin [47, 180]. Applying this method to resorufin and cresyl violet as guests in various amorphous hosts at 4.2 K, values of S were obtained between ~ 1.3 and ~ 2 [187]. As one may expect from these charged molecules, the value of S depends on the polarity of the host, Z. The parameter Z is a measure of the microscopic solvent-solute interaction based on spectroscopic parameters, and represents the ionizing power of the solvent [188]. In contrast to these charged molecules, the values of S for the neutral molecule H_2P are lower, between ~ 0.5 and 1.0, and vary randomly with Z [187].

Since it is difficult to obtain the vibrational frequencies of the first excited singlet state S_1 of these charged molecules by means of site-selection spectroscopy, due to overlapping spectral bands, a different method based on hole-burning was

used in ref. [187]. A hole was burnt into the broad absorption band of resorufin in
ethanol, and the spectral distances to satellite holes which appeared on its low
energy side were measured, as proposed in ref. [189]. Small and coworkers have
reported similar satellite hole structures for cresyl violet, nile blue, and oxazine
720 in polyvinylalcohol [185].

The hole-burning mechanism taking place in ionic dye systems has mostly been
attributed in the literature to a non-photochemical reaction (NPHB) [59, 93, 96,
185, 190] (see also Sections 1.2 and 3.3), although it has not been properly
justified. It is usually assumed that in a NPHB process a slight reorientation of a
photostable molecule relative to its host takes place [20, 26]. A criterion to
distinguish between a photochemical and a non-photochemical reaction has been
suggested by Hayes and Small [20]: if the spectral position of the absorption band
of the photoproduct lies very close (< 2 cm^{-1}) to that of the originally burnt hole,
the mechanism is NPHB. "Hole-filling" experiments, similar to those described in
refs. [20, 59, 190], have been performed on cresyl violet and resorufin at 4.2 K.
The results indicate that the spectral distance of the photoproduct to the original
absorption band is ~ 80 to 400 cm^{-1} [49, 187]. This distance increases approximately in a linear fashion with host polarity Z for both ionic dyes in alcoholic
glasses (see Fig. 33) [23, 187]. In contrast, no dependence on Z has been found for
the neutral molecule H$_2$P. The results suggested that *intermolecular* photochemical

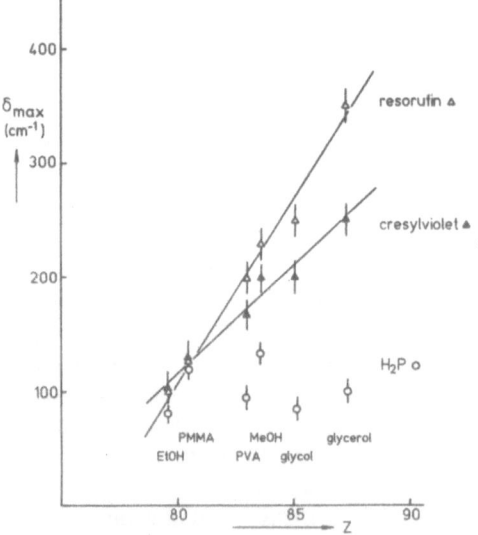

Fig. 33. Spectral distance of the photoproduct absorption maximum to the originally burnt hole,
δ_{max}, as a function of host polarity, Z (kcal/mol) [188], for resorufin, cresyl violet and H$_2$P in ethanol
(EtOH), polymethylmethacrylate (PMMA), polyvinylalcohol (PVA), methanol (MeOH), glycol and
glycerol. Notice the linear dependence on Z for the two ionic dyes, whereas no dependence on Z is
observed for the neutral molecule H$_2$P [187].

hole-burning takes place in these ionic dye systems [49]. The mechanism is most probably a charge redistribution in the excited state of the dye molecule, which induces a reorganization of the solvent cage surrounding the guest molecule [187]. As might be expected, this process is markedly dependent on the interaction of the ion with its environment. Furthermore, the hole-burning efficiency of these ionic dyes is apparently much larger than that of typical NPHB-systems like pentacene in PMMA [86].

Despite the large inhomogeneous linewidths and the appreciable electron-phonon coupling strength of these ionic dye systems, very small homogeneous linewidths, in the MHz-regime, have been observed [49, 86, 186, 187, 191]. The temperature dependence of these small widths, as well as that of larger linewidths reported earlier, obtained by hole-burning [49, 59, 86, 93, 96, 186, 190, 191] and photon echoes [96, 191], will be compared and discussed in Sections 4.2.2 and 4.2.3.

3.2.2. Two-Photon and Photon-Gated PHB Systems

a. *Dimethyl-s-tetrazine (DMST)*. This molecule (Fig. 34), incorporated as guest in durene at liquid helium temperature, was one of the first in which PHB was observed [17]. The hole-burning mechanism in DMST, similar to that in s-tetrazine [40], occurs by a sequential absorption of *two photons* of the same wavelength. The first photon excites the molecule into a selective energy level, whereas the second photon is responsible for the photodissociation of the molecule into N_2 and $2(CH_3)CN$ [192]. The photodissociation seems to involve several intermediate states at low temperature [50, 193]. From the quadratic dependence of the photochemistry on light intensity, the non-linear character of the reaction was demonstrated [50].

DMST also undergoes hole-burning when embedded in amorphous hosts like polyvinylcarbazole (PVCa) [45, 69], PMMA [45, 49, 88], polyethyleneglycol (PEG) [45], polyethylene (PE) [88] and ethanol-methanol glass [194]. The temperature dependence of the holewidth in these amorphous systems will be dis-

DMST

Fig. 34. Structure of dimethyl-s-tetrazine (DMST). This molecule undergoes intramolecular two-photon PHB [17, 50, 192].

cussed in Section 4.2.2. It is interesting to notice that two-level saturation hole-burning has also been reported for s-tetrazine in the gas phase [195].

b. *Photon-gated systems*. For optical storage applications in the frequency-domain, it is very inconvenient to work with materials that undergo a one-photon PHB-mechanism because they bleach during the probe process and, thus, do not fulfill the requirement of non-destructive reading [196] (see Section 4.1.5). Molecular systems which undergo PHB via two-photon *"photon-gated"* processes [197] do not suffer from this limitation. By this mechanism, a hole is burnt at a specific laser frequency v_h in the presence of a second gating light source (at frequency v_g) only. Persistent spectral holes can then be probed in the absence of the gating light without degradation of the signal.

Photon-gated spectral hole-burning was first observed in an inorganic system, Sm^{2+} *in BaClF* [198]. Here, the mechanism responsible is a two-step photoionization of Sm^{2+}, which leads to persistent holes in the first-step transition at ~688 nm. The ratio of the gated hole depth to the one-photon hole depth is greater than 10^4. Holewidths of 25 MHz at 2 K were observed in a 13 GHz wide inhomogeneous absorption band. A remarkable feature of this system is that the holes are stable (they broaden by a factor of less than 2) after thermal cycling to room temperature for several days [198].

The second system for which "photon gated" hole-burning was reported is *carbazole in boric acid glass* [51]. In this case, the first photon excites a $S_1 \leftarrow S_0$ transition of carbazole at ~335 nm. Subsequently, the molecules undergo inter-system crossing into the lowest triplet state T_1 with a high yield. In the presence of a second photon tuned at a wavelength between ~350 and 514 nm, which induces a $T_n \leftarrow T_1$ transition, holes are burnt at the first singlet excitation wavelength. This reaction is due to photoionization of the molecule during the second step, by which the ejected electron is trapped in the boric acid glass. Holes were stable after cycling from 1.4 to 77 K for several hours, although they become broader and shallower. Holewidths were not reported in ref. [51].

After these two first "photon-gated" hole-burning experiments, a few other inorganic systems have been found that show the same effect: Sm^{2+} *in CaF$_2$* at ~696 nm [199], Co^{2+} *in LiGa$_5$O$_8$* at ~660 nm [200] and Cr^{3+} *in SrTiO$_3$* at ~790 nm [201]. The PHB-mechanism in all of them is a two-step photo-ionization. Also some *organic* photon-gated systems have been found, like the *anthracene-tetracene photoadduct in PMMA* which shows hole-burning in the $S_1 \leftarrow S_0$ transition at 326 nm, with holewidths of ≤ 0.07 cm^{-1}, in the presence of a second $T_n \leftarrow T_1$ excitation at 442 nm only. The result of the two-photon photochemistry here is a substantial change in the molecular structure. A *Zn-tetrabenzoporphyrin derivative in PMMA*, in the presence of a halomethane acceptor molecule at high concentration, has been found to undergo a donor-acceptor electron transfer "photon-gated" hole-burning mechanism [71, 203]. "Gated" holes were burnt in this system by simultaneous excitation of the 0—0 singlet transition of the por-phyrin donor at ~630 nm and its triplet-triplet absorption at ~488 nm. The electron transfer process occurs from the excited triplet state of the porphyrin donor to the chloroform acceptor, with a gating ratio greater than 30 [203]. Holewidths of ~ 3.3 GHz at 1.4 K were reported. They are much broader than the fluorescence lifetime-limited value of Zn-tetrabenzoporphyrin of $\simeq 40$ MHz, expected from extrapolation to $T \to 0$. Very fast burning, in a time of $\simeq 30$ ns at ~ 630 nm with hole depths of $\simeq 1\%$ could be achieved, when a simultaneous gating pulse of 200 ns was present at ~ 488 nm [71].

3.3. NON-PHOTOCHEMICAL HOLE-BURNING (NPHB) SYSTEMS

NPHB, also called "photophysical" hole-burning (see Section 1.2), has first been observed in the *visible* spectral region of photostable organic molecules, like perylene [16] and tetracene [20] embedded in organic *glassy* hosts. Afterwards, it has also been reported for organic *crystalline* systems, like pentacene [61, 63] and thioindigo [204] both incorporated in benzoic acid, at liquid He temperature. The hole-burning mechanism in the latter case has been interpreted as a hydrogen-bond rearrangement in the host crystal after laser irradiation, which leads to a modification of the local environment of the guest molecule and a shift in absorption. Very narrow holes, of widths limited by the fluorescence lifetime at 1.8 K, and "antiholes" absorbing at an energy of about 7 to 130 cm^{-1} from the hole, have been reported for the 0—0 transition of pentacene in benzoic acid [61]. Further details on the NPHB-dynamics of these crystalline systems can be found in the literature (see Chapter 5 of ref. [21], and refs. [61, 63, 204]).

Another class of *crystalline* systems that has been studied undergoes "photophysical" hole-burning in the *infrared* spectral region. HB occurs here by excitation of a molecule into a vibrational state of the ground state with an IR-laser. During this process a molecular reorientation in the host seems to take place. Systems belonging to this class are *1,2-difluoroethane* isolated in matrices of Ar, Kr and N_2 [205, 206], ReO_4^- substitutionally doped into alkali halide single crystals [207], and CN^- in alkali halides [208]. A review on *infrared hole-burning* of vibrational transitions of impurity molecules in solids is given in Chapter 6 of ref. [21].

Let us now turn to non-photochemical hole-burning (NPHB) in *organic glasses*. In the first system, *perylene* in ethanol at 4.2 K, studied by Personov and coworkers [16], the guest as well as the host are photostable molecules at low temperature. The holes were reported to last for hours, and the sample could be restored by heating it up to 20—30 K with white light. From the holewidth, the homogeneous linewidth was estimated to be Γ_{hom} ~ 0.4 cm^{-1} [16]. From a subsequent study of the kinetics of this system, it was concluded that hole-burning occurs by a one-photon process [209]. The mechanism was attributed to a possible reorientation of the impurity molecule in the glassy host accompanied by a shift of the frequency of the 0—0 transition. This type of NPHB effect was later observed by the same group for other photostable molecules like: *tetracene, dibenzpyrene* and *Zn-tetraphenylporphin* in ethanol glass [189]. In these experiments not only a hole was observed at the laser frequency, but satellite holes over a wide spectral region appeared simultaneously with it. As mentioned above (Section 3.2.1), the vibronic frequencies of the first excited singlet state S_1 of the guest molecule can be obtained in this way by measuring the spectral distances of the resonant hole to the satellite holes [185, 187, 189].

Many NPHB studies on *tetracene* and other photostable guests embedded *in various alcohol matrices* have been carried out by Small and coworkers. Detailed discussions on their findings can be found in Chapter 5 of ref. [21] and ref. [26a, b]. A mechanism for NPHB, called phonon assisted tunneling by two-level systems (TLS) of the glass, was proposed by this group [20, 52]. A brief description of this model was given in Section 1.2 by means of an energy level diagram (see Fig. 3). It has been pointed out in ref. [52] that an appreciable guest-host interaction is necessary in order for a local rearrangement of the matrix around the excited guest molecules to occur. Quantum yields for NPHB are generally low ($\leqq 10^{-5}$) [210a, b] (see also Chapter 5 of ref. [21]), as compared to those for PHB ($\simeq 10^{-2}-10^{-3}$) [18, 86, 123a, b].

As already mentioned in Sections 1.2 and 3.2.1., a characteristic feature of NPHB in glasses is that the photoproduct, or "antihole", is expected to absorb at a frequency very close ($<$ a few cm^{-1}) to that of the burnt hole [20]. By means of "hole filling" experiments [20, 187, 190], it is possible to distinguish between PHB and NPHB processes. This is illustrated in Fig. 35, which shows a hole burnt into the $S_1 \leftarrow S_0$ 0—0 transition of *pentacene in PMMA* at 1.2 K (top). By burning a second hole at a distance of less than 50 GHz (<2 cm^{-1}) to either the higher or lower energy side from the original hole, the latter is filled in, whereas no hole-filling was observed at larger spectral distances [56]. These results indicate that the "antihole" of pentacene in PMMA lies very close to the originally burnt hole. Thus, the NPHB process most probably changes the molecule-matrix configuration only slightly.

It has been reported that NPHB in polar hydrogen-bonding glasses is significantly more efficient than in other glasses or polymers [210a]. NPHB, however, has also been observed for organic molecules in non-hydrogen bonding amorphous hosts, like tetracene in anthracene-derivatives glasses [54, 211a, b, 212], pentacene in PMMA [55, 56], and tetracene and pentacene in methyltetrahydrofuran (MTHF) [213]. *Inorganic systems* undergoing NPHB are, for example, *rare earth ions* doped in many *inorganic glasses* [30, 53] and *organic glasses*, like Pr^{3+} and Nd^{3+} in polyvinylalcohol (PVA) [190], and Eu^{3+} in ethanol, H_2O, ether, PMMA and PVA [214a, b]. In these systems, the NPHB-mechanism is due to either population redistribution between the nuclear hyperfine levels of the ground state of the rare earth ion or rearrangement of the local environment [30, 214b] (see also Section 4.2.7).

As will be seen in Sections 4.2.3 and 4.2.4, there is still a controversy regarding holewidths and homogeneous linewidths, and their temperature dependence in organic amorphous systems [24, 30]. In fact, various hole-burning and photo-echo experiments recently performed on PHB and NPHB systems have yielded contradictory results. The purpose of these experiments was to separate the contributions to the holewidth due to dephasing and to "spectral diffusion" processes [55, 56, 86, 96, 186, 191].

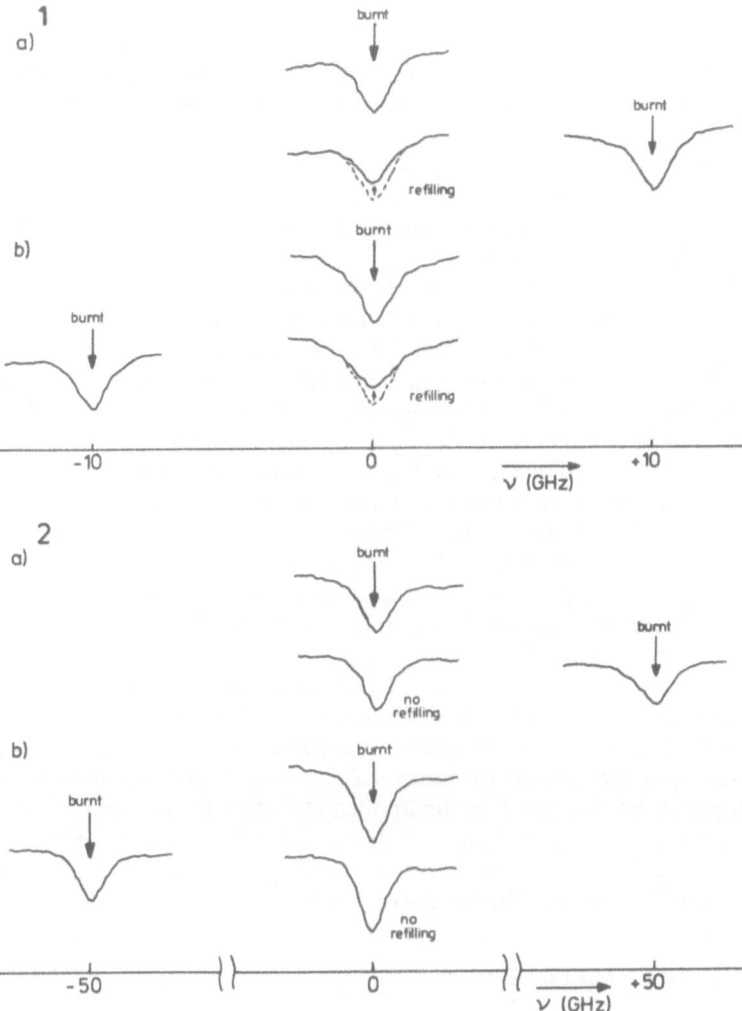

Fig. 35. "Hole-filling" experiment on the NPHB-system pentacene in PMMA at 1.2 K. From top to bottom: (1a) A hole burnt at $\nu = 0$ GHz is filled in when a second hole is burnt at $\nu = +10$ GHz. (1b) The same is valid as in (1a) for a second hole burnt at $\nu = -10$ GHz. If a second hole is burnt at $\nu \geq +50$ GHz (2a), or $\nu \geq -50$ GHz (2b), the originally burnt hole is not filled in [56].

4. Applications

The high spectral resolution of the order of MHz attainable with hole-burning (HB) makes this technique attractive for many applications. In particular, since *permanent* holes obtained either by *PHB* or *NPHB* at liquid helium temperature represent sharp frequency makers in the absorption spectrum, it is possible to detect frequency shifts, in addition to hole broadening, as a function of temperature and applied external fields that are many orders of magnitude smaller than the inhomogeneously broadened absorption band. Thus, optical dephasing processes and very accurate Zeeman and Stark effects can be studied with HB, which are not possible with any other technique. Also strain fields, like hydrostatic pressure, can be applied to holes which yields information on matrix compressibilities. In contrast to conventional methods, HB has the advantage that very small fields and pressures suffice for this purpose.

Another potential technological application of HB is data storage in the frequency domain. The presence or absence of a hole at a specific frequency in the absorption band can be used as one "bit" of information. This permits to increase the storage density in comparison to conventional optical techniques by a factor $\Gamma_{inhom}/\Gamma_{hole}$, the ratio of the inhomogeneous width to the holewidth, which can be as large as 10^5 in polymer hosts doped with organic molecules (see Section 4.1.5). Since permanent HB allows to burn away arbitrary portions of an inhomogeneously broadened absorption band, it is possible, in principle, to make almost any spectral profile. By combining this HB-property with the application of external fields that modulate or shift such a profile, very interesting devices like time domain laser pulse shaping, and frequency-multiplexed optical spatial filters could be made. The reader interested in this type of technological applications should consult Chapter 7 of ref. [21].

We will present only a few selected HB-applications. As principal subject, we have chosen the study of optical dephasing and relaxation processes of electronically excited states of organic molecules incorporated in either crystalline (Sections 4.1.1 to 4.1.3) or amorphous (Sections 4.2.2 to 4.2.7) hosts at low temperature. A further topic to be discussed is the application of magnetic and electric fields to holes burnt in zero-field in crystalline (Section 4.1.4) and in disordered materials (Section 4.2.8). The effect of pressure applied to holes burnt in amorphous systems will be presented in the last part of Section 4.2.8.

4.1. MOLECULAR MIXED CRYSTALS

Optical dephasing experiments on electronic $S_1 \leftarrow S_0$ 0—0 transitions of dilute molecular mixed crystals will be discussed and compared with theoretical models in the next two sections. It will further be shown in Section 4.1.3, that hole-burning is also a useful technique to measure relaxation times from excited S_1-vibronic states, and may serve to assign accurate vibrational frequencies in multiple-site spectra. Most of the Zeeman and Stark experiments studied by HB have been performed on *inorganic* crystals doped with rare earth ions and colour centers [166, 169, 171, 172b, 175, 215—217]. Only a few hole-burning experiments in external fields have been reported for *organic* crystals, which will be presented in Section 4.1.4. The systems studied were free-base porphin [42] and chlorin [114—116] in *n*-alkane hosts. Finally, the potential use of permanent hole-burning as a method to increase the density of information in optical storage devices [218] will be discussed in Section 4.1.5.

4.1.1. *Optical Dephasing in Organic Crystalline Systems: Experimental Results*

a. *Photochemical hole-burning (PHB) systems*. It was shown above (Section 1.1), that a hole burnt in the inhomogeneously broadened absorption band of the $S_1 \leftarrow S_0$ transition of an isolated molecule embedded in a crystalline host lattice may represent a negative replica of the homogeneous spectral line. The holewidth, if correctly extrapolated to negligible hole depth (for zero laser power and zero burning time, see Section 2.2) is then determined by the homogeneous linewidth, Γ_{hom} [19, 219]. The first MHz-resolution HB-experiment performed on an organic system in which the holewidth was close to the true value of Γ_{hom} was reported in 1976 by de Vries and Wiersma for dimethyl-s-tetrazine (DMST) in a durene crystal [17]. It was mentioned in Section 3.2.2 that DMST photodissociates through a two-photon process, and undergoes intramolecular photochemical hole-burning when excited with a narrow-band laser into the 0—0 transition at 587.5 nm. A hole width of 120 MHz was observed in this absorption band at 2 K, which is still significantly larger than the width of ~25 MHz, that corresponds to the 6 ns fluorescence decay time of DMST. The extra contribution, assumed to be due to phase-destructive events (T_2^*, see eq. (1)), was attributed in ref. [17] to low temperature electron-phonon coupling.

In a crystalline system at the lowest temperatures one would, however, expect that the population decay time T_1 (see eq. (1)) is ultimately the factor limiting the holewidth, because phonon processes are frozen out. This was, indeed, observed for holes burnt into the $S_1 \leftarrow S_0$ 0—0 transition of free-base porphin (H_2P) in an n-octane (n-C_8) crystal (see also Section 3.1.1). The holewidth, $\frac{1}{2}\Gamma_{hole} = \Gamma_{hom} \simeq$ 9 MHz for $T \lesssim 2$ K, was entirely determined by the fluorescence lifetime of H_2P, $T_1 = 17$ ns, and thus, $\Gamma_{hom} = (2\pi T_1)^{-1}$ when $T \to 0$ [19, 58]. The same result was found for other porphin molecules in n-alkane hosts: ZnP in n-C_8 [38], chlorin in n-C_6, n-C_8 and n-C_{10} [44], H_2P in n-C_{10} [60] and MgP in n-C_8 [39]. By contrast, photon echo decay times, $\frac{1}{2}T_2$, obtained for the $S_1 \leftarrow S_0$ 0—0 transitions of tetracene and pentacene in p-terphenyl [220], naphthalene in durene and per-deuteronaphthalene [221], and pentacene in naphthalene [222] at the lowest temperature of ≈ 1.4 K were not found to be determined by T_1. It was suggested in ref. [220] that even at 1.2 K not all relaxation processes in these organic crystals were frozen out. Also energy transfer [221] and quadratic electron-phonon coupling [222] processes were held responsible for these discrepancies. Subsequent hole-burning experiments on DMST in durene and s-tetrazine in benzene, carried out with a better stabilized laser, yielded homogeneous linewidths at 2 K which were entirely determined by T_1 [40]. The same result, namely $T_2 = 2T_1$, was later found by photon echoes in dilute mixed crystals of pentacene in p-terphenyl [223], pentacene in bezoic acid [61, 224] and xanthione in xanthone [225].

As the temperature increases, thermally induced dephasing processes rapidly set in. *Permanent hole-burning* (PHB) has the unique possibility, compared to coherent transient techniques, that it can measure simultaneously the homogeneous *linewidth* and its *frequency shift*. It is, therefore, a powerful tool for studying optical dephasing processes at low temperatures, because it yields more experimental parameters for the interpretation of the data [58]. Since most of the PHB studies in crystals have been performed on porphins and their derivatives in *n*-alkane hosts, we will use the results obtained for these systems as a framework for the understanding of optical dephasing processes in dilute molecular mixed crystals. Specific experiments will now be discussed in more detail.

Fig. 36 shows the temperature dependence of the holewidth in the $S_1 \leftarrow S_0$ 0—0 transition of free-base porphin (H_2P) in the A- and B-sites of n-C_8 [19, 41, 58] between 1.6 and 4.2 K. Below 5 K, little broadening occurs for the A-site, and Γ_{hom} is almost entirely determined by the fluorescence decay time of the S_1-state [19]. This is not the case for the B-site, where Γ_{hom} increases strongly with temperature [58]. In addition to line broadening, also frequency shifts were observed in this system. The latter, illustrated in Fig. 37, were measured by using

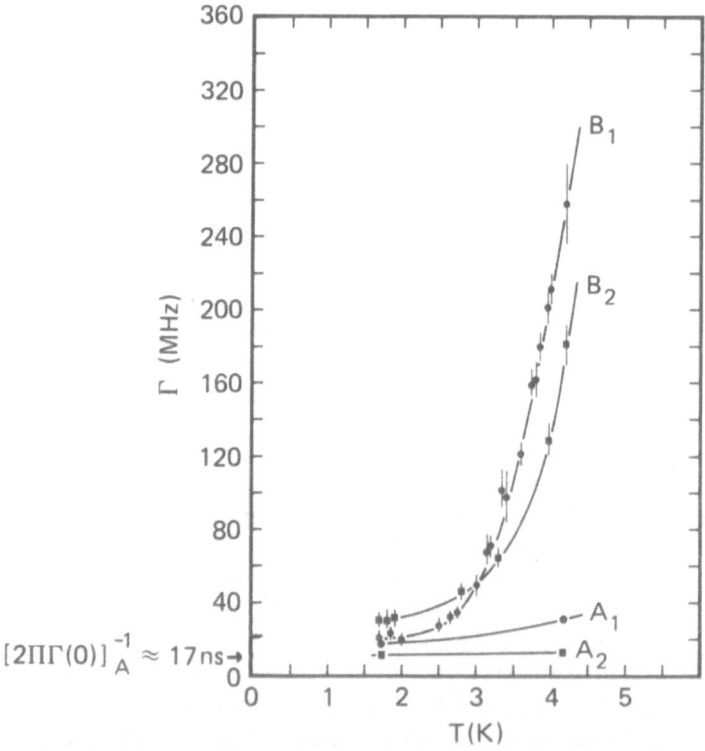

Fig. 36. Temperature dependence of the homogeneous linewidth derived from holes burnt into the $S_1 \leftarrow S_0$ 0—0 transition of H_2P in the A- and B-sites of *n*-octane [58]. The linewidth of the A-sites below 5 K is almost entirely determined by the fluorescence decay time of 17 ns [19, 41].

(a)

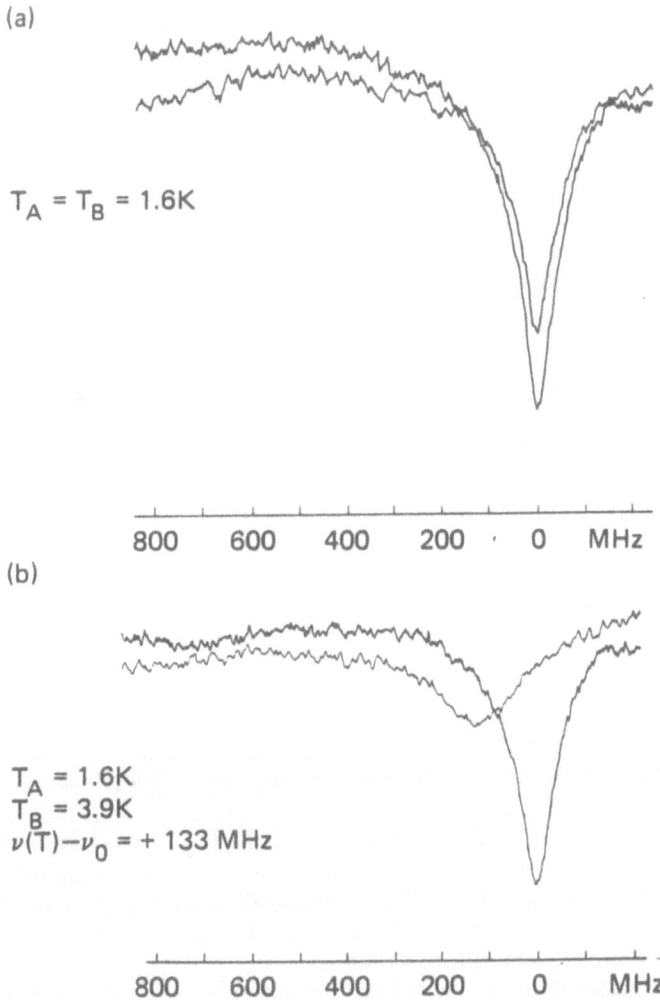

$T_A = T_B = 1.6K$

800   600   400   200   0   MHz

(b)

$T_A = 1.6K$
$T_B = 3.9K$
$\nu(T) - \nu_0 = + 133$ MHz

800   600   400   200   0   MHz

Fig. 37. Frequency shift and broadening of a hole burnt in the B_1 0—0 line of H_2P in n-C_8. (a) Excitation spectra of two holes burnt at the same frequency at 1.6 K in identical samples in two cryostats, A and B. (b) Excitation spectra of the same hole after raising the temperature of cryostat B to 3.9 K. The temperature behaviour of the hole is reversible [58].

two identical samples in separate cryostats, and increasing the temperature in one of them [58]. This temperature behaviour was found to be reversible. Both the width and the shift have an exponential dependence on the inverse temperature, $\Gamma_{hom} - \Gamma_0 \propto \exp(-E/kT)$, with the same activation energy, $E \sim 15$ cm^{-1} for the B-site (see Fig. 38), and $E \sim 35$ cm^{-1} for the A-site [58].

These low activation energies have been interpreted in refs. [19, 58] in terms of pseudo-local phonon modes [220, 226]. Such a local phonon in a dilute molecular

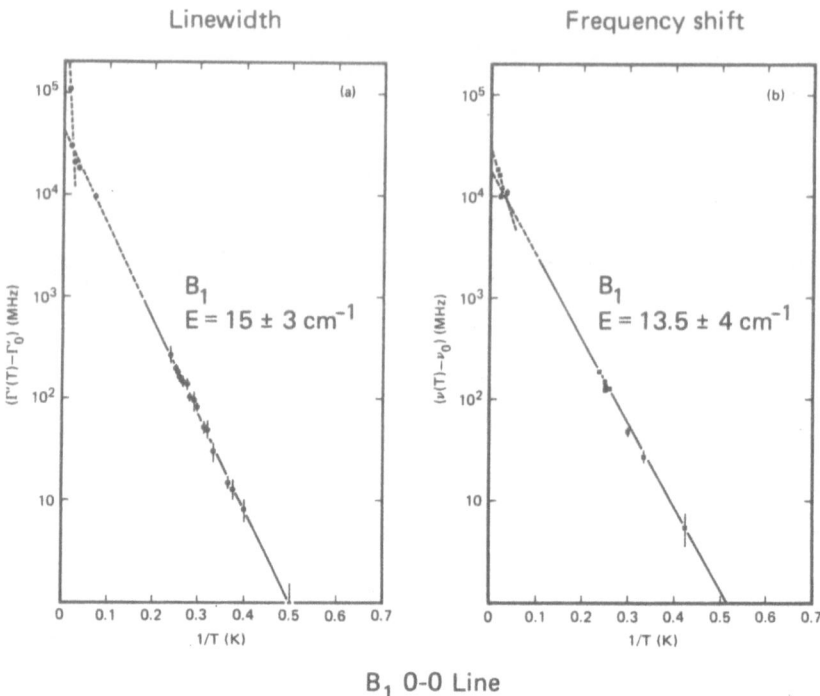

Fig. 38. Logarithmic increase of the homogeneous linewidth (left), and logarithmic frequency shift (right) of the B_1 0—0 transition of H_2P in n-C_8 as a function of the inverse of the temperature. Notice that the same activation energy is responsible for both processes [58].

mixed crystal arises from a strong distorsion of the host crystal due to the presence of the guest molecule with very different properties. As a consequence, the phonon density of states sharply peaks at the frequency of these pseudo-local modes [227]. In the case of porphin molecules in n-alkane crystals, they are assumed to correspond to librational motions of the guest in the host [19, 58, 107]. Dephasing by such low frequency modes have been proposed by Aartsma and Wiersma as the explanation for the temperature dependent photon echo results obtained for tetracene and pentacene in p-terphenyl [220]. Localized modes have been observed as phonon side-bands in excitation and fluorescence spectra of mixed molecular crystals like pentacene in naphthalene [228] and benzoic acid [229], and for various porphin molecules and their derivatives (Zn-porphin [107], H_2-chlorin [44], H_2-bacteriochlorin [118], H_2-tetra-ter-butylphthalocyanine [98]) in n-alkane crystals. An example of such low frequency localized librational modes is given in Fig. 39 for H_2P in anthracene at 4.2 K [85]. It has been observed that these phonon side-bands burn together with their corresponding 0—0 lines and that their frequencies coincide with the activation energy, $E \simeq 17 \pm 2$ cm^{-1}, found from the temperature dependent broadening of the holewidths [85].

Phosphorescence spectra, in which low frequency localized modes have been

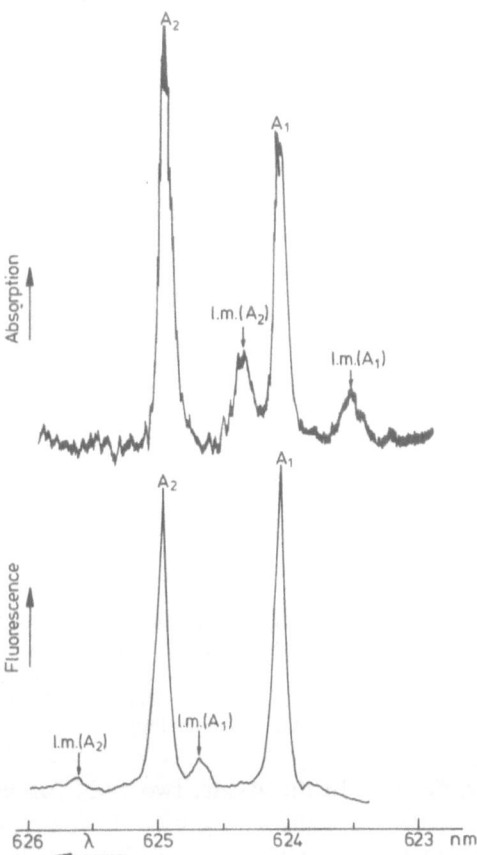

Fig. 39. Excitation (a), and fluorescence (b) spectra of the $S_1 \leftarrow S_0$ 0—0 region of H_2P in anthracene at 4.2 K. The two strong bands (A_1 and A_2) correspond to 0—0 transitions of H_2P in two different orientations. Localized librational modes (l.m.) are present as phonon side-bands at the high energy side (~ 15 cm^{-1}) of the 0—0 transitions in the absorption (a), and at the low energy side (~ 16 cm^{-1}) in the fluorescence spectrum (b) (see ref. [85], p. 33).

observed as hot-bands of the $T_1 \rightarrow S_0$ transition, have also been reported for naphthalene in durene [230a], quinoxaline in durene [230b] and napthalene [230c]. These localized phonons are assumed to be responsible for the activation energies observed in ESR and electron-spin-echo signals as a function of temperature of excited triplet states (see chapter on spin-lattice relaxation by J. Schmidt in this book).

The frequencies and lifetimes of pseudo-local phonon modes depend on the tightness of the fit of the guest in the host [58, 60, 231]. This effect has very convincingly been shown for H_2P in the A- and B-sites of n-octane (n-C_8) and

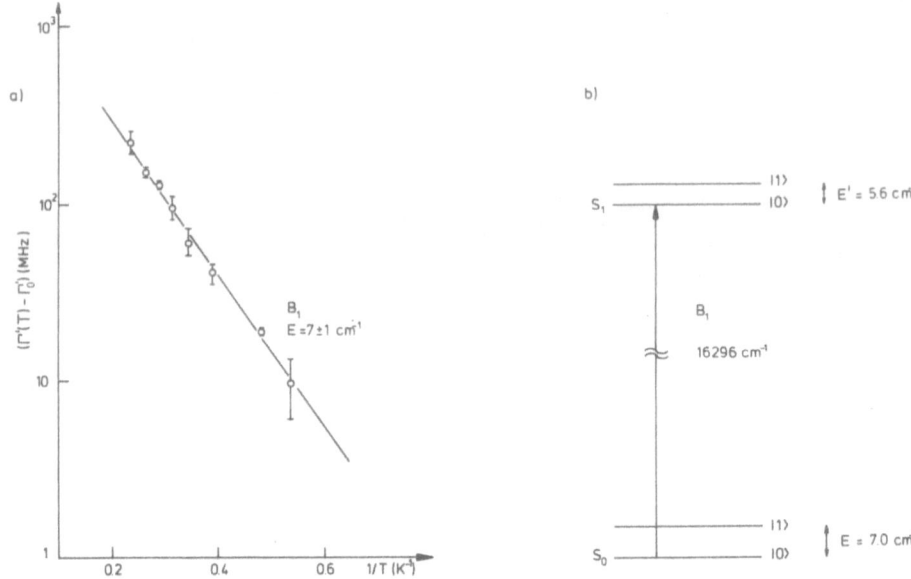

Fig. 40. Left: Logarithmic increase of the homogeneous linewidth, $\Gamma_{hom} - \Gamma_0$, of the $S_1 \leftarrow S_0$ 0—0 transition of H_2P in the B_1-site of n-C_{10} as a function of the inverse of the temperature [60]. Right: Diagram of the energy levels involved in the "exchange model" [236, 237]. The activation energy, $E \simeq 7 \pm 1$ cm^{-1}, is consistent with the frequency of the localized mode observed as phonon side-band in the spectrum [60].

n-decane (n-C_{10}) [19, 58, 60]. In the A-site, two n-C_8 (or n-C_{10}) molecules are replaced by one H_2P molecule; in the B-site, three n-C_8 (or n-C_{10}) molecules are replaced by one H_2P [110, 231]. Whereas the tautomer splitting of the B-site in n-C_8 is about 25 cm^{-1}, that of the same site in n-C_{10} is much smaller (≤ 2 cm^{-1}), which suggests a looser fit of the guest in the n-C_{10} host [46]. Furthermore, from the temperature dependence of the holewidths, a very low activation energy of about 6—7 cm^{-1} was, in fact, found for the B-site in n-C_{10} (see Fig. 40). In addition, from a fit of the exchange model for optical dephasing (see Section 4.1.2) to the data, a very long lifetime of ~ 120 ns was estimated for the librational mode in the S_1-state [60]. The same lifetime was independently obtained from the width of a hole burnt into the phonon side-band at 5.6 cm^{-1} above the 0—0 transition, at 1.2 K. Such a librational phonon hole ($\Gamma_{hom} \sim 1.5$ GHz) is illustrated in Fig. 41b. Notice that its width is two orders of magnitude larger than that of a hole burnt into the 0—0 band ($\Gamma_{hom} \sim 15$ MHz), as shown in Fig. 41a [60].

In contrast to the B-site, the A-site of n-C_{10} has a much larger tautomer splitting of 63 cm^{-1} [46], a larger activation energy of ~ 37 cm^{-1} [85] (see Fig. 42), and the librational mode has a much shorter lifetime, $\tau \sim 0.5$—1 ps [60]. These values, which are very similar to those of the A-site of n-C_8 [58], suggest that the A-site in both n-C_8 and n-C_{10} provides a much tighter fit for the guest molecule in

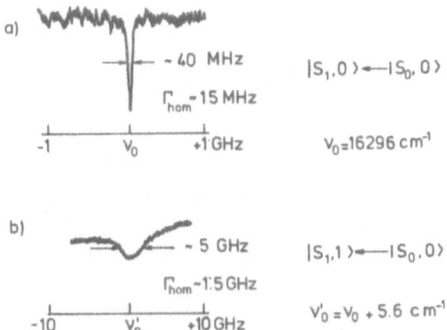

Fig. 41. (a) Hole burnt in the $S_1 \leftarrow S_0$ 0—0 transition of H_2P in the B_1-site of n-C_{10} at 1.2 K. (b) Hole burnt in the localized phonon mode at 5.6 cm^{-1} from the B_1 0—0 transition. Notice the difference of about two orders of magnitude in Γ_{hom} [60].

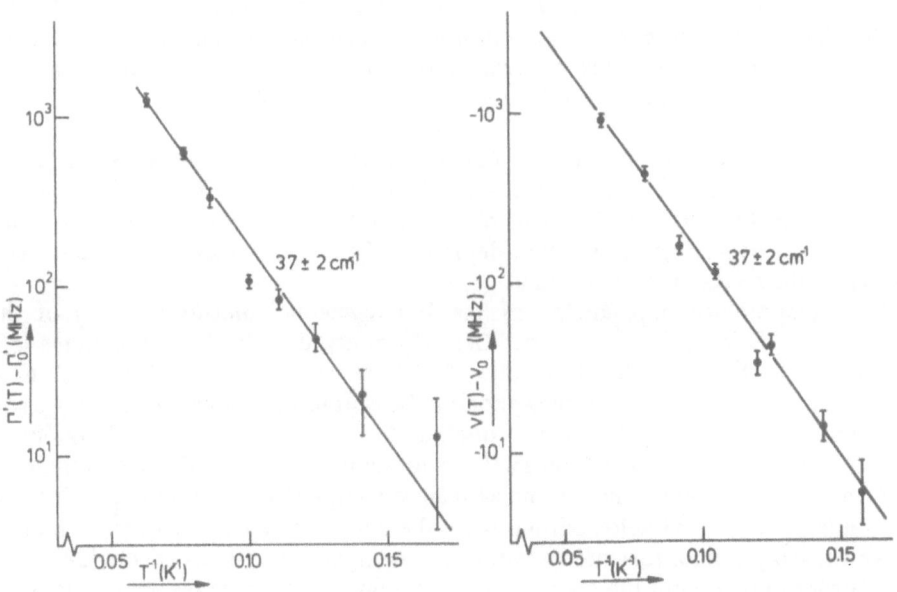

Fig. 42. Logarithmic increase of the homogeneous linewidth (left) and frequency shift (right) of the $S_1 \leftarrow S_0$ 0—0 transition of H_2P in the A_2 site of n-C_{10} as a function of the inverse of the temperature (see ref. [85], p. 29). Notice that the activation energy, $E \simeq 37$ cm^{-1}, in the A-site is much larger than that in the B-site ($E \simeq 7$ cm^{-1}) of n-C_{10}.

the n-alkane host than the B-site. Monte-Carlo calculations for H_2P in n-C_8 support these conclusions [231].

Chlorin (H_2Ch) in n-alkane hosts (n-C_6, n-C_8 and n-C_{10}) exhibits a very similar dephasing behaviour as the A-site of H_2P. Homogeneous linewidths of ~ 40 MHz at 1.6 K were obtained in these systems, which correspond to the measured fluorescence decay time of chlorin, $T_1 = (2\pi\Gamma_{\text{hom}})^{-1} \simeq 8$ ns [41, 44]. This value is approximately half of that for H_2P (17 ns) [19], and is due to the stronger allowed $S_1 \leftrightarrow S_0$ transition of H_2Ch, the symmetry of which (C_{2v}) is lower than that of H_2P (D_{2h}). From the very small increase of the holewidth with temperature between 1.6 and 4.2 K, it was inferred that H_2Ch occupies A-like sites in the n-alkanes studied [44]. The same conclusion was reached from Zeeman [115, 116] and Stark [114] effect measurements (see Section 4.1.4.).

For $T > 50$ K the temperature dependence of the homogeneous linewidth of the $S_1 \leftarrow S_0$ 0—0 transition of porphin molecules in n-alkanes deviates from that at low T. The activation energy of about $E = 150$ cm^{-1} becomes independent of the n-alkane host, and corresponds to the frequency of the lowest internal vibration of H_2P [41, 232].

A theory for optical dephasing that gives a consistent explanation for the hole broadening and frequency shifts of the $S_1 \leftarrow S_0$ 0—0 transitions of porphin molecules (free-base and metal-porphins) and their derivatives in n-alkane crystalline hosts at low temperatures ($T \simeq 1.2$—50 K) is the "exchange model" [233—237], which will be discussed in Section 4.1.2. The idea of this model is that the electronic oscillator becomes modulated by absorption and emission of a low-frequency phonon mode localized at the guest molecule [58].

b. *Transient hole-burning (THB) systems.* The mechanism of transient hole-burning (THB) has already been discussed in Section 3.1.4. Two metal porphin systems, ZnP [38] and MgP [39] in n-octane (n-C$_8$), were given as examples for hole-burning occurring by selective depletion of the ground state population and storage in the metastable triplet state.

The temperature dependence of the homogeneous linewidth (Γ_{hom}) of the $S_{1x} \leftarrow S_0$ 0—0 transition has been differently interpreted in the two systems [38, 39]. From the hole broadening results of ZnP in n-C$_8$ between 1.6 and 4.2 K, it was suggested that a resonant one-phonon absorption occurs between the S_{1x} and S_{1y} components [38], rather than coupling to a low-frequency local mode, as observed in PHB-experiments on H_2P in various hosts [58, 60, 85]. The argument was based on the coincidence of the activation energy derived from dephasing and the value of the crystal field splitting δ for the particular site studied. No evidence, however, was given for this interpretation by a comparison with other sites [38].

In order to figure out the correct interpretation, experiments on the temperature dependence of holes burnt in the 0—0 transition were performed on four sites of MgP in n-C$_8$ between 1.2 and 4.2 K. The homogeneous width Γ_{hom}^x of the $S_{1x} \leftarrow S_0$ 0—0 transition at 1.2 K proved to be determined by the fluorescence lifetime of Mg ($T_1 \sim 10$ ns), independent of the site, as previously found for H_2P [19, 58, 60], ZnP [38], and chlorin [44]. Thermally induced hole broadening up to 4.2 K was observed for only two of the four sites. Activation energies of $E \simeq 14$ cm^{-1} suggest that these sites are of the B-type [39]. By comparing the E-values with the frequencies of the librational modes identified as phonon side-bands in the spectra (see Fig. 43), it was concluded that optical dephasing of the $S_{1x} \leftarrow S_0$ 0—0 transition of MgP in n-C$_8$ is caused by thermal excitation and de-excitation of local

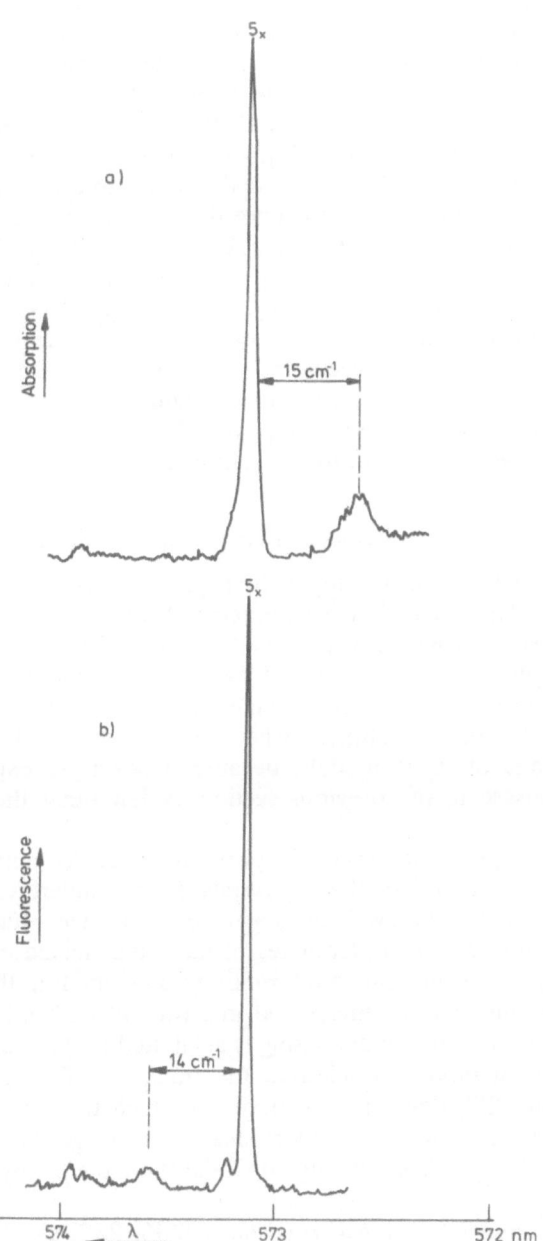

Fig. 43. Site-selected (site 5) excitation spectrum (a), and fluorescence spectrum (b) of the $S_1 \leftarrow S_0$ 0—0 region of MgP in n-C_8 at 4.2 K detected and excited, respectively, via the 1187 cm^{-1} vibrational band and the 714 cm^{-1} vibronic band. The 0—0 5_x line and its local mode at 15 cm^{-1} higher energy (a) and 14 cm^{-1} lower energy (b) are shown [39].

phonon modes, just as for H_2P in n-alkanes, and not by one-phonon scattering between the S_{1x} and S_{1y} states. The latter mechanism can be disregarded, because the crystal-field splittings, $\delta = E_{1x} - E_{1y}$, for the various sites of MgP in n-C_8 differ markedly from the activation energies measured [39].

It was mentioned earlier (Section 3.1.4), that two of the sites of MgP in n-C_8 exhibit photochemical hole-burning [39], which may be attributed to a reorientation of the ethanol ligand with respect to the MgP molecule in the MgP-ethanol complex [135, 136]. From the temperature dependence of these photochemically burnt holes, it was concluded that these sites belong to the A-type [39].

Holes have also been burnt in the $S_{1y} \leftarrow S_0$ 0—0 transitions of MgP in n-C_8, from which the relaxation time τ_y from $S_{1y} \rightarrow S_{1x}$ could be determined. It was found that the value of τ_y decreases with an increase of the crystal-field splitting δ: $\tau_y \simeq 40$ ps for $\delta \sim 20$ cm^{-1}, whereas $\tau_y \sim 2$ ps for $\delta \sim 200$ cm^{-1} [39]. It would be interesting to study the role of vibronic coupling between S_{1x} and S_{1y}, by means of the application of a magnetic field, and check whether such a mechanism introduces additional pathways for the relaxation from S_y to S_x.

4.1.2. *Models for Optical Dephasing in Mixed Molecular Crystals*

Various theories have been developed to explain optical dephasing of impurity molecules in *crystalline* hosts. In these models it is usually assumed that the optical transition is represented by a two-level system coupled to either acoustic, optical or pseudo-local phonons. For a detailed description of these theoretical models, the reader may consult various review articles [36, 239—241] (see also the chapter by R. Silbey in this book [238]). In what follows, we will discuss the physical implications of one of these models, because it seems to explain most of the experiments discussed in the previous section. A few other theories will just be mentioned.

The results on optical dephasing of porphin molecules and their derivatives incorporated as guests in n-alkane crystals have similarities with dephasing phenomena previously observed in magnetic resonance studies between the sublevels of phosphorescent triplet states of molecular mixed crystals [242, 243] (see also the chapter on spin-lattice relaxation by J. Schmidt in this book [244]). In both cases, the results can be interpreted in terms of the "exchange mechanism" [233—237, 242, 243], in which dephasing is attributed to thermal excitation of low frequency librational modes coupled to the transition. For the optical regime, Harris *et al.* [236, 237] developed a model in which the idea of low frequency local phonons was combined with the stochastic exchange theory of Kubo [233] and Anderson [234] (see also chapter on relaxation theory by R. Silbey in this book [238]).

The *exchange model for optical transitions* [236, 237] can be summarized as follows (see Fig. 44). It is assumed that dephasing is predominantly due to coupling of the electronic oscillator to a specific local mode, which may be thermally excited. In a four level scheme, the transition between the zero-phonon levels is given by $\langle S_1, 0|\ \leftarrow\ \langle S_0, 0|$ with frequency ν_0, and between the one-phonon-levels by $\langle S_1, 1|\ \leftarrow\ \langle S_0, 1|$ with frequency $\nu_0 + \Delta/2\pi$ (Δ can be >0 or <0). The latter represents the transition to which the electronic oscillator makes short stochastic jumps by absorption and re-emission of a phonon (at $T > 0$) [58]. A non-zero value of $\Delta/2\pi = E' - E$ implies that the librational mode frequency in the excited state, E', is different from that in the ground state, E. Another quantity involved in the model is the lifetime τ of the local phonon, which is assumed to be same in both phonon states, but much shorter (ps) than that of the electronic excitation (ns) [237]. If $kT \ll E$, then the broadening of Γ_{hom}, and the frequency shift are given, respectively, by [58, 237]:

$$\Gamma_{\text{hom}}(T) - \Gamma_0 = 2\Delta\tau\ \frac{\Delta/2\pi}{1 + \Delta^2\tau^2}\ \exp(-E/kT). \tag{2}$$

$$\nu(T) - \nu_0 = \frac{\Delta/2\pi}{1 + \Delta^2\tau^2}\ \exp(-E/kT). \tag{3}$$

It should be noticed that the parameters Δ and τ can be independently determined, because information is available on both frequency shifts and widths from PHB experiments [58]. By analogy to NMR, three different regimes are to be distinguished: $\Delta\tau \gg 1$ which corresponds to slow exchange, $\Delta\tau \approx 1$ to intermediate, and $\Delta\tau \ll 1$ to fast exchange [235]. The exchange model was successfully applied to interpret the results of H_2P [58], H_2Ch [44] and MgP [39] in A- and B-sites of n-C_8. In all these cases one is dealing with a situation of intermediate to fast exchange, i.e. $\Delta\tau \lesssim 1$, which implies that the $|S_1, 1\rangle \leftrightarrow |S_0, 1\rangle$ transition is not separately observable as "hot" band in the spectrum.

The first example of slow exchange for optical transitions seems to be the

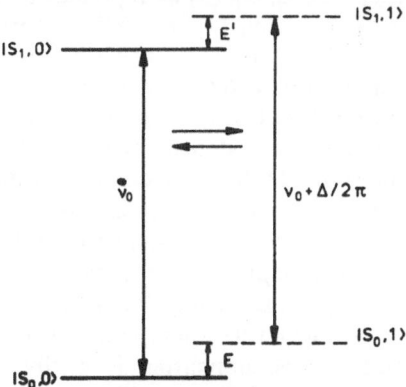

Fig. 44. Diagram of the transitions involved in the "exchange model" [236, 237]. The energies of the low-frequency modes are assumed to be somewhat different in the ground state S_0 and the electronically excited state S_1, resulting in a frequency difference $\Delta/2\pi$ between the $\langle S_1, 1|\ \leftarrow\ \langle S_0, 1|$ and $\langle S_1, 0|\ \leftarrow\ \langle S_0, 0|$ transitions. The lifetime of the local modes is of the order of ps, while that of the S_1 state is about ns [58].

system H_2P in the B-site of n-C_{10} [60]. In this case, $\Delta\tau \approx 30 \gg 1$, and eq. (2) can be approximated by

$$\Gamma_{hom}(T) - \Gamma_0 \approx (\tau\pi)^{-1} \exp(-E/kT). \tag{4}$$

The parameter $\Delta \approx 1.4$ cm^{-1} was independently obtained from the difference in frequency of phonon side-bands observed in excitation and emission spectra, and $\tau \approx 120$ ps was derived from the width of holes burnt in the phonon side-band of the S_1-state (see Section 4.1.1). The value of τ is consistent with that obtained from a fit of eq. (4) to the data of ref. [60]. As already discussed above, the long lifetime of the librational mode implies that H_2P has a very loose fit in the B-site of n-C_{10}.

It will be shown below (Section 4.2.5) that one of the theoretical models for optical dephasing in amorphous systems also assumes that local phonon modes are responsible for the broadening of the homogeneous linewidth, in addition to two-level systems of the glass [238, 245, 246]. These local phonon modes are supposed to have a Gaussian distribution of frequencies in glasses, as opposed to a single frequency like in crystals [245].

An exponential temperature dependence of Γ_{hom}, as found experimentally, can also be interpreted in terms of resonant scattering of a vibrational mode. The expression for the line broadening then becomes [19, 98]:

$$\Gamma_{hom}(T) - \Gamma_0 = \Gamma'\bar{n}(E)[\bar{n}(E) + 1] \tag{5}$$

where Γ' is the coupling constant for the scattering process, and $\bar{n}(E) = [\exp(E/kT) - 1]^{-1}$ is the average number of thermally excited phonons of energy E. For low temperature, $kT \ll E$, eq. (5) extrapolates to an exponential function $\exp(-E/kT)$, whereas for $kT \gg E$, eq. (5) becomes proportional to T^2.

The first theory that explained optical linewidths and temperature shifts in crystalline systems was reported in 1963 by McCumber and Sturge [247]. Refinements of this theory have subsequently been proposed [240, 248]. All these models predict different temperature dependences for the linewidth and for the frequency shift. The exchange model [237], in contrast, predicts the same exponential dependence on temperature for both, which is consistent with the experimental results obtained for porphin molecules in n-alkane crystals [58, 60, 85].

Another theory for optical dephasing in molecular mixed crystals at low temperatures was presented by De Bree and Wiersma [249], who applied Redfield relaxation theory [250] to a four-level system coupled to an anharmonic phonon bath. As in the exchange model, dephasing is attributed to coupling of the electronic transition to specific low frequency librations active in the ground and excited state. But contrary to the exchange model, the phonon scattering in this case is "uncorrelated", i.e. allowance is made for a difference in lifetime of the local mode in the two electronic states [249]. In this theory, the homogeneous linewidth Γ_{hom} increases bi-exponentially with temperature, and for diagonal quadratic electron-libration coupling the expression for the line broadening becomes [9, 229]:

$$\Gamma_{\text{hom}}(T) - \Gamma_0 = \frac{1}{2\pi} [\tau^{-1}(T) \exp(-E/kT) + \tau'^{-1}(T) \exp(-E'/kT)] \tag{6}$$

where $\tau(T)$, $\tau'(T)$ are the temperature dependent lifetimes of the librational modes in the ground and excited state, respectively, and E and E' are the frequencies of the local phonon modes in these two states. De Bree and Wiersma [249] showed that eq. (6) reduces to the single exponential activation energy (eq. (2)) of the exchange model, when the values of E and E' are very similar. In that case $\bar{E} = \frac{1}{2}(E + \bar{E})$, and τ is identical in the S_0 and S_1-states. The photon echo results of refs. [225, 228, 229] were interpreted with this model. From these experiments, it was concluded that librational lifetimes in the ground and electronically excited states of organic mixed crystals are rather different and, therefore, only consistent with the "uncorrelated phonon scattering" model [249]. The optical Redfield theory [250] as worked out in ref. [249], was recently also adapted to interpret optical dephasing data in glasses [55], as will be shown in Section 4.2.5.

The fact that optical dephasing of porphin molecules in n-alkane crystalline hosts can nicely be interpreted in terms of the exchange theory; that this model does not seem to work for aromatic-type mixed crystals was tentatively explained in ref. [229]. In order to observe coherence exchange effects, the librational potential must be little affected by optical excitation. This leads to similar librational frequencies in ground and electronically excited states, which seems to be fulfilled in Shpolskii and noble-gas matrices. By contrast, aromatic-type mixed crystals apparently have a much stronger guest-host interaction, and therefore, very different E and E'-values.

A third, non-perturbative, theory for optical dephasing in crystals has recently been proposed by Hsu and Skinner [227], who assumed a harmonic approximation and quadratic electron-phonon coupling. The homogeneous linewidth in this model depends only on three of the four parameters τ, τ', E and E'. Thus, τ' can be expressed in terms of the others. For low temperatures, and sufficiently long lived librational phonon modes, a bi-exponential expression results:

$$\Gamma_{\text{hom}}(T) - \Gamma_0 = \frac{(\Delta\tau^*)^2}{1 + (\Delta\tau^*)^2} \frac{1}{2\pi} [\tau^{-1} \exp(-E/kT) +$$
$$+ \tau'^{-1} \exp(-E'/kT)], \tag{7}$$

where $\Delta/2\pi = E' - E$, $\tau^{*-1} = \frac{1}{2}(\tau^{-1} + \tau'^{-1})$. Eq. (7) is valid for τ^{-1}, $\tau'^{-1} \ll kT/\hbar$ and $\exp(-E/kT)$, $\exp(-E'/kT) \ll 1$. It reduces to eq. (2) of the exchange theory [58, 237], if $\tau = \tau'$ and $\exp(-\Delta/kT) \simeq 1$, and to the "uncorrelated phonon scattering" result of ref. [249] for $\Delta\tau^* \gg 1$ [251]. Several experimental results of temperature dependent linewidths have been tried to be fitted with this theory [36, 227, 239, 251].

A review of Redfield theory and its applications to microscopic models that describe optical homogeneous line shapes of localized excitations in crystals and glasses has been given by Silbey in another chapter of this book [238].

4.1.3. *Vibronic Relaxation in Organic Crystalline Systems*

So far, we have discussed hole-burning experiments on the electronic origin of the $S_1 \leftarrow S_0$ transition of organic molecules in crystalline hosts. Let us turn now to vibronic spectral lines which arise from transitions involving both electronic and vibrational excitations. The mechanism by which the vibrational energy is dissipated in large organic molecules in solids has not been completely understood yet [23, 25, 41, 156, 228, 252—258], and hole-burning has helped to further unravel this problem.

At very low temperatures, where the contribution to the homogeneous linewidth, Γ_{hom}, of the pure dephasing term, $(\pi T_2^*)^{-1}$, is negligible (see eq. (1)), one can determine vibrational relaxation times in the first excited singlet states S_1, directly from the widths of the holes burnt into vibronic bands of the $S_1 \leftarrow S_0$ transition. These relaxation times are often very fast ($\leq$ ps) and difficult to measure in the time domain. Initial hole-burning experiments on single vibronic lines [17, 19, 40, 92, 259] have confirmed that these energy levels relax much faster than the electronic origin. In many of these cases [17, 19, 40] no holes were observed, because the homogeneous width, Γ_{hom}, was larger than the inhomogeneously broadened band.

The first systematic study of vibrational relaxation using PHB was performed on the $S_1 \leftarrow S_0$ transition of free-base porphin (H_2P) embedded in an n-octane (n-C_8) crystal at 4.2 K [156, 260]. Fig. 45 shows an example of two vibronic lines in the A-site of this system. A 10 GHz broad hole was burnt in the 710 cm^{-1} vibrational band, which corresponds to a relaxation time of 32 ps. On the other hand, it was not possible to burn a hole in the 155 cm^{-1} band, since it bleached entirely. This vibronic band is essentially homogeneously broadened, which corresponds to a relaxation time of about 2 ps. A systematic study of 14 vibronic levels between the origin and 1600 cm^{-1} above it revealed large variations (by a factor of 40) in the relaxation rates. This is illustrated in Fig. 46, where the homogeneous linewidths and their corresponding vibrational relaxation times have been plotted as a function of the excess energy above the $S_1 \leftarrow S_0$ 0—0 transition. No correlation was found between these parameters or between the relaxation times and the symmetry species of the vibrations involved [156]. A crucial question arouse from these experiments: are the large differences in vibronic relaxation times due to intramolecular effects characteristic of the guest molecule, or do they result from coupling of the guest to the host lattice?

Similar unexpected results as those obtained for $S_1 \leftarrow S_0$ vibronic bands of H_2P in n-octane, were subsequently observed for pentacene in naphthalene and p-terphenyl by accumulated picosecond photon echoes [228, 257].

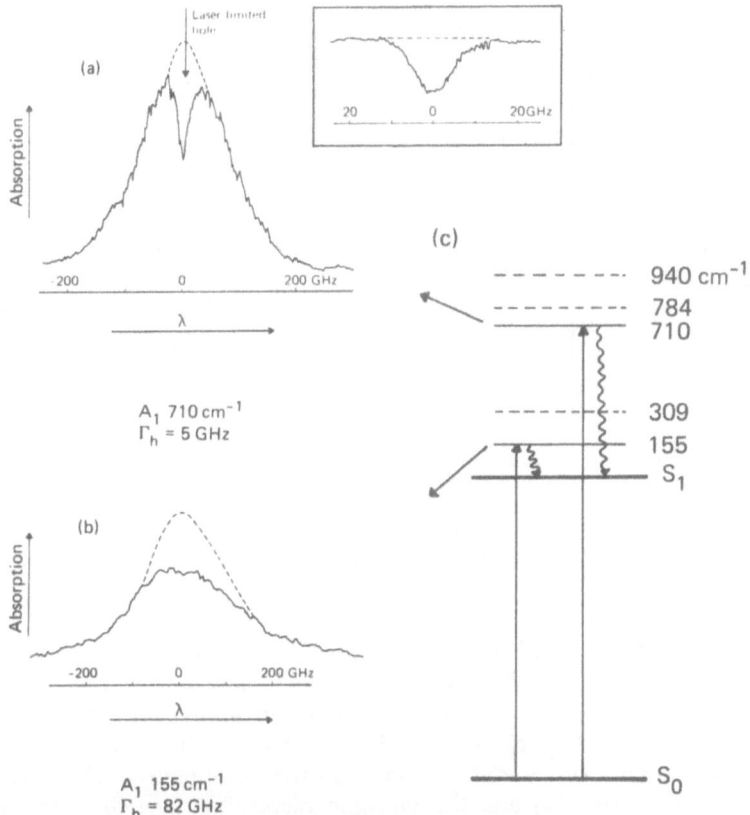

Fig. 45. Photochemical holes burnt in two vibronic bands of the $S_1 \leftarrow S_0$ transition of H_2P in the A_1-site of n-C_8 at 4.2 K. (a) 20 GHz-hole burnt in the 710 cm^{-1} band. The width is limited by the resolution of the probe laser. Insert: hole burnt at the same frequency and probed with a high-resolution cw dye laser (≈ 2 MHz). The holewidth is determined by $\Gamma_{hom} \approx 5$ GHz. (b) The 155 cm^{-1} vibronic band is essentially homogeneously broadened, with $\Gamma_{hom} = 82$ GHz. The larger width is due to rapid vibrational relaxation of ≈ 2 ps to the electronic origin of S_1. No hole is observed in this band. (c) Energy level diagram with the S_0 and S_1 electronic states and some of the vibronic levels below 1000 cm^{-1} [41, 156].

The assumption that the homogeneous linewidth of a vibronic transition is exclusively determined by vibrational relaxation for H_2P in n-alkanes, was confirmed by measuring the widths of narrow holes (a few GHz wide) of several vibronic bands at different temperatures [258]. Since no broadening was observed between 1.2 K and 4.2 K, it was concluded that dephasing in these bands is only a minor effect, probably in the MHz regime at 4.2 K, as for the 0—0 band [19, 58, 60]. This was independently proven by means of a three-pulse stimulated photon echo experiment on pentacene in naphtalene [257].

The influence of the host on vibronic relaxation of the $S_1 \leftarrow S_0$ transition was

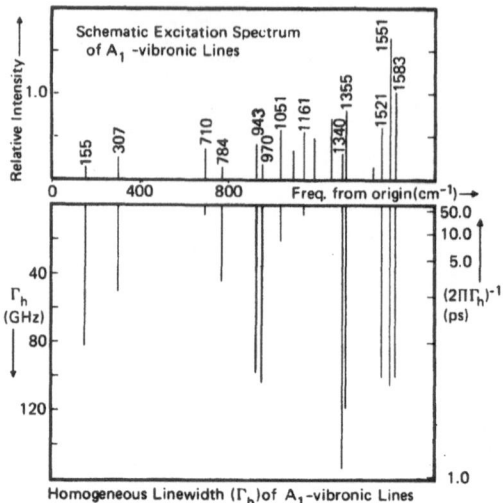

Fig. 46. Top: Schematic excitation spectrum of vibronic bands of the $S_1 \leftarrow S_0$ transition of H_2P in the A_1-site of n-C_8. Bottom: Homogeneous widths of the same vibronic bands obtained from hole-burning experiments [156].

investigated in various n-alkanes doped with H_2P at 1.6 K: the A-sites of n-C_6, n-C_7 and n-C_{10}, and the B-sites of n-C_7, n-C_8 and n-C_{10} [258]. The results demonstrated that the local environment plays an important role in the energy dissipation of free-base porphin, and that the vibronic relaxation time depends strongly on site and host. A correlation was observed between the "cage" size of the n-alkane (see Fig. 47) and the vibronic relaxation time: the latter becomes longer for molecules having a looser fit in their cage (see, for example, the 710 cm^{-1} vibration of H_2P in the B-site of n-C_{10} as compared to that in n-C_7 and n-C_8 in Fig. 48). Furthermore, a tight fit induces fast relaxation, which is the case of the A-site in any n-alkane crystal [258].

The very long relaxation times observed for many vibrations of H_2P in the B-site of n-C_{10} ($10 \leq \tau \leq 120$ ps) suggest that this site provides a very large cage for the guest molecule with weak guest-host interaction [258]. This conclusion is supported by the long lifetime and the low frequency of the localized librational mode of H_2P in the B-site of n-C_{10} [60] (see also Section 4.1.1).

A practical application of hole-burning, which is worthwhile mentioning here, is its use as a technique for excited state vibrational assignments. In multi-site system, like porphin molecules in n-alkane hosts, it is often difficult to correctly identify higher vibrational modes, because they overlap with vibrations from other sites. By burning away all 0—0 lines, with the exception of one site, the number of spectral lines can be reduced considerably, and an unambiguous vibrational assignment is usually possible [41].

A similar method has also been used for molecules embedded in glassy hosts, as first demonstrated by Personov and coworkers [189] (see also Section 3.2.1).

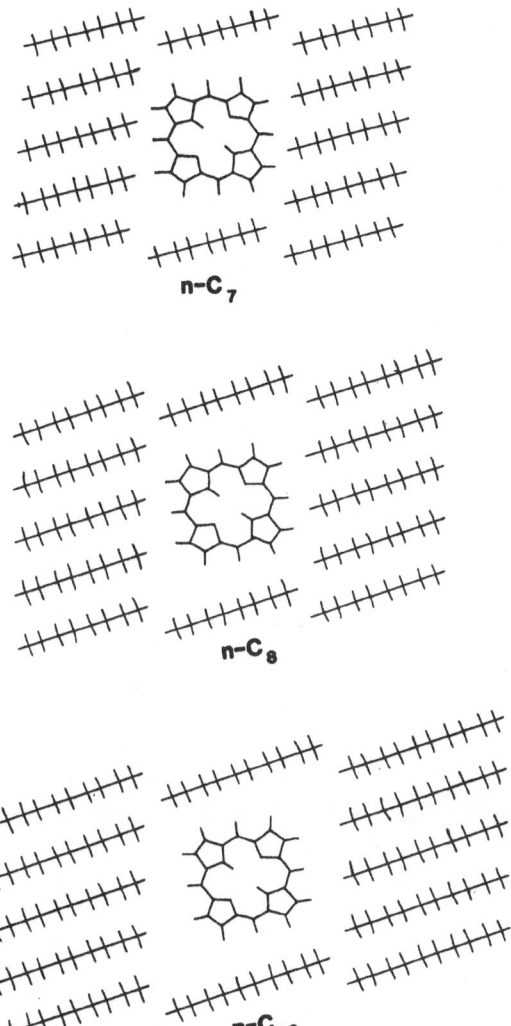

Fig. 47. H_2P in the B-site of n-C_7, n-C_8 and n-C_{10}. The longer the n-alkane chain, the looser the fit of the guest molecule in the substitutional n-alkane cage [258].

Vibronic transitions of tetracene in ethanol glass [20, 26], and various ionic dyes in alcoholic glasses and polymer hosts [185, 187] were assigned in this way. An illustration of this method is given in Fig. 49, which shows a hole burnt in the $S_1 \leftarrow S_0$ 0—0 band of H_2P in MTHF glass, and satellite holes appearing over the entire spectrum at the frequency of each vibronic transition [1]. An extensive discussion on this subject, and site-selection spectroscopy of organic molecules in glassy hosts can be found in ref. [4].

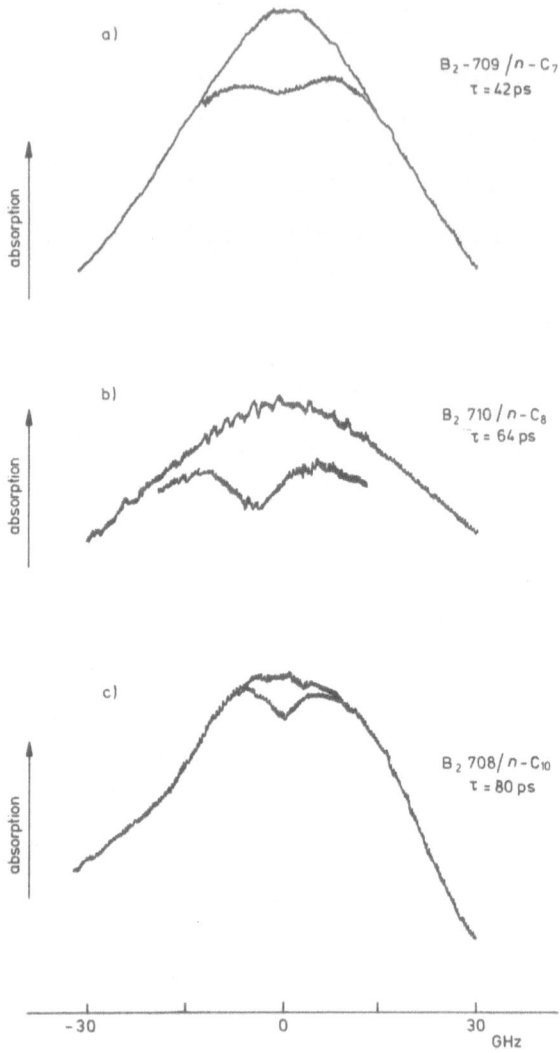

Fig. 48. Holes burnt in the 710 cm^{-1} vibronic band of the $S_1 \leftarrow S_0$ transition of H_2P in the B_2-site of (a) n-C_7, (b) n-C_8, (c) n-C_{10} at 1.6 K. The holes become narrower, i.e. the vibronic relaxation time τ becomes longer, the looser the fit of the guest in the host [258].

4.1.4. *External Field Effects in Mixed Molecular Single Crystals*

A quite different application of PHB is to use it as a method to measure very accurate frequency shifts of spectral lines with a resolution of a few MHz. Such frequency shifts are usually the result of the application of an external field (e.g.

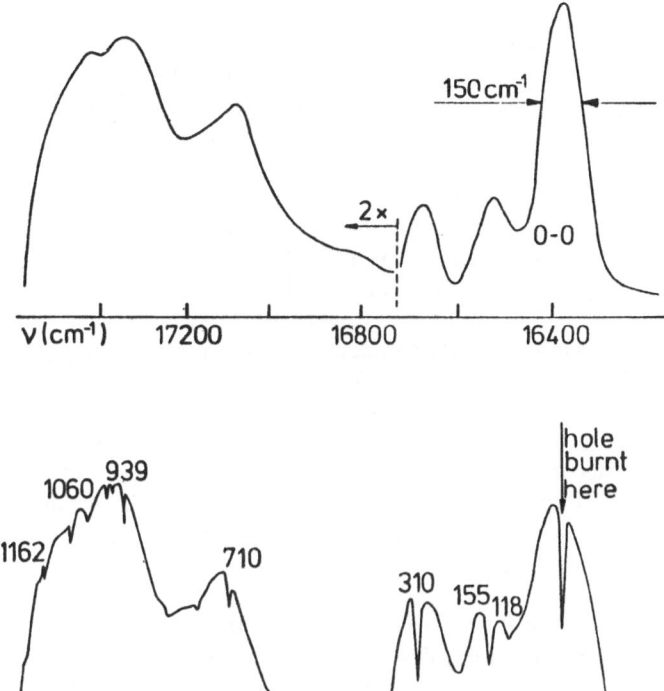

Fig. 49. Top: Excitation spectrum of H_2P in MTHF glass at 4.2 K, before burning. Bottom: The same spectrum after a hole was burnt in the $S_1 \leftarrow S_0$ 0—0 transition. Satellite holes at the frequency of each vibronic transition appear over the entire spectrum [1]. This is an accurate method for vibrational assignment in broad spectra of organic molecules in glasses at low temperature [185, 187, 189].

electric, magnetic, temperature, pressure). With conventional spectroscopic techniques it is possible to detect shifts of the order of the width of the absorption band only, which in case of organic crystals amounts to ~ 1 cm^{-1} (30 GHz). Zeeman and Stark effects are often two to three orders of magnitude smaller, and therefore, inaccessible to measure by traditional methods. Since hole-burning increases the spectral resolution by a factor of 10^3—10^5, a permanent hole may serve as an accurate frequency marker when burnt into a broad inhomogeneous band at low temperature. The shift of such a hole caused by an external field can be measured with respect to a hole burnt in a reference sample in zero-field in another cryostat with an accuracy of a few MHz [42, 58]. The resolution of this method is given by the holewidth, which in an organic crystalline system at $T \leqslant 2$ K is limited by the population decay time T_1 of the guest molecule, $\Gamma_{hole} \approx 2\Gamma_{hom} \approx (\pi T_1)^{-1} \approx 10$—50 MHz.

In this section, the effects of magnetic and electric fields on holes burnt in zero field in *organic crystalline* systems will be discussed, in particular, studies performed on the $S_1 \leftarrow S_0$ 0—0 and some vibronic transitions of porphin molecules (free-base porphin and chlorin) in *single crystals* of *n*-alkanes [42, 85, 114—116]. Zeeman and Stark effects in organic amorphous systems will be treated in Section 4.2.8.

Hole-burning has extensively been used to study the *Zeeman* effect in *inorganic crystals* doped with rare earth ions. We will not discuss these systems here, but only mentioned a few of them: Pr^{3+} in LaF_3 [172b] and CaF_2 [171], Er^{3+} in LaF_3 [215], Sm^{2+} in CaF_2 [261], SrF_2 [216] and $BaClF$ [217].

Also electric fields have been combined with hole-burning to study the *Stark* effect in many *inorganic crystalline* systems. For instance, colour centers like F_3^+ in NaF [166] at 5456 Å, the 6070 Å center in NaF [262], the 5770 Å center in NaF [263] and the R' center in LiF [264], and some rare earth ions in the crystals mentioned above [169, 216, 217] have been investigated in this way. To our knowledge, electric fields in combination with hole burning have been applied to only one type of *organic crystalline* system so far, chlorin in single crystals of *n*-hexane and *n*-octane [114]. A review on external electric field effects on spectral holes can be found in ref. [35].

a. *Zeeman effect in free-base porphin* (H_2P) *and chlorin* (H_2Ch). The Zeeman effect in metal porphin molecules incorporated as guests in single crystals of *n*-alkanes at low temperature has been studied in great detail by means of conventional spectroscopy in magnetic fields up to $\sim 8T$ by van der Waals and coworkers (for a review, see ref. [131]). The first excited singlet state of PdP [265], ZnP [107, 111] and MgP [136], the triplet state of PdP [266], ZnP [111] and PtP [267], and the lowest quartet state of CuP [268, 269] are some of the systems investigated.

In order to understand the Zeeman effect in the excited singlet state of free-base porphins and derivatives, it is helpful to consider first a metal porphin (MeP), like MgP or ZnP (see also ref. [131]). A MeP can be described as a highly conjugated planar system with 26 delocalized π-electrons. Owing to its relatively high symmetry (approximately D_{4h}) the lower excited states are doubly degenerate. In the ground state the molecule possesses zero angular momentum around the out-of-plane axis z, $L_z = 0$, but in the first excited singlet state the isolated molecule possesses an orbital angular momentum Λ (if one takes $h = 1$). In the presence of a crystal field, however, the degeneracy of the S_1 state is lifted and the $S_1 \leftarrow S_0$ transition splits into two components, $S_{1x} \leftarrow S_0$ and $S_{1y} \leftarrow S_0$, usually referred to as the Q_x and Q_y bands, respectively. The energy separation between the two corresponding 0—0 transitions is the "crystal field splitting", δ, which varies between ~ 10 and 100 cm^{-1} in *n*-alkanes, depending on site and host [131]. In such a situation the angular momentum is quenched, but the operator L_z still gives rise to an off-diagonal matrix element between the two orbital states S_{1x} and S_{1y},

$$\Lambda = |\langle S_{1y}| L_z |S_{1x}\rangle|. \tag{8}$$

As a consequence, a second order Zeeman effect arises, which causes a shift of the $0-0$ Q_x transition to lower energy, given by:

$$\Delta E = -(\beta \Lambda H_z)^2/\delta, \tag{9}$$

where β is the Bohr magneton and H_z is the strength of the magnetic field along the z axis. For metal-porphins values of Λ between 4.2 and 4.4 have been obtained with conventional spectroscopy in fields up to ~ 8 Tesla [107, 131, 178].

The approximate symmetry of H_2P (D_{2h}) is lower than that of a MeP, and the first two excited singlet states S_1 and S_2, which stem from the two degenerate S_1 states of a MeP, are separated by ~ 3000 cm^{-1} [270]. This means that, if Λ $(H_2P) \simeq \Lambda$ (MeP), one would expect an energy shift of the S_1 and S_2 levels for H_2P about 100 times smaller than that for a MeP, for a given field strength. Such small shifts are not detectable with a conventional resolution of ~ 1 cm^{-1}.

By means of photochemical hole-burning, however, it has been possible to measure very small Zeeman shifts of the $S_1 \leftarrow S_0$ $0-0$ transition of H_2P and H_2Ch in oriented *single crystals* of n-C_8 at 4.2 K [42, 115—116]. The method was similar to that previously used to detect frequency shifts of holes caused by an increase in temperature [58] (see Section 4.1.1). Two identical samples were mounted in two cryostats, one equipped with a superconducting magnet and another one as a reference. A hole was simultaneously burnt in both samples in zero field at 4.2 K. Subsequently, the magnetic field, applied parallel to the molecular out-of-plane axis z, was increased in the Zeeman cryostat and the excitation spectra of both holes were probed. Relative frequency shifts of holes could be measured in this way with an accuracy of about 10 MHz (see Fig. 50). The holes burnt in the A-site of H_2P in n-C_8 shifted quadratically as a function of the magnetic field with -60.6 MHz/T^2 (see Fig. 51, full line) [42]. By assuming a value of ~ 2940 cm^{-1} for the energy separation between the S_1 and the S_2 states of H_2P in n-C_8, a value of $\Lambda = 5.6$ was obtained [42], which is only in fair agreement with the theoretical prediction of $\Lambda = 4.29$ [271], and is higher than for metal complexes investigated earlier [131]. Although the origin of this discrepancy is not understood, it is probably related to vibronic coupling, which was not taken into consideration in this study.

Zeeman experiments performed on holes burnt in two vibronic bands of the $S_1 \leftarrow S_0$ transition of H_2P in n-C_8 yielded Zeeman slopes of -38.7 MHz/T^2 for the 710 cm^{-1} vibration, and -32.3 MHz/T^2 for the 1160 cm^{-1} band [85]. No satisfactory explanation has yet been provided for this reduced Zeeman effect.

PHB experiments in combination with a magnetic field carried out on the $S_1 \leftarrow S_0$ $0-0$ transition of *chlorin* (7,8-dihydroporphin, H_2Ch) and its photoproduct (H_2Ch^*) incorporated as guest in single crystals of n-hexane at 4.2 K yielded surprising results [115, 116]. As mentioned in Section 3.1.1, the stable isomer, H_2Ch, has the axis through its central protons parallel to the reduced bond, whereas the photoproduct, H_2Ch^*, has the axis perpendicular to this bond (see Fig. 14, right). The $0-0$ transition of H_2Ch^* lies ~ 1500 cm^{-1} above that of H_2Ch [41, 44]. Holes burnt in zero field in one of the four sites in n-C_6 (site 1 and 1' of Fig. 15) shifted with a quadratic field dependence of -39.0 MHz/T^2 for H_2Ch, and -245.3 MHz/T^2 for H_2Ch^* [115]. These slopes are also plotted in Fig. 51

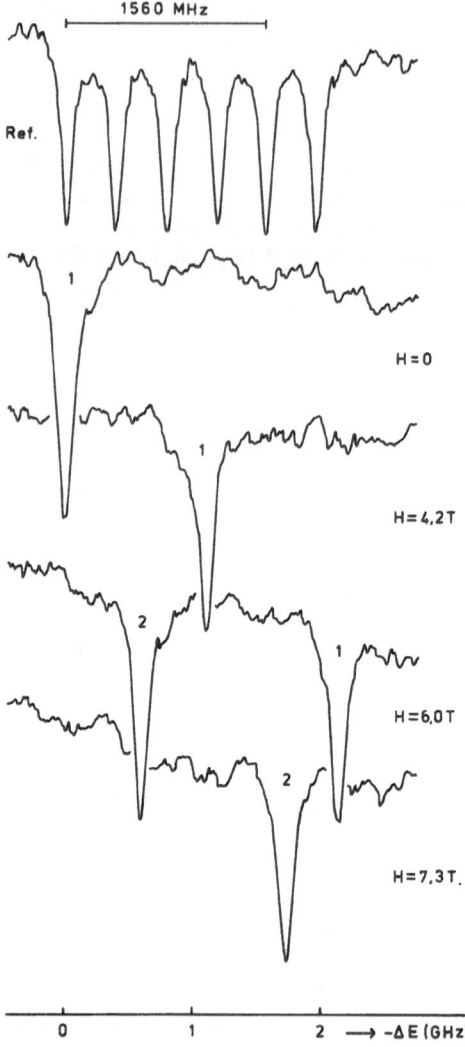

Fig. 50. Zeeman shifts of holes burnt in the A_1 0—0 line of H_2P in n-C_8 at 4.2 K. Top: Reference pattern of six holes burnt and kept at zero field. Bottom: Zeeman shifts of two holes burnt at zero field. The first hole(1) was burnt coincident with the hole at the high-frequency end of the reference pattern, the second hole(2) at 1560 MHz higher frequency. By increasing the magnetic field, hole 2 shifts into the reference pattern towards lower frequencies, whereas hole 1 has moved out of the laser scan range [42].

(dashed lines), for comparison with the slope obtained for the A-site of H_2P in n-C_8 [42]. From the results, a value of $\Lambda = 4.5$ for H_2Ch [115] was estimated, which is somewhat smaller than that for H_2P ($\Lambda = 5.6$) [42]. The reason for this difference is probably due to the hydrogenation of a double bond in one of the

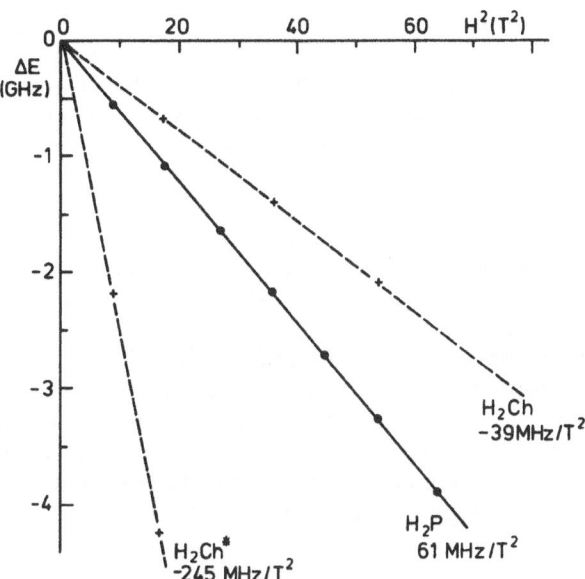

Fig. 51. Zeeman shifts of holes burnt in the $S_1 \leftarrow S_0$ 0—0 transition of H_2P in n-C_8 (full line), and H_2Ch and its photoproduct H_2Ch^* in n-C_6 (dashed lines), as a function of the squared value of the strength of the field H applied along the out-of-plane molecular axis z [42, 115].

pyrrole rings of chlorin. The energy separation between the states S_1 and S_2 for H_2Ch is ~ 2900 cm^{-1} [115], whereas that for H_2Ch^* is only about 400—700 cm^{-1}, and varies from site to site [116]. The latter frequencies are comparable in magnitude to the crystal field splitting of a few 100 cm^{-1}. Assuming that the values of Λ for H_2Ch and H_2Ch^* were similar, one would expect much larger Zeeman shifts for H_2Ch^* than for H_2Ch. In fact this has been observed, because the Zeeman slopes for the four different sites of H_2Ch^* in n-C_6 varied between -136 MHz/T^2 and -245 MHz/T^2. From the results, a common value of $\Lambda = 3.9$ for H_2Ch^* was obtained [116]. These experiments have demonstrated that a description of the Zeeman effect in a basis of only two purely electronic states, S_1 and S_2, seems to hold for H_2Ch, but not for H_2Ch^*. For the latter, the $E_2 - E_1$ energy separation is comparable to the strength of the vibronic coupling and crystal field effects. These have therefore to be taken into account [115, 116].

b. *Stark effect in chlorin* (H_2Ch). The Stark effect expected for optical transitions in solids may also be orders of magnitude smaller than the inhomogeneous spectral linewidths. If conventional techniques are used to get information on line splittings and shifts, large modulated electric fields of the order of 10^5 V cm^{-1} are necessary. However, the problem can be solved in an easier and more accurate way by means of hole-burning. Due to the intrinsic high resolution achieved with this technique, much smaller fields can be used by which electric breakdown and fast switching difficulties are eliminated.

Since electric fields are sensitive to molecular symmetry, it is possible to determine dipole moment changes and polarizations of molecules in excited electronic states with Stark experiments. A molecule without center of inversion possesses a permanent dipole moment, and will exhibit a linear Stark effect on application of an electric field. In contrast, molecules with inversion symmetry will yield a quadratic Stark shift.

Only a few *organic crystalline* systems have been studied so far with high resolution Stark spectroscopy in the MHz-regime: *free-base chlorin* (H_2Ch) as guest in single crystals of *n*-alkanes [114]. The symmetry of this molecule (C_{2v}) is lower than that of H_2P (D_{2h}). Thus, unlike the latter, H_2Ch possesses a permanent ground state dipole moment, μ, from which a first order Stark effect is expected.

Hole-burning experiments were performed on the $S_1 \leftarrow S_0$ 0—0 transition of H_2Ch and its photoproduct H_2Ch^* incorporated in different sites of *n*-C_6 and *n*-C_8 single crystals at 1.2 K [114]. Under the influence of a weak electric field (< 3 kV/cm), a permanent photochemical hole burnt at zero field splits into two components with a separation linearly proportional to the applied field. This effect is illustrated in Fig. 52 (right). A splitting, instead of a shift, is observed because

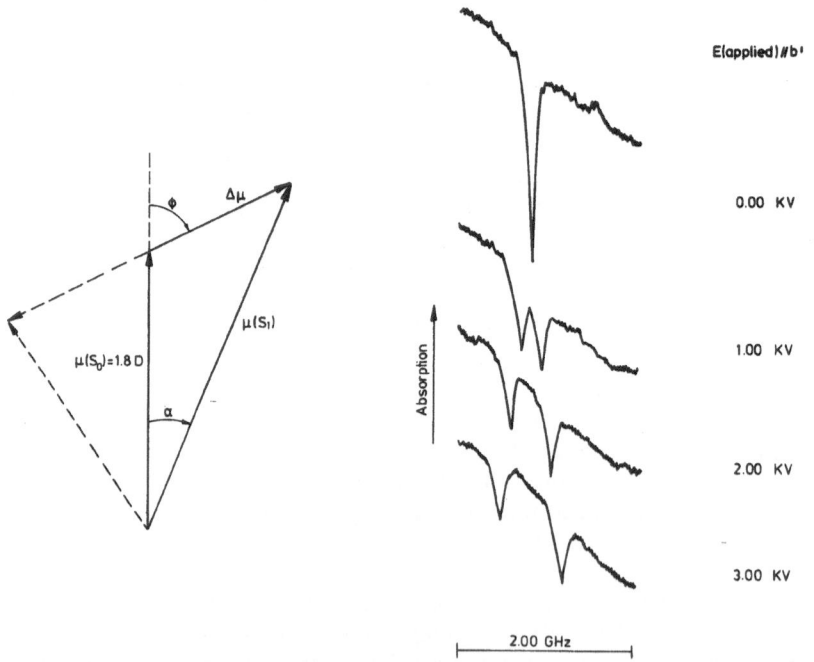

Fig. 52. Right: Stark splitting of a photochemical hole burnt in line 2 of the $S_1 \leftarrow S_0$ 0—0 transition of H_2Ch in *n*-C_8 at 1.2 K. The hole is first burnt and probed at zero electric field E. In a series of fields, the hole splits linearly with E. The width of the hole is $\approx$ 100 MHz. Left: change in magnitude and direction of the electric dipole moment, $\Delta\mu$, when exciting H_2Ch^* from the S_0 to the S_1-state. The ground state dipole moment, $\mu(S_0)$, is assumed to be parallel to $\Delta\mu(H_2Ch)$ and equal to 1.8 D. The value of $\Delta\mu$ for H_2Ch was found to be 0.23 D. The value of $\mu(S_1)$ of H_2Ch^* is $\approx$ 2.2 D with a site-dependent angle between $\mu(S_0)$ and $\mu(S_1)$ of up to 24° [114].

the dipoles of the guest molecules in a given site of the host crystal may be oriented in two opposite directions with approximately equal probability. In this way the inversion symmetry of the crystal is maintained.

From the Stark shifts, the magnitude and direction of the change in dipole moment $\Delta\mu$ on excitation to the first excited singlet state S_1 for both molecules were obtained. While the magnitude $|\Delta\mu|$ is directly related to the charge redistribution in the molecule on excitation to S_1, the direction of $\Delta\mu$ gives insight into the orientation of the H_2Ch guest in the host lattice. The results indicate that H_2Ch occupies A-sites in both n-C_6 and n-C_8 n-alkane crystals [114]. It was found that $|\Delta\mu| = 0.21 \pm 0.02$ D and 0.255 ± 0.005 D for H_2Ch in n-C_6 in n-C_8, respectively, in good agreement with the theoretical prediction of $\Delta\mu = 0.26$ D [272]. Furthermore, $\Delta\mu \parallel \mu(S_0)$ for H_2Ch, whereas the S_0- and S_1-dipole moments for H_2Ch^* are no longer parallel, but their relative orientation, $\Delta\mu$, depends on the site measured. The value of $|\mu(S_1)|$ for H_2Ch^* was estimated to be 2.2 D, as compared to $|\mu(S_0)| \sim 1.8$ D, with an angle between $\mu(S_1)$ and $\mu(S_0)$ that varies up to 24° (see Fig. 52, left).

The fact that $\mu(S_1)$ for H_2Ch^* changes as a function of the site is probably related to the near degeneracy of the S_1 and S_2 states in this molecule. As discussed in the previous section, the energy separation $E_2 - E_1$ seems to vary between 400 and 700 cm^{-1} for the four sites of H_2Ch^* in n-C_6 [116]; thus the two orbital states are probably mixed by the crystal field and vibronic coupling. In such a situation, the change in dipole moment, $\Delta\mu$, no longer is an intrinsic property of the guest, but also depends on the guest-host interaction [114].

4.1.5. *Frequency Domain Optical Storage*

Permanent hole-burning, whether photochemical or non-photochemical, has potential technological applications for optical data storage in the frequency domain [218]. The presence or absence of a hole at a particular frequency can be used to store one "bit" of information. Fig. 53 shows a schematic diagram of an inhomogeneously broadened absorption band with information in binary code.

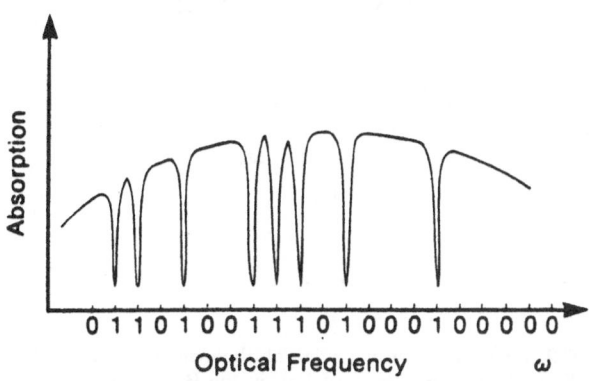

Fig. 53. Schematic diagram of an inhomogeneously broadened absorption band with data storage in the frequency domain. Information is obtained in binary code [218].

Conventional optical recording can approach storage densities of $\sim 10^8$ bits/cm^2, which is fundamentally limited by optical diffraction (minimal spot size of laser $\simeq \lambda^2 \simeq 1\ \mu m^2$) [273]. Hole-burning adds the frequency dimension to the two-dimensional spatial storage of conventional optical devices [218]. The crucial parameter in this case is the number of holes, N, that can be burnt in the inhomogeneously broadened absorption band, Γ_{inhom}, at a single spatial spot, i.e. $N \simeq \Gamma_{inhom}/\Gamma_{hole}$, where Γ_{hole} represents the holewidth. For example, for H$_2$P in an n-octane *crystalline* host at 1.2 K, $N \geq 10^3$ [19, 58], which would allow the storage capacity in this case to be increased by more than 3 orders of magnitude with respect to conventional optical storage. At 77 K, however, $N \simeq 1$ [18, 19]. These results imply that high storage densities can probably be achieved at very low temperatures only. Temperature cycling in the above mentioned *organic crystalline* system is limited to $T \lesssim 30$ K, because the information stored in the holes is destroyed above this temperature [41] (see Fig. 54). Very recently, however, an *inorganic crystalline* system was found in which holes could be recovered after cycling the temperature to 300 K [198].

An alternative approach to use hole-burning at higher temperatures is to increase Γ_{inh}, such that $\Gamma_{inh} \gg \Gamma_{hom}$. It has been shown that *organic amorphous* systems like H$_2$P, resorufin and pentacene in various polymer hosts at 1.2 K have

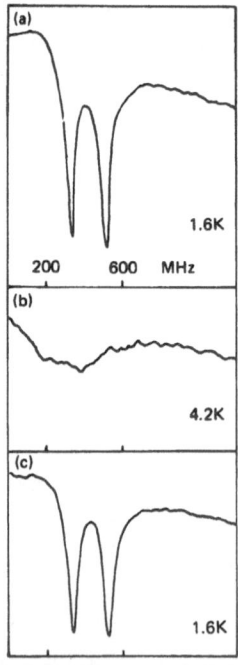

Fig. 54. Example of thermal cycling of two adjacent photochemical holes. (a) Two holes burnt 180 MHz apart in the $S_1 \leftarrow S_0$ 0—0 transition of H$_2$P in the B$_1$-site of n-C$_8$ at 1.6 K. (b) The same spectrum heated to 4.2 K. (c) After cooling to 1.6 K the information is recovered [41].

a ratio of $\Gamma_{inh}/\Gamma_{hom} \simeq 10^5$ [1, 23, 24, 43, 45, 49, 56, 186], which is already 100 times higher than in organic *crystalline* systems. Moreover, a study of the influence of the structure of polymer hosts on Γ_{hom} has shown that very narrow holes (tens of MHz at 1.2 K) can be burnt into organic guest molecules incorporated in polymers with short sidegroups, like polyethylene [43], which also leads to a potentially higher storage capacity.

Most of the hole-burning systems studied so far undergo a single-photon mechanism (see upper part of Fig. 55 [196]), with the disadvantage that the signal-to-noise ratio degrades during the data-reading process. This occurs because photo-induced bleaching takes place simultaneously with the reading. *Non-destructive* hole detection, however, is possible by means of a *"gated"* mechanism. Such "gating" requires, in addition to optical excitation, a second step to initiate the reaction leading to hole-burning. A few systems have been found until now

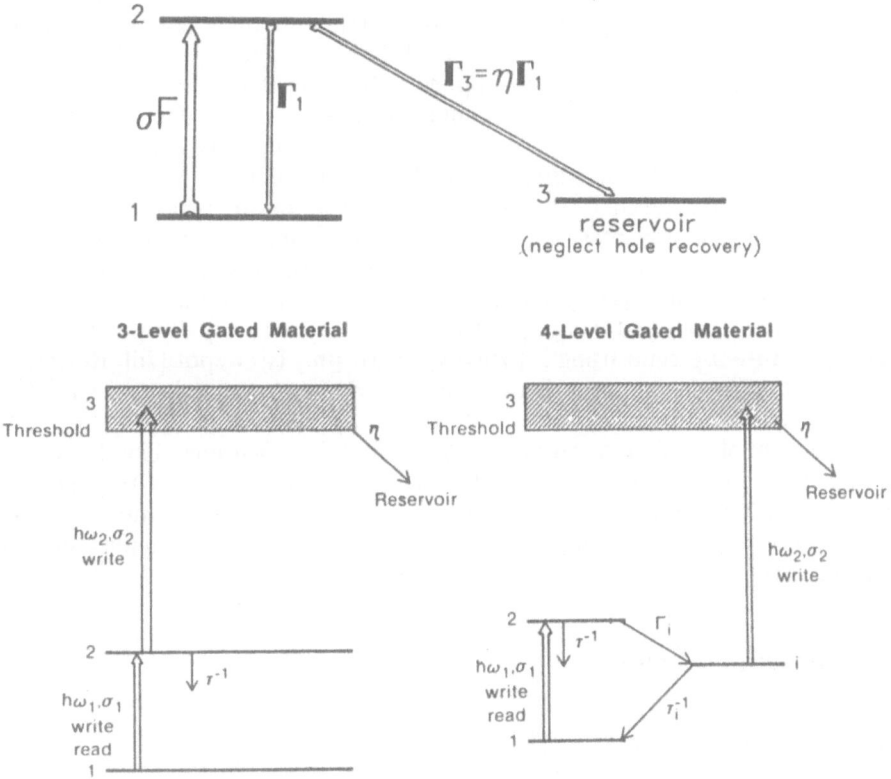

Fig. 55. Top: Schematic energy level diagram representing a single-photon hole-burning mechanism. Level 1 is the ground state, level 2 the excited state, and level 3 the photoproduct after hole-burning. Γ_1 represents the excited-state decay rate, η the hole-burning quantum efficiency, F the writing or reading flux, and σ the peak absorption cross section (from ref. [196]). Bottom: Energy level schemes for three-level (left) and four-level (right) photon-gated PHB mechanisms (see text) (from ref. [197]).

that undergo "gated" hole-burning: Sm^{2+} in BaClF [198] and CaF_2 [199], Co^{2+} in $LiGa_5O_8$ [200], Cr^{3+} in $SrTiO_3$ [201], carbazole in boric acid glass [51], anthracene-tetracene photoadducts in PMMA [202] and Zn-tetrabenzoporphyrin derivatives with halomethanes in PMMA [71, 203]. The HB-mechanisms in these systems have already been discussed in detail in Section 3.2.2. In order to burn a hole, a first photon is needed for frequency selective excitation, whereas absorption of a second photon of different energy triggers the hole-burning process. This process may either be a photoionization [51, 198—201], photodissociation [202], or donor-acceptor electron transfer reaction [71, 203]. Because of the difference in frequency of these two steps, the reading and writing processes are decoupled, and non-destructive reading is achieved.

A *photon-gated* mechanism will only be attractive for storage applications, if the overall efficiency and the gating ratio are high [76]. Fig. 55 (lower part) illustrates some of the requirements to be satisfied in 3-level (left) and 4-level (right) photon-gated systems [197]. For example, a 3-level system should have the wavelength $\lambda_1 > \lambda_2$, and the lifetime τ long enough to allow population to build up in level 2 [197]. The transition from level 2 to 3 does not need to be frequency selective. This type of level scheme applies to Sm^{2+} in BaClF [198]. For a 4-level system, the rate Γ_i (from $2 \rightarrow i$) should be as large as possible, and the intermediate state lifetime τ_i should be long to achieve a large population in level i. Further, for efficient photon gating the yield η should be large. A 4-level scheme is appropriate for materials like donor-acceptor electron transfer processes [71, 203].

In spite of the advantages of photon-gated systems, none of the present materials fulfills all the conditions required for a successful frequency domain optical storage medium [197]. The principal problem is burning a hole in a very short time (~ 30 ns) and, subsequently, detecting it very rapidly (~ 30 ns) with high signal-to-noise ratio using a focussed (~ 10 μm) laser spot [76]. Reviews on this subject, including an analysis of the engineering and material requirements for practical optical storage systems with frequency domain read-out, and other possible technological applications of persistent hole-burning, like laser pulse shaping and multiplexed optical spatial filters, can be found in refs. [22, 76]. Further, a study of the parameters that permit to evaluate the usefulness of one-photon and photon-gated hole-burning materials for storage applications are reported in refs. [196, 197].

4.2. AMORPHOUS SOLIDS

4.2.1. *Introduction*

Glasses can be regarded as frozen liquids. They do not have long-range order, but a high degree of local correlation because their atoms or molecules are held together, as opposed to a dilute gas in which they are randomly distributed.

At low temperature, one might expect the physical properties of amorphous solids to be similar to those of crystals, since the structure of the material should become unimportant when long wavelength phonons are excited. In 1971, however, Zeller and Pohl presented experimental evidence that the thermal properties are quite different from those of crystalline materials [273]. The specific heat in glasses revealed to be much larger and to depend linearly on temperature, while the thermal conductivity was shown to be much smaller than in crystals, and to vary quadratically with temperature. This is in contrast to the T^3-dependence observed for both parameters in crystalline solids. Further, these physical properties are rather insensitive to the chemical composition of the glass. The experimental results suggested that amorphous solids at low temperature are dominated by low-frequency excitations characteristic of the glasses state [274].

Many theories have been proposed to explain these anomalies, the most successful of which has been the tunneling model by Phillips, and Anderson *et al.* [275]. It is based on the assumption that atoms or groups of atoms occupy two energetically inequivalent configurations of the glass, the so-called *"two-level-systems"* (*TLS*), and may tunnel between them at very low temperature. Owing to the random distribution of atoms or molecules in glasses, there will be a wide distribution of such TLS which can be represented by a nearly constant density of states. Detailed predictions of this model depend on assumptions made about the distribution of tunneling parameters. Most of the experiments performed on glasses at low temperatures since 1972 have been interpreted in the framework of this model. However, the nature of the low frequency excitations is still unknown. In fact, one of the main questions in the physics of amorphous solids is whether the TLS are intrinsic to the absence of long-range order [276].

Not only thermal, acoustic and dielectric properties, but also *optical properties* of glasses at low temperature have been found to differ from those of crystals. In the last ten years much work has been devoted to the study of linewidths of impurity ions and molecules incorporated as guests in amorphous materials.

In the next sections we will present experimental results on the temperature dependence of optical linewidths in glasses measured by different techniques. A brief account will be given on *inorganic glasses* (Section 4.2.2a), and the reader interested in a detail review of the subject may consult refs. [30, 32]. Optical dephasing in *organic glasses* will be presented chronologically. We have included an actual controversy between the results obtained from hole-burning and photon-echo experiments, because it is a problem not solved yet and a matter of lively discussions (Section 4.2.3). Another subject closely related to it is "spectral diffusion", or slow relaxation processes in glasses, which is discussed in Section 4.2.4. Theoretical models developed to explain optical dephasing in amorphous solids will be summarized in Section 4.2.5. More formal approaches to the theories can be found in R. Silbey's chapter in this book [238] and refs. [246, 277]. The physical implications of some of these models have led to further experimental work. Results obtained from these checks on *semi-crystalline* polymers and *mixed organic-inorganic* amorphous systems will be shown in Sections 4.2.6 and 4.2.7. Finally, we will discuss experimental results on the influence of external magnetic and electric fields and pressure on holes burnt in zero field in the $S_1 \leftarrow S_0$ 0—0 transitions of organic molecules in glasses (Section 4.2.8).

4.2.2. *Optical Dephasing in Glasses*

Due to the structural disorder of amorphous solids, a large range of local environments exists around an impurity ion or guest molecule embedded in a glassy host. As a consequence, optical linewidths are strongly inhomogeneously broadened ($\Gamma_{inh} \simeq$ a few hundred cm^{-1} in glasses, as compared to a few GHz to tens of GHz in crystals). Techniques such as fluorescence line-narrowing (FLN) [2—4, 14], spectral hole-burning [15—20] and photon echoes [5—9] (see also section 1) have been used in amorphous systems to eliminate the influence of the inhomogeneous linewidth and get information on the homogeneous linewidth, Γ_{hom}. The first experiments on the temperature dependence of Γ_{hom} on amorphous systems at low temperature were performed on rare earth-doped *inorganic glasses*. In the next section we will discuss some of the results on these systems in order to be able to compare them to *organic glasses*.

a. *Inorganic glasses*. The first results on optical dephasing in glasses were reported in 1976. The system studied was Eu^{3+} in a silicate glass ($^5D_0 \leftrightarrow {}^7F_0$ transition) between 8 and 90 K by means of FLN [278]. The linewidths measured were one to two order of magnitude larger than those previously found in crystals and, in contrast to the exponential temperature dependence characteristic of the latter, a $T^{1.8}$ temperature dependence was observed. It was subsequently checked by means of hole-burning that the same T-dependence seems to extend down to 1.6 K [53].

From FLN experiments on Eu^{3+} in other inorganic glasses at high temperature (up to 800 K), it was concluded that $\Gamma_{hom} \propto T^n$, where $n = 2.0 \pm 0.3$ depends slightly on the composition of the glass [279, 280]. The T^2 dependence at high temperatures was attributed to a two-phonon Raman process, as in crystals. The quadratic low temperature dependence, however, could not be explained in a satisfactory manner. Theories developed to interpret these results were based on the "tunneling model" [275], i.e. the optical transition is coupled to TLS of the glass which, on their hand, interact with phonons of the bath [278, 281—283]. Recent FLN-experiments on the same $^5D_0 \leftrightarrow {}^7F_0$ transition of Eu^{3+} in lithium and sodium silicate, and potassium germanate glasses performed between 10 and 300 K have again shown a T^2-dependence of the linewidth [284], but the results were interpreted in terms of "fractons" which are localized vibrational excitations [285]. In this model, the impurity ion is assumed to interact with these fractons, instead of coupling to TLS-modes of the glass [284].

A similar T^2-temperature dependence of Γ_{hom} has also been observed for the $^3P_0 \leftrightarrow {}^3H_4$ transition of Pr^{3+} in BeF$_2$ and GeO$_2$ glasses between 10 and 300 K by FLN [286]. Since only very few dephasing data exist for inorganic glasses doped with Eu^{3+} and Pr^{3+} at temperature below 10 K, it has strongly been recommended in the literature [30] to perform hole-burning together with very high-resolution (~ 1 MHz) FLN experiments between ~ 0.4 and 10 K, in order to check the validity of the T^2-dependence of Γ_{hom} for these systems. In contrast to the results mentioned above, the $^1D_2 \leftrightarrow {}^3H_4$ transition of Pr^{3+} in silicate glass yields a linear T-dependence of Γ_{hom} between 2 and 20 K, when measured by both hole-burning and accumulated photon echoes [53]. It would be very informative to extend these experiments to lower temperatures (at least down to 0.3 K) to be able to strictly test the theoretical models.

Another rare earth impurity ion that has extensively been studied in inorganic glasses is Nd^{3+}. At temperatures between 80 and 220 K, FLN experiments on the $^4F_{3/2} \leftrightarrow {}^4I_{9/2}$ transition have shown that Γ_{hom} broadens with an approximately T^2-dependence [287]. A similar temperature dependence was observed for the $^4G_{5/2} \leftrightarrow {}^4I_{9/2}$ transition between 10 and 80 K by means of accumulated photon echoes [288]. However, ns two-pulse photon echo studies on the $^4F_{3/2} \leftrightarrow {}^4I_{9/2}$ transition of Nd^{3+} in a silicate glass fiber (fused silica) have shown that $\Gamma_{hom} \propto T^{1.3}$ between 0.1 and 1 K [289a, b]. A model developed to explain these results will be discussed in Section 4.2.5 [289c]. This experiment is the only one so far performed on a pure inorganic glassy system below 1 K. Unfortunately, these photon echo experiments have neither been extended to temperature above 1 K nor to another inorganic glass doped with a rare earth ion. Thus, it is still unknown whether the $T^{1.3}$-dependence of Γ_{hom}, observed for many organic glasses [24], is perhaps also valid for *inorganic glasses* but in a much lower temperature regime than the former. Furthermore, it should be checked whether there is a cross-over to a T^2 dependence at higher temperatures.

There has been one other system, Yb^{3+} in a phosphate glass ($^2F_{5/2} \leftrightarrow {}^2F_{7/2}$), studied by FLN which was reported to follow a $T^{1.3}$ law between 7 and 40 K, and a $T^{1.8}$ law between 40 and 70 K [95]. The results, however, should be critically re-examined with better instrumental resolution, because the bandwidth (~ 4 GHz) of the pulsed colour center laser in combination with the Fabry-Pérot etalon used was much larger than the deconvoluted value of Γ_{hom} (< 1 GHz).

The only optical dephasing experiment reported so far for a transition metal ion has been on Cr^{3+} in ED-2 silicate glass ($2E \leftrightarrow {}^4A_2$) between 5 and 150 K by means of FLN by Imbusch and coworkers [290]. The data were interpreted as the sum of two processes, a linear T dependence of Γ_{hom} below 80 K, and an approximately T^2-dependence above 80 K, the general behaviour being similar to that of rare earth-doped glasses. As in the previous example, the data at low temperature are not reliable, because of the poor instrumental resolution (of ~ 1 cm^{-1}).

In summary, the majority of the optical dephasing experiments on *inorganic glasses* (silicate, phosphate, borate, germanate) have been performed on rare earth impurity ions (Eu^{3+}, Pr^{3+}, Nd^{3+}, Yb^{3+}) by means of FLN, in the temperature range between ~ 7 and 300 K. In most cases, it was found that $\Gamma_{hom} \propto T^{2.0 \pm 0.3}$, the only exception being Pr^{3+} in silicate glass (measured by hole-burning and accumulated photon echoes between 2 and 20 K, for which $\Gamma_{hom} \propto T^{1.0 \pm 0.2}$ [53]). Only one inorganic amorphous system has been investigated below 1 K, Nd^{3+} in fused silica. Two-pulse photon echoes used in this case yielded a $T^{1.3}$ dependence of Γ_{hom} between 0.1 and 1 K [289].

b. *Organic glasses.* The first hole-burning experiments on optical linewidths in *organic glassy* systems were reported by Personov and coworkers in 1974 [16] for the $S_1 \leftarrow S_0$ 0—0 transition of perylene in ethanol. Since then most of the optical dephasing results on organic glasses have been obtained with this technique [33]. There have been only a few organic amorphous systems that have been investigated by other methods, like phosphorescence line-narrowing [291], accumulated photon echoes [55], and two pulse ps-photon echoes [96, 191].

Homogeneous linewidths, Γ_{hom}, of $S_1 \leftarrow S_0$ 0—0 transitions of organic molecules embedded in organic glasses (mostly alcohols) were initially reported to be a few cm^{-1} broad at temperatures of about 2 K [16, 20, 26, 47, 48, 52, 54, 69]. They were orders of magnitude larger than the homogeneous linewidths in crystalline hosts [17, 19, 38—42, 44, 58, 60, 61, 85]. Furthermore, it was found that $\Gamma_{hom} \propto T^n$, with $n = 1$ [26, 52, 69] or 2 [48, 54, 69], depending on the system studied.

In none of these experiments the fluorescence lifetime-limited value of the transition, $\Gamma_0 = (2\pi T_1)^{-1}$ of a few MHz to tens of MHz, was reached, neither at low temperature nor by extrapolation to zero temperature. From eq. (1) (Section 1.1) one would expect, in principle, that the second term, $(\pi T_2^*)^{-1}$, which represents the "pure" dephasing contribution and accounts for thermally induced fluctuations of the optical transition frequency, should go to zero when $T \rightarrow 0$. For organic glasses, residual linewidths of $\Gamma_0 \simeq 0.4$—1 cm^{-1} (for $T \rightarrow 0$) were reported [26, 48, 52, 69], which were about three order of magnitude larger than the expected values. It was suggested that picosecond dephasing processes would still take place at $T = 0$ [26, 48, 52]. These striking results were subsequently proven to be incorrect. Many of the holewidths reported were much too large, and did not represent the true homogeneous linewidth. The reason for these excessively broadened holes is that in most cases the laser fluences used for burning were too high, often combined with an instrumental bandwidth in the probing step that was too large [43] (see also Section 2.2).

It has been argued in the literature that the large differences in homogeneous linewidths, Γ_{hom}, and their T-dependences, between different organic glassy systems were due to the hole-burning mechanisms involved. This statement is incorrect, as we will see below. In Section 1.2, it was mentioned that hole-burning mechanisms can be classified according to their *photochemical* or *non-photochemical* nature. In the first category, we have further distinguished between *intramolecular* photochemical reactions which take place within the guest molecule [1, 15, 17, 18, 43—45, 62, 87, 88, 102], and *intermolecular* reactions which occur between the guest and the host [47—49, 183, 184, 185, 187]. In the case of *non-photochemical hole-burning* (NPHB), a relative reorientation of the guest molecule with respect to its environment is assumed to take place after excitation [16, 20, 26, 55, 56]. Thus, potentially many photostable organic molecules incorporated in organic glasses could be studied by means of NPHB. In this section we will present the results on optical dephasing in organic glasses, and compare the temperature dependence of Γ_{hom} of systems undergoing different hole-burning mechanisms. For a discussion of the individual HB-mechanisms, see Section 3.

Most of the organic amorphous *hosts* used in hole-burning experiments so far have been either *alcohols* (methanol, ethanol, glycol, glycerol, and mixtures), or *polymers* (polyvinylcarbazole, polymethylmethacrylate, polystyrene, polyethylene, etc.). A schematic representation of the structure of polymers is given in Fig. 56,

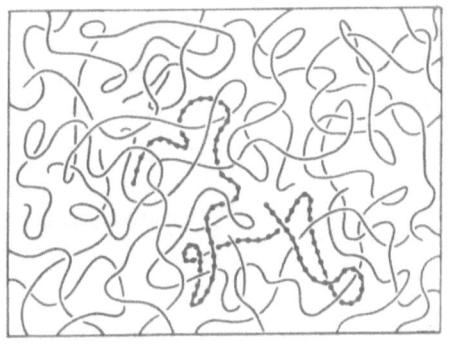

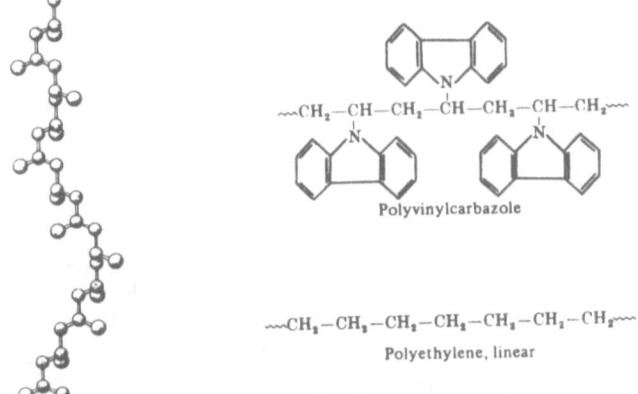

Polyvinylcarbazole

$\sim CH_2-CH_2-CH_2-CH_2-CH_2-CH_2-CH_2\sim$

Polyethylene, linear

Fig. 56. Top: Schematic representation of the structure of polymers, according to the random coil model of Flory [292a]) (from ref. [292b], p. 112). Bottom, left: Conformation of an organic polymer chain (from ref. [292b], p. 110). Bottom, right: chemical structure of polymers with bulky side-groups (polyvinyl-carbazole) and no side-groups (polyethylene).

which shows the random coil model of Flory (top) [292a, b], and the general conformation of an organic polymer chain (bottom, left), together with the chemical structure of a polymer with bulky side-groups (polyvinylcarbazole) and no side-groups (polyethylene) (bottom, right).

Organic molecules studied as *guests* in organic glassy hosts are illustrated in Fig. 57. They have been ordered according to their HB-mechanism: the top row shows molecules that undergo *intramolecular* PHB, the middle one, molecules that undergo *intermolecular* PHB with the host, and the bottom one, photostable molecules for which *NPHB* has been observed. Of these molecules, only two have also been investigated by other techniques: pentacene by accumulated phonon echoes [55], and resorufin by two pulse ps-photon echoes [96, 191] (see Section 4.2.3).

A systematic hole-burning study of organic amorphous systems was started in

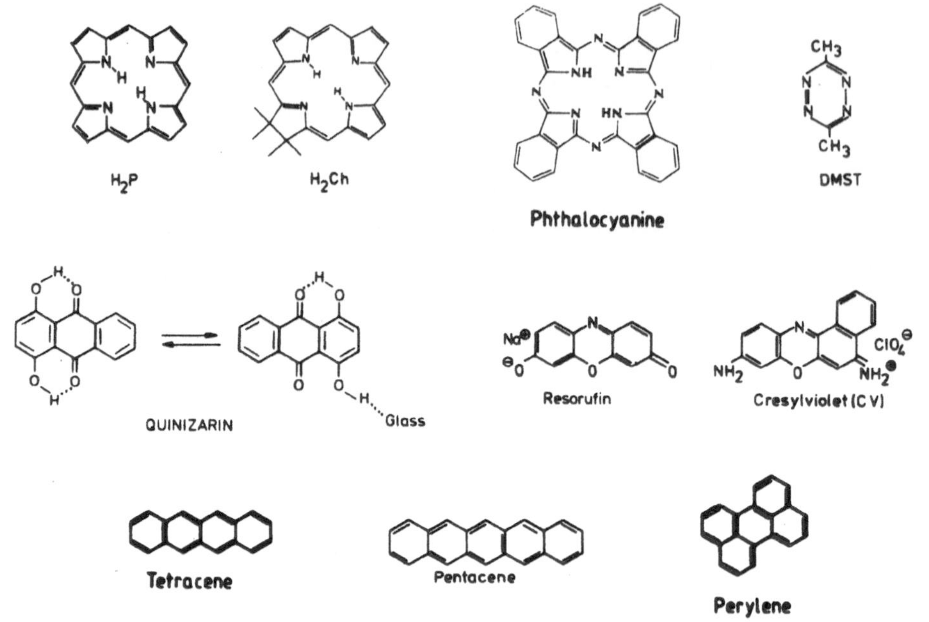

Fig. 57. Organic molecules studied as guests in organic amorphous hosts. Top: free-base porphin (H_2P), free-base chlorin (H_2Ch), phthalocyanine, and dimethyl-s-tetrazine (DMST). They all undergo intramolecular-PHB. Middle: quinizarin, and the ionic dyes resorufin and cresylviolet (CV). They undergo intermolecular-PHB. Bottom: tetracene, pentacene and perylene. They undergo non-photo-chemical-HB.

1982 by the group in Leiden [1]. The aim was to clear up the contradictions between different hole-burning results on glasses reported until then in the literature. As a first step, the molecule free-base porphin (H_2P) was chosen, because of its well known one-phonon *intramolecular* PHB-reaction [18] (see also Section 3.1). Hole-burning experiments were performed on the $S_1 \leftarrow S_0$ 0—0 transition of H_2P embedded in a large variety of glasses and polymers between 1.2 and 4.2 K [1, 43, 45]; some of the systems were also studied over an extended temperature range, down to 0.3 K [24, 49, 72, 86, 87, 293]. The results were surprising, and in strong contrast to the data reported by other groups: (1) very narrow holes of about hundreds of MHz to a few GHz at $T \simeq$ 2 K were found, (2) the value of Γ_{hom}, defined as $\Gamma_{hom} = \frac{1}{2}\Gamma_{hole}$ for very low burning fluences (see also Section 2), extrapolated to the fluorescence lifetime-limited value, $\Gamma_0 \simeq$ 10 MHz of H_2P, when $T \rightarrow 0$ [1, 43, 45, 86, 87, 293], and (3) $\Gamma_{hom} - \Gamma_0$ followed a $T^{1.3}$-temperature law, independent of the organic amorphous host used [1, 45]. The latter result is illustrated in Fig. 58, where a log-log plot of $\Gamma_{hom} - \Gamma_0$ versus T for H_2P in various polymers between 1.2 and 4.2 K [43, 45] is shown. Two of the samples were also measured between ~0.3 and ~20 K [86] (see comment below). Notice that the value of Γ_{hom} at a given temperature is larger for polymers with bulky side-groups (PVCa) than with small side-groups (PE). A systematic study of the

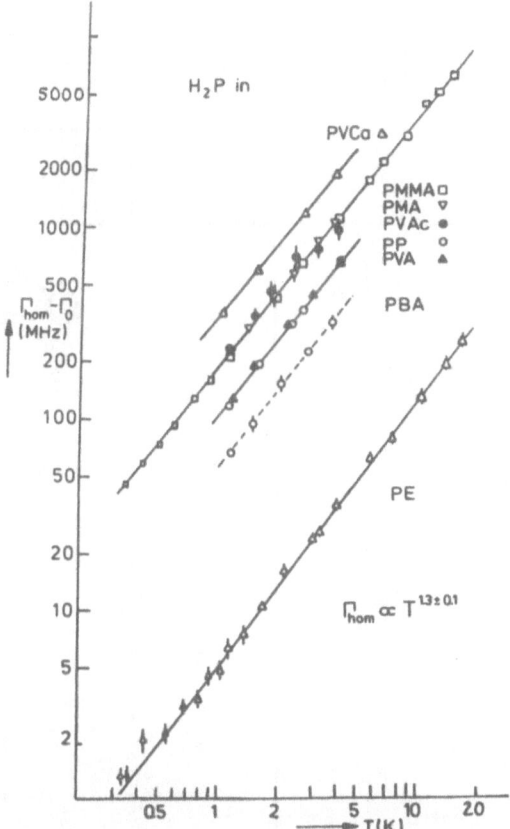

Fig. 58. Log-log plot of $\Gamma_{hom} - \Gamma_0$ versus temperature for H_2P in several polymers between 1.2 and 4.2 K (PVCa: polyvinylcarbazole, PMA: polymethylacrylate, PVAc: polyvinylacetate, PP: polypropylene, PVA: polyvinylalcohol, PBA: polybutylacrylate) [43, 45], and for H_2P in PMMA (polymethylmethacrylate) and PE (polyethylene) between 0.3 and 20 K [86]. Notice that $\Gamma_{hom} - \Gamma_0 \propto T^{1.3 \pm 0.1}$, independent of the host. The value of Γ_{hom} is related to the size of the side-group in the polymer chain (see also fig. 59) [43, 45].

influence of the size of the side-groups on Γ_{hom} has yielded a correlation between these two parameters [43]. The structure of the polymers used in ref. [43] are reproduced in Fig. 59. From these results, it was concluded that the larger the voids (empty spaces) of the polymer in which the guest molecule has freedom to move, the faster the optical dephasing seems to be [43], in analogy to what occurs in n-alkane crystals of varying chain length [58, 60, 231] (see also Section 4.1.1).

A similar behaviour of Γ_{hom} was observed when H_2P was incorporated in a series of alcoholic glasses: Γ_{hom} becomes narrower the larger the number of hydroxylic groups per host molecule (from one OH-group in ethanol to four in diglycerol) [45]. This effect has been attributed to the increase in hydrogen bond

H_2P in polymers $\left[-CH_2 - \overset{R_1}{\underset{R_2}{C}} - \right]_n$

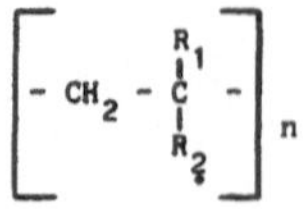

Name	R_1	R_2
Polyvinylcarbazole : PVCa	– H	
Polymethylmethacrylate: PMMA	– CH_3	$-\overset{O}{\overset{\|}{C}} - O - CH_3$
Polymethylacrylate : PMA	– H	$-\overset{O}{\overset{\|}{C}} - O - CH_3$
Polyvinylacetate : PVAc	– H	$- O - \overset{O}{\overset{\|}{C}} - CH_3$
Polypropylene : PP	– H	– CH_3
Polyvinylalcohol : PVA	– H	– OH
Polyethylene : PE	– H	– H
Polybutylacrylate : PBA	– H	$-\overset{O}{\overset{\|}{C}} - O - (CH_2)_3 - CH_3$

Fig. 59. Polymers studied as hosts for free-base porphin (H_2P). The side-groups of the polymer chain are represented by R_1 and R_2 in decreasing size from top to bottom (see also Fig. 58). An exception is PBA, for which a rather small homogeneously linewidth was found. This polymer has a long paraffin side chain with high mobility, able to surround each individual polymer chain as a swelling agent or solvent. The effective size of the side-groups in this polymer is reduced, and thus a smaller value of Γ_{hom} would be expected, as observed [43].

strength, since the alcoholic host tends to form a stiff polymeric aggregated network that inhibits thermal motion of the CH_2OH groups [45]. These results demonstrate that hole-burning is a highly sensitive technique, because it is able to distinguish very small structural changes in the direct vicinity of the guest molecule [43]. In Section 4.2.6, this HB-property will be used again to distinguish samples of various degrees of crystallinity [88].

In relation to Fig. 58 it should be remarked that the $T^{1.3}$-dependence of $\Gamma_{hom} - \Gamma_0$ has recently found to be valid between 0.3 and, at least, ~ 20 K for various organic amorphous systems [86], two of which are given in the figure.

Another interesting result is that holes burnt in the $S_1 \leftarrow S_0$ 0—0 transition of the dimer of H_2P, which absorbs at ~ 300 cm^{-1} to the red of the monomer 0—0 transition, show the same $T^{1.3}$-dependence as holes burnt in the monomer band of H_2P [1]. At a given temperature the holewidths in the two absorption bands are however different. The results for H_2P in glycerol: ethanol (10 : 1) glass and MTHF between 1.2 and 4.2 K are given in Fig. 60 [1]. These experiments suggested that the $T^{1.3}$ law is independent, not only of the glassy host, but also of the chemical structure of the guest.

In order to clear up this point, the temperature dependence of Γ_{hom} for other guest molecules was investigated: free-base chlorin (H_2Ch), which undergoes a *one-photon intramolecular PHB*-reaction [44] (see Section 3.1.1), dimethyl-s-tetrazine (DMST), which undergoes a *two-photon intramolecular* photodissociation [17, 50] (see Section 3.2.2a), and the ionic dyes resorufin and cresyl violet (CV), which undergo an *intermolecular* PHB-reaction with the host [49, 86, 187] (see Section 3.2.1b). In all cases, again very narrow holes were observed with $\Gamma_{hom} - \Gamma_0 \propto T^{1.3}$, independent of the guest molecule, and thus of the hole-burning

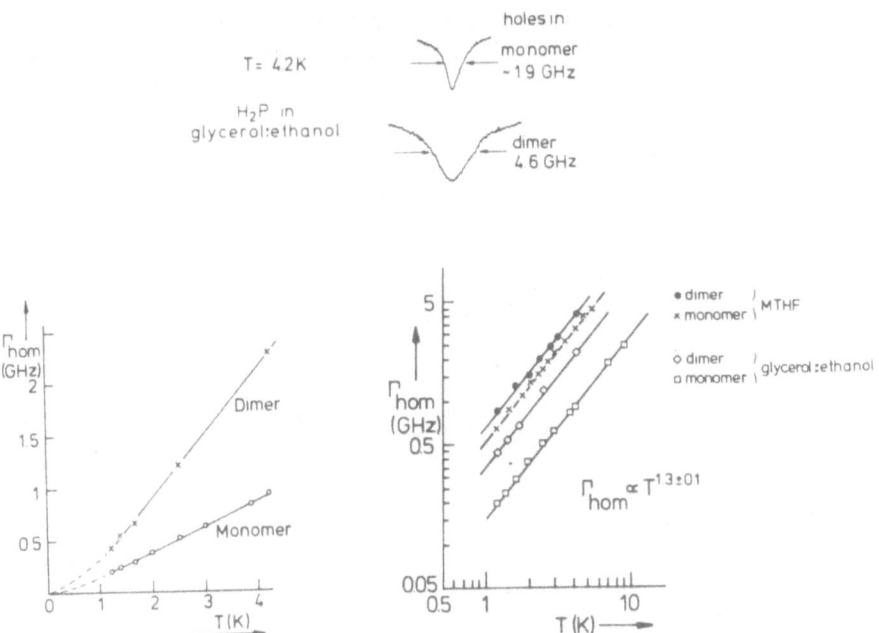

Fig. 60. Top: Holes burnt in the $S_1 \leftarrow S_0$ 0—0 transition of the monomer and dimer of H_2P in glycerol: ethanol (10:1) glass at 4.2 K (see text). Bottom, left: Γ_{hom} versus T for the same transition and sample between 1.2 and 4.2 K. Bottom, right: Log-log plot of Γ_{hom} versus T for the monomer and dimer of H_2P in MTHF and glycerol: ethanol (10:1) glass between 1.2 and 4.2 K. Notice the $T^{1.3}$ dependence of Γ_{hom} for all samples [1].

mechanism [45, 49, 86]. Fig. 61 illustrates the results for the ionic dyes. Notice that resorufin in PMMA follows the $T^{1.3}$ law between 0.3 and, at least, ~ 10 K [86], as seen above for H_2P in PMMA and PE [86]. This T-dependence is also valid for systems undergoing *NPHB* [56] (see Section 4.2.3). For all organic amorphous systems mentioned and investigated down to 0.3 K, Γ_{hom} extrapolates smoothly to $\Gamma_0 = (2\pi T_1)^{-1}$ when $T \to 0$ [49, 86–88, 186, 293]. This is summarized in Fig. 62, where Γ_{hom} is plotted as a function of T for various samples undergoing different HB-mechanisms [86].

Hole-burning experiments reported by other groups indicate that a similar power law is obeyed in a variety of organic amorphous systems, at least, in a restricted temperature range [62, 67, 102, 134, 141, 191, 212, 294, 295]. Recently a study was reported on the dependence of the holewidth on concentration of the monomer band of tetracene in ethanol: methanol (3 : 1) glass between 0.4 and 4 K. For a highly concentrated sample a residual linewidth Γ_0 was observed which was more than one order of magnitude larger than the value expected from the fluorescence lifetime [295]. This effect was attributed to fast energy transfer processes to lower lying aggregate states. The temperature dependence of $\Gamma_{hom} - \Gamma_0$, however, was not affected by the presence of aggregates. No other HB-experiment on organic glasses has been reported so far, in which the value of Γ_0 was observed to vary with concentration.

Only one system, to our knowledge, octaethylporphyrin in polystyrene, has been investigated below 0.3 K [102]. At temperatures between 0.05 K and ~ 0.08 K, a $T^{1.6} - T^{1.7}$ dependence on $\Gamma_{hom} - \Gamma_0$ was reported, whereas between

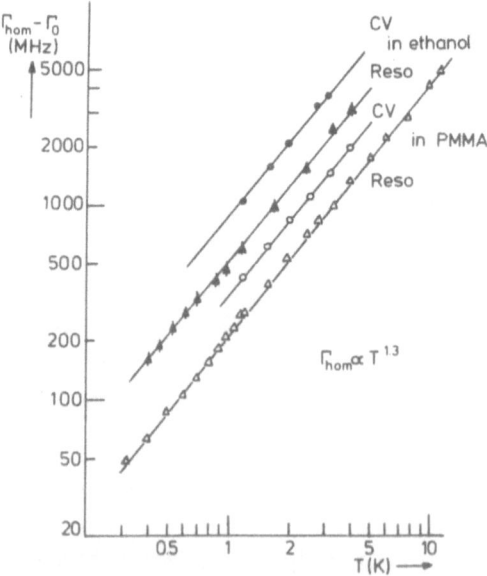

Fig. 61. Log-log plot of $\Gamma_{hom} - \Gamma_0$ versus T for the ionic dyes cresyl violet and resorufin in ethanol and PMMA between 0.3 and 10 K. In all cases, $\Gamma_{hom} - \Gamma_0 \propto T^{1.3 \pm 0.1}$.

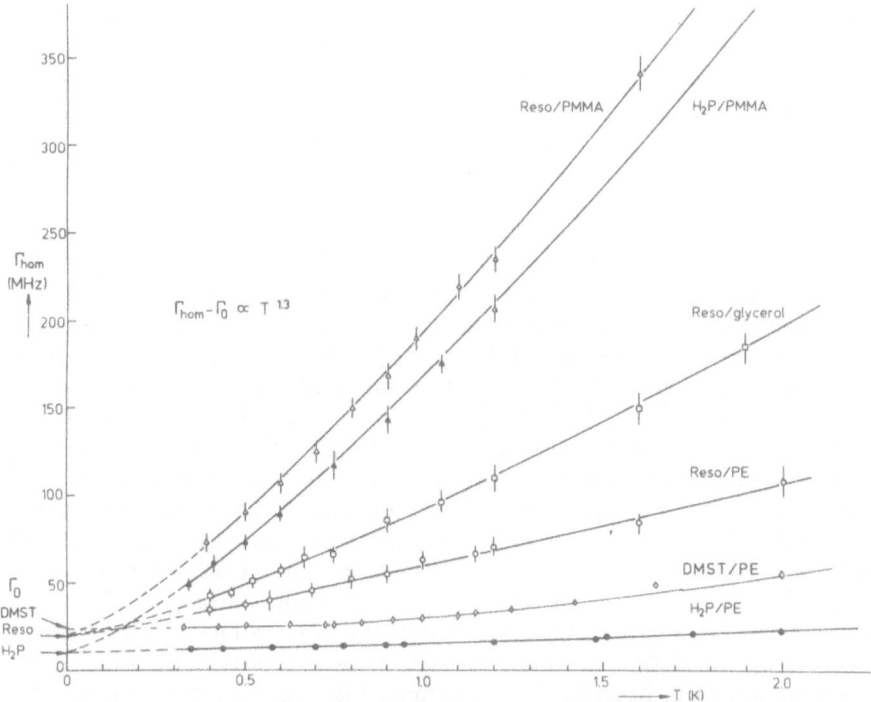

Fig. 62. Temperature dependence of the homogeneous linewidth, Γ_{hom}, of resorufin in PMMA, glycerol and PE, H_2P in PMMA and PE, and dimethyl-s-tetrazine (DMST) in PE at $T \leqq 2$ K. For $T \rightarrow 0$, Γ_{hom} extrapolates to $\Gamma_0 = (2\pi T_1)^{-1}$ of each guest molecule. In PE, the value of Γ_0 is almost reached at 0.3 K for all three guest molecules. The lines through the data were traced according to

$$\Gamma_{hom} = \Gamma_0 + a\, T^{1.3}\ [86].$$

0.08 and 0.3 K, a $T^{0.6} - T^{0.8}$ dependence was observed, which crossed over into $T^{1.2} - T^{1.3}$ above 0.3 K. A $T^{1.3}$-dependence was also observed for *organic* glasses by means of phosphorescence line-narrowing [291] and accumulated photon echoes [55] (Section 4.2.3), and for an *inorganic* glass between 0.1 and 1 K by means of two pulse ns-photon echoes [289] (Section 4.2.2a). Furthermore, very recent HB experiments performed on a number of *mixed organic* guest-*inorganic* glassy host systems between 0.3 and 4.2 K have shown the same T-dependence [296, 297] (Section 4.2.7). In the last few years, theoretical models have been developed to explain this non-integral power law. Some of them will be discussed and related to the experimental results (see Section 4.2.5).

A comment on optical dephasing experiments performed with femtosecond lasers is interesting at this point. Cresyl violet in PMMA between 15 and 290 K was studied by Ippen and coworkers by means of a three-pulse scattering technique [298a]. Dephasing times of the order of 100 fs at 15 K were reported, and attributed to fast intramolecular vibrational relaxation. But, one has to keep in mind that fs-pulses are very broad and excite a multi-level structure. Similar

results, obtained by incoherent ns-photon echoes, were reported for cresyl fast
violet in a cellulose film, for which T_2-values between 30 and 70 fs at 10 K were
derived [298b]. Recently, optical dephasing times of 75 fs were deduced from fs
spectral hole-burning experiments on cresyl violet in ethylene glycol at room
temperature by Shank and coworkers [298c]. The recovery time of the holes, of
the order of 50 fs, was attributed to thermalization of a non-equilibrium distribution of excited states.

4.2.3. *Apparent Contradictions Between Hole-Burning and Photon Echo Results*

It has been stressed in the recent literature that *non-photochemical hole-burning*
(*NPHB*) does not reflect the homogeneous linewidth, Γ_{hom}, but only measures slow
relaxation processes in the glass [55, 96, 191]. This statement was first used by
Molenkamp and Wiersma [55] to explain contradictory results they had obtained
for pentacene in PMMA by means of accumulated photon-echo and NPHB
experiments. The linewidths measured by NPHB were reported to be much larger
than those derived from photon echoes. Moreover, the temperature dependence of
the holewidths was T^2, whereas that of linewidths derived from photon echoes was
$T^{1.3}$ between 1.5 and 12 K. Similar differences in linewidths were reported by
Fayer and coworkers for data obtained for resorufin in ethanol [191] and glycerol
[96] by means of two pulse ps-photon echoes and hole-burning. In all cases, the
difference between the linewidths was attributed to long-time "spectral diffusion"
processes. In principle, hole-burning and photon echoes may yield different
physical information, because these techniques measure on different time scales
(ps to ms for photon echoes, s to min for permanent HB), and glasses, as is well
known, have a broad range of relaxation rates [274].

  These results led to a NPHB-reinvestigation of pentacene in PMMA [56].
Subsequently, also resorufin in ethanol [186] and glycerol [86] were rechecked by
PHB. In particular, the influences on the holewidth of laser power, burning time,
sample preparation, optical density and sample heating effects were carefully
studied. The NPHB-results obtained in ref. [56] for the homogeneous linewidth of
pentacene in PMMA as a function of temperature are given on the lowest curve of
Fig. 63. The NPHB and accumulated photon echo data of ref. [55] are also
reproduced on this log-log plot, for comparison (dashed lines). In the figure, Γ'_{hom}
has been defined as $\Gamma'_{hom} = \frac{1}{2}\Gamma_{hole}$ for HB-data, and $\Gamma'_{hom} = (\pi T_2)^{-1}$ for photon
echo data [55]. Notice that the lowest NPHB-curve again follows a $T^{1.3}$ depen-
dence between 1.2 and 4.2 K [56]. The holes on this set of data were burnt with
very low laser fluences ($Pt \lesssim 7.5 \times 10^{-3}$ J cm^{-2}) and probed by fluorescence
excitation. In contrast, the holes of ref. [55] were burnt with a laser fluence more
than 10^3 times higher, and probed in transmission through the glassy sample
(thickness 2.2 mm). As can be seen from Fig. 63, the results of the accumulated
photon echoes [55] are very similar to those of the lower NPHB curve [56]. The
small discrepancy between these data has been attributed to temperature effects
[56]. Since the photon echoes were measured in transmission through a glassy
sample, known to have poor thermal conductivity, the energy dissipated in the
bulk of this sample may have produced local heating, and the temperature was

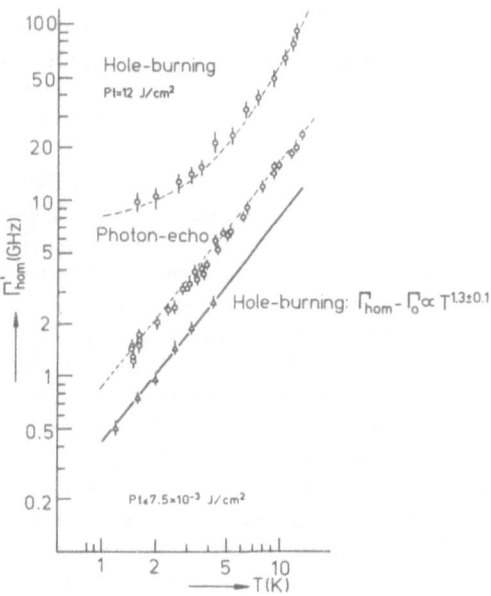

Fig. 63. Log-log plot of Γ'_{hom} versus T for pentacene in PMMA. Γ'_{hom} is defined as $\Gamma'_{hom} = \frac{1}{2}\Gamma_{hole}$ for hole-burning data, and $\Gamma'_{hom} = (\pi T_2)^{-1}$ for photon echo data. The two dashed lines for hole-burning and photon echo results are reproduced from ref. [55]. The lowest full line represents hole-burning data from ref. [56] obtained with a laser burning fluence $Pt \leq 7.5 \times 10^{-3}$ J cm^{-2}. The data follow $\Gamma_{hom} - \Gamma_0 \propto T^{1.3}$. The hole-burning results of ref. [55] follow a T^2-dependence, whereas the photon echo data lie on a $T^{1.3}$-curve [55].

probably higher than assumed. In addition, the experiments of ref. [55] were performed in a He-gas cryostat instead of an immersion cryostat [56].

The large discrepancy between the two NPHB results [55, 56] can be attributed to the same origin. In ref. [55] orders of magnitude more energy was used to burn the holes than to produce the photon echoes. Furthermore, the power broadened holes were probed with an instrument of insufficient resolution for detecting the true homogeneous linewidth. A detailed discussion on the dependence of Γ_{hole} on burning time and burning power for samples probed by *fluorescence excitation* and by *transmission* through the sample — either by scanning a narrow-band laser or a monochromator — has already been presented in Section 2.2. From this study it was concluded that: (1) the true value of Γ_{hom} at a given temperature is only reached at very low laser fluences; (2) for a given laser burning power (above a threshold value), holes detected in *transmission* through the sample are broader than holes detected in *fluorescence*. The holewidths only coincide at very short burning times [56].

The other two organic amorphous systems for which hole-burning and photon-echoes have yielded different results are resorufin in ethanol glass [191] and resorufin in glycerol [96]. Fayer and coworkers obtained linewidths for the former system that were a factor of about four smaller when measured by two-pulse

picosecond photon echoes than when measured by hole-burning [49, 191]. Moreover, the T-dependence of linewidths derived from photon-echoes in both samples
strongly deviated from $T^{1.3}$. From these results, it was concluded that the holes do
not reflect the homogeneous linewidth, but are broadened by additional processes
[96, 191].

Both systems have been re-investigated by means of hole-burning, resorufin in
ethanol between ~ 0.3 and 4.2 K [186], and resorufin in glycerol between ~ 0.3
and 22 K [86, 214]. For the first system it was found that the sample preparation,
i.e. its cooling rate, is of crucial importance in the determination of the value of
Γ_{hom} [186], since ethanol can exist in more than one solid phase [299]. The
samples used in ref. [186] were cooled in two ways, quickly to obtain phase I, and
slowly to obtain phase II. Thus, great care must be taken when ethanol is used as a
host matrix. This is not the case for resorufin in glycerol, which was cooled in the
same two ways as resorufin in ethanol, but only yielded one phase [86]. The top and
bottom curves of Fig. 64 represent the values of Γ_{hom} for resorufin in phases I and
II of ethanol as a function of temperature between ~ 0.3 and 4.2 K obtained by
hole-burning [186], whereas the middle (dashed) curve represents results obtained
by photon echoes [191]. It should be noticed that, at a given temperature, the hole-
burning values of Γ_{hom} for the two samples of resorufin in ethanol differ by a factor
of ~ 8. On the other hand, the photon echo linewidths [191] are ~ 4 times smaller
than the HB-linewidths for resorufin in phase I [49, 186], but ~ 2 times larger than
those for resorufin in phase II of ethanol [186]. From these experiments, it was

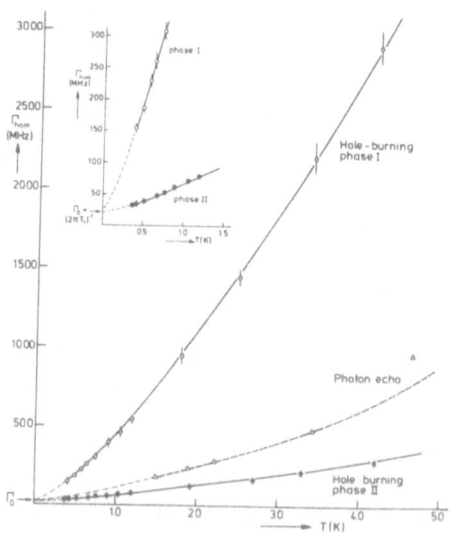

Fig. 64. Γ_{hom} versus T for resorufin in ethanol glass between 0.4 and 4.2 K. The highest and lowest
curves correspond to hole-burning data for phase I and II of ethanol, respectively [186]. The middle
curve (dashed) is derived from two-pulse photon echo data [191]. Insert: Detail of the data for
$T \to 0$. The value of Γ_{hom} extrapolates to the fluorescence lifetime-limited value of resorufin, $\Gamma_0 =$
$$(2\pi T_1)^{-1} \approx 20 \text{ MHz } [49, 186].$$

concluded in ref. [186] that (a) the holewidths measured for resorufin in the two phases of ethanol glass are determined by their homogeneous linewidths, and (b) in order to prove the contrary, photon echoes should be measured in both phases of resorufin in ethanol, and yield linewidths still smaller than the smallest ones measured by hole-burning in phase II. Thus, the discrepancies between hole-burning and photon echoes for resorufin in ethanol can probably be attributed to differences in the sample preparation.

It should be remarked that the values of Γ_{hom} for both resorufin in phases I and II of ethanol extrapolate to ~ 20 MHz (see insert of Fig. 64), which corresponds to the fluorescence lifetime-limited value of the guest, $\Gamma_0 = (2\pi T_1)^{-1}$, when $T \to 0$ [49]. At 0.4 K the value of Γ_{hom} for resorufin in phase II is very close to the value of Γ_0 [186].

A log-log plot of $\Gamma_{hom} - \Gamma_0$ versus temperature for resorufin in the two phases of ethanol is given in Fig. 65. Both curves follow a $T^{1.3}$ dependence over a temperature range of, at least, one decade between ≈ 0.3 and 4.2 K [186]. From a fit of Jackson and Silbey's model [245] (see Section 4.2.5) to the experimental

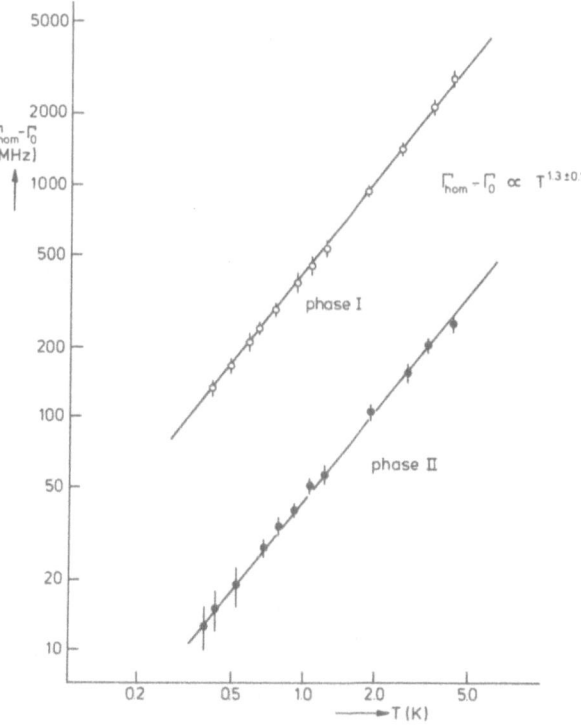

Fig. 65. Log-log plot of $\Gamma_{hom} - \Gamma_0$ versus T between 0.3 and 4.2 K for resorufin in the two phases of ethanol. The value of $\Gamma_{hom} - \Gamma_0$ for phase II is about 10 times smaller than for phase I. Both curves follow $\Gamma_{hom} - \Gamma_0 \propto T^{1.3}$ [186].

data, it appears that very low frequency modes ($E \sim 2$ cm^{-1}), in addition to two-level systems (TLS) [275], may be responsible for the optical dephasing in these systems. These very low frequency modes are consistent with those derived for other organic amorphous systems measured down to 0.3 K [88], but are inconsistent with a value of $E \simeq 19$ cm^{-1} deduced from photon echo data between 1.5 and 11.5 K [191]. In the latter case, deviations from a $T^{1.3}$ law were reported for $T \gtrsim 5$ K.

Hole-burning experiments performed at very low burning fluences on resorufin in glycerol between 0.3 and 22 K also yielded a $T^{1.3}$ dependence of $\Gamma_{hom} - \Gamma_0$ [86]. Very different results, however, have been obtained by Fayer et al. on the same system [96]. The differences are illustrated in Fig. 66. The linewidths derived from photon echoes and hole-burning between ~ 1.5 and 25 K (dashed curves) show a strong deviation from this power law [96]. The discrepancy between the hole-burning data of the two groups could be explained in terms of differences in burning fluences, Pt [86]. More than ten times larger Pt-values have been used in

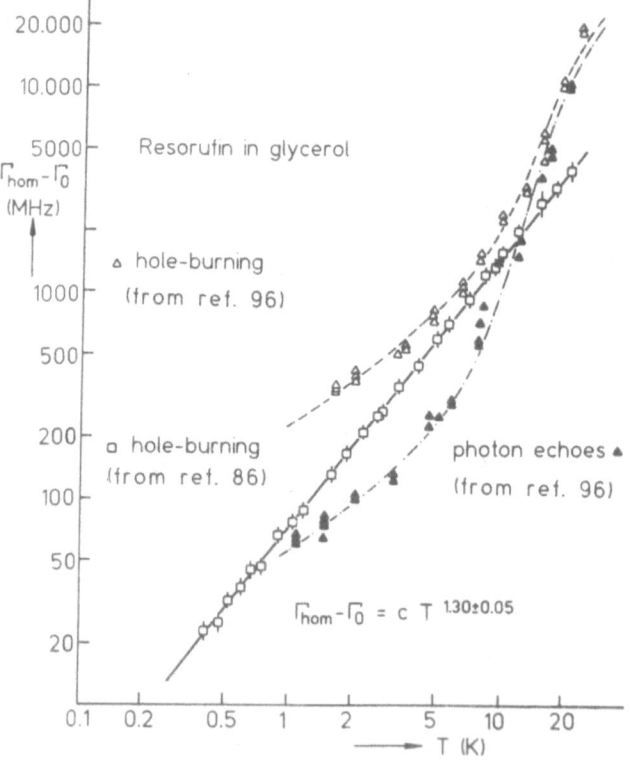

Fig. 66. Log ($\Gamma_{hom} - \Gamma_0$) versus log T between ≈ 0.3 and ≈ 25 K for resorufin in glycerol. The dashed curves have been obtained from hole-burning and photon echo data of ref. [96]. The full line, following a $T^{1.3}$ power law over two decades in T, is the result of hole-burning experiments from ref. [86]. All data were plotted with $\Gamma_0 = 20$ MHz.

ref. [96], as compared to ref. [86], over the temperature range measured. The log-log plot of $\frac{1}{2}\Gamma_{hole}$ versus burning fluence for resorufin in glycerol at 1.2 K, given in Fig. 67, illustrates this point. The curve obtained for holes probed in transmission is parallel to that for holes probed in fluorescence excitation. Both curves follow $\frac{1}{2}\Gamma_{hole} = c(Pt)^{0.23\pm0.01}$ over more than three decades in Pt-values [86]. Notice that a very similar dependence of $\frac{1}{2}\Gamma_{hole}$ on Pt has been obtained for pentacene in PMMA (see Fig. 7b). The two plots, however, differ by 4 orders of magnitude in the Pt-values. In fact, much less burning fluence was needed for resorufin than for pentacene to obtain the same hole depth. This, together with a few other arguments, led to the conclusion that the ionic dye molecules resorufin and cresyl violet undergo an *intermolecular PHB* reaction [86, 187], as opposed to a *NPHB* reaction typical of pentacene in PMMA [56]. A $\sim (Pt)^{0.23\pm0.01}$ dependence of Γ_{hole} has been observed for many organic glassy systems at 1.2 K, and more experiments are currently in progress to understand this behaviour [300].

The discrepancy between the hole-burning data of ref. [86] and the photon-echo data of ref. [96] has not been explained in a quantitative manner yet. It should be commented, however, that photon-echo experiments on resorufin in glycerol is not straightforward: the data have to be corrected for the simultaneous hole-burning by the laser pulses. In addition, the high peak powers needed to observe an echo may disturb the local thermal equilibrium around the guest molecules due to the low thermal conductivity of the glassy host. The authors of ref. [96] suggested that the extra broadening of their hole-burning data with respect to the photon-echo data, $\Gamma_{SD} = \Gamma_{HB} - \Gamma_{ph.echo}$, which in their case followed a power law dependence T^α with $\alpha \leq 1$, arises from "spectral diffusion". It is obvious from Fig. 66 that the difference between the linewidth determined by hole-burning from ref. [86] and that determined by photon echoes [96] does not obey such a power law. In fact, the data cross each other at $T \sim 12$ K and probably at $T < 1.1$ K. Thus, the

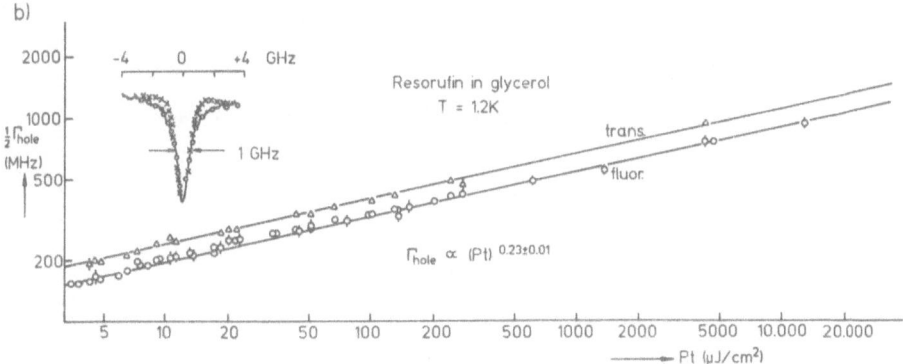

Fig. 67. Log-log plot of $\frac{1}{2}\Gamma_{hole}$ versus burning fluence, Pt, for resorufin in glycerol at 1.2 K. Holes were probed simultaneously in transmission (upper curve) and fluorescence excitation (lower curve). The data follow $\frac{1}{2}\Gamma_{hole} = c(Pt)^{0.23\pm0.01}$ over, at least, three orders of magnitude in Pt-values ($Pt \geq 3 \mu$ J cm^{-2}). Insert: Hole burnt with $Pt = 600 \mu$ J cm^{-2} at 1.2 K, $\frac{1}{2}\Gamma_{hole} = 500$ MHz. The hole shape is Lorentzian (circles). For comparison, a Gaussian curve (crosses) has also been drawn [86].

temperature dependence of "spectral diffusion" processes, if defined as the difference between holewidths and photon echo linewidths [96], cannot be described by a T^α-function, if one assumes that the photon echo data of ref. [96] are correct [86]. This leads us to the subject of the next section.

4.2.4. Long-Time "Spectral Diffusion" Processes

Slow relaxation processes ("spectral diffusion") in organic glasses have been observed by means of hole-burning on a time scale of hours, and even days [55, 90, 94, 301a, b, 302]. As already mentioned in the last section, the source of the discrepancies between optical linewidths obtained by HB and photon-echoes [55, 96, 191] has been attributed to such long-time "spectral diffusion" processes.

A systematic hole-burning study of pentacene in PMMA as a function of burning fluence and detection method has shown that this "spectral diffusion" effect is only observed under special experimental conditions, as will be discussed

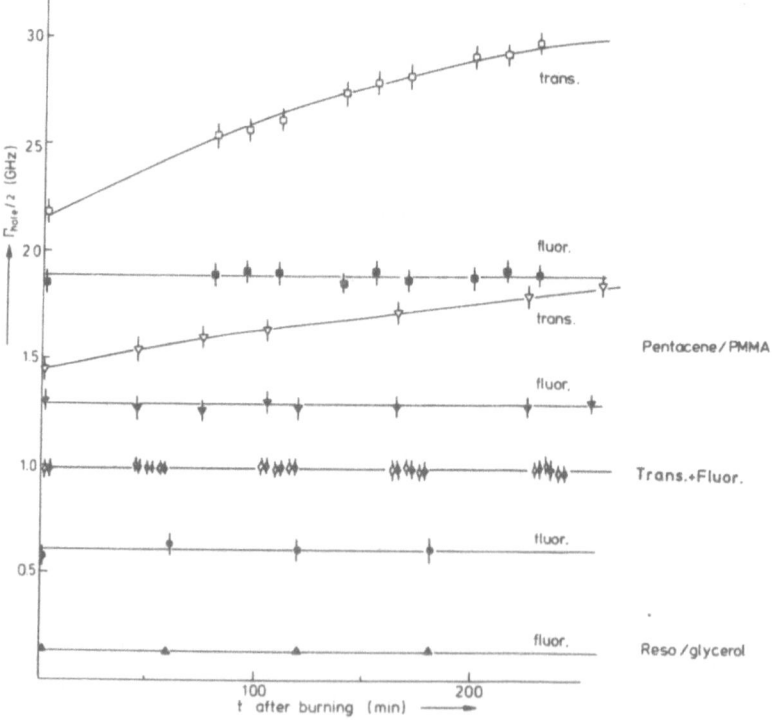

Fig. 68. $\tfrac{1}{2}\Gamma_{hole}$ as a function of time after burning, for resorufin in glycerol (lowest curve), and pentacene in PMMA (other curves) at 1.2 K. Holes were probed simultaneously in fluorescence (closed symbols) and in transmission (open symbols). Burning fluence (P: power, t: burning time) used for resorufin in glycerol: $Pt = 1.8 \ \mu\mathrm{J\,cm^{-2}}$; for pentacene in PMMA (bottom to top): $Pt = 0.035 \pm 0.005, 1.4 \pm 0.1, 6.0 \pm 0.5 \ \mathrm{J\,cm^{-2}}$ [56, 214].

below [56]. Similar conclusions were reached from HB-experiments on resorufin in glycerol [86, 214] and other organic amorphous systems [300]. Fig. 68 illustrates some of the results. The holewidth, $\frac{1}{2}\Gamma_{hole}$, as a function of time after burning has been plotted for pentacene in PMMA for different burning fluences, at 1.2 K. The time after burning is defined here as the time elapsed in the dark between burning a hole and probing it. The holes were probed simultaneously by fluorescence excitation, and in transmission through the sample. One clearly sees that the holes *detected in fluorescence* do *not broaden* on a time scale of hours, not even at large burning fluences, although the holewidth at $t = 0$ increases with burning fluence (see also Fig. 7a) [56]. A similar effect is observed for resorufin in glycerol (lowest horizontal line) [214]. These results indicate that, as long as the molecules are probed in fluorescence, i.e. they sit close to the surface of the sample and are in thermal contact with liquid helium, *no "spectral diffusion"* occurs, at least not on a time scale between a few seconds and 10^4 s [56, 86, 214].

However, when holes are *probed in transmission* through a rather thick amorphous sample (for example, pentacene in PMMA, $d = 2.2$ mm, OD $\approx$ 0.9 [56]), they *do broaden* on a time scale of minutes to hours (in this system, $\tau \approx$ 220 min) [55, 56], the effect being more pronounced for higher burning fluences (see upper curves in Fig. 68). At burning fluences below a certain threshold, the holes do not broaden which suggests that heat is responsible for this effect. In the bulk of the glass which has poor thermal conductivity, the heat of the laser probably induces strain causing slow structural relaxation of the host. Thus, the inhomogeneous linewidth is affected during this process, which in turn is reflected as a *long-time* (hours) *broadening* of the hole. This interpretation is supported by time-persistent thermal relaxation observed over hours and days in pure PMMA at very low temperatures [303], which was attributed to slowly relaxing TLS excitations in the amorphous material.

It should be remarked that *short-time "spectral diffusion"* processes occurring on a time scale between that determined by photon echoes (ps to ms) and that of conventional hole-burning experiments (s to min) have *not* been unambiguously *demonstrated* yet for an organic amorphous system (see also Section 4.2.3). Preliminary transient hole-burning experiments on H_2P in PE between 1.2 and 4.2 K performed on a time scale of ms [304] have yielded results that were identical to those obtained by conventional HB on a time scale of minutes [43, 86]. Furthermore, holes burnt in H_2P in PE at 1.2 K probed in fluorescence did not broaden as a function of time after burning, not even after many hours, just as for pentacene in PMMA [56] and resorufin in glycerol [86, 214]. These results suggest that *"spectral diffusion"* does *not take place* in a time span of seven orders of magnitude (from 10^{-3} to 10^4 s), at least, for the system H_2P in PE [86], if the temperature of the sample is kept constant.

It is worth to remark here that accumulated photon echo and hole-burning experiments on the inorganic amorphous system Pr^{3+} in a silicate glass between 1.6 and 20 K yielded identical results, despite the very different time scales of these two experiments of $\sim 10^{-4}$ s and 10^2 s, respectively [53]. Thus, no slow rearrangement of the glass network seems to contribute to the holewidth in this inorganic system. However, three-pulse stimulated ns-photon echo experiments on Nd^{3+} doped-silica optical fibers at $T \lesssim 0.1$ K [289b] revealed a non-exponential decay which was much faster than the fluorescence decay time, $T_1 = 490$ μs. This was interpreted as an indication of the presence of short-time "spectral diffusion" processes in this system.

Further, long-time "spectral diffusion" hole-burning experiments on organic glasses have been reported by Haarer and coworkers (for a review, see ref. [90]). Holes were burnt into the $S_1 \leftarrow S_0$ 0—0 transition of organic amorphous systems at rather *high laser powers* ($\Gamma_{hole} \simeq 1$ cm^{-1}) at $\sim 1.3 - 4$ K, and probed in *transmission* through the sample. Hole areas and widths were measured as a function of time after burning on a very long time period of up to 10^4 min. Experiments performed on quinizarin [90, 94, 301a, b, 302], tetracene [90, 302], and dimethyl-s-tetrazine [194] in ethanol/methanol mixed glasses have shown that both the hole area and the holewidth change logarithmically with time. The results were interpreted in terms of a model for slow structural relaxation processes in glasses based on the concept of spectral diffusion of a dilute spin system [305]. By assuming a wide distribution of TLS rates in the glass [275], a time dependent expression for the "spectral diffusion" holewidth was obtained [90, 301], similar to that known for the time dependence of the specific heat [306].

Recent temperature dependent hole-burning experiments on phthalocyanine in various polymers were combined with thermal cycling of holes between 4 and 25 K [68]. The results yielded reversible and irreversible line broadening contributions, which were found linear in temperature. The very small irreversible part of the linewidth, of about $1 - 20 \times 10^{-3}$ cm^{-1}/K, was attributed to "spectral diffusion" caused by TLS-flips occurring on a slow time scale. From the results, it was concluded that temperature-induced "spectral diffusion" at these temperatures is much smaller than pure dephasing contributions [68].

4.2.5. *Models for Optical Dephasing in Glasses*

Soon after the first experiments on optical dephasing in amorphous materials at low temperature were reported, a number of theories were proposed to explain the findings [52, 278, 281—283]. Since then, many more models have been developed to account for subsequent experimental results [55, 245, 289c, 307—314]. All these theories are based on the same physical concept, i.e. the optical transition of the impurity ion, or guest molecule, interacts with the two-level-systems (TLS) of the glass [275], and the total system interacts with the phonons of the bath. The interaction with the TLS is assumed to be different in the ground and electronically excited state. Furthermore, the density of the TLS-states is assumed to be nearly constant as a function of energy. The interacting systems and bath are schematically depicted in Fig. 69 (taken from ref. [277]). Since the TLS absorb and re-emit phonons of the bath, they modulate the energy levels of the electronic states causing dephasing of the optical transition. The various models differ from each other in the way the coupling impurity-TLS-phonons is introduced, and in the approximations made to evaluate the linewidth. Furthermore, the procedures

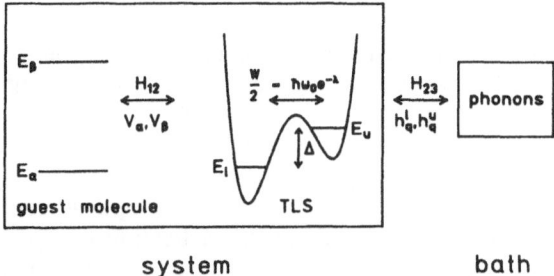

Fig. 69. The components of optical dephasing models for amorphous systems and their interactions (from ref. [277]). The system includes the optical transition and the TLS. The latter is represented in configuration space by a double well potential. The phonons are included in the bath.

used to average over the TLS parameters are different. Some of the models include additional processes in order to be able to explain the experimental results. Thus, small changes in the calculated temperature dependence of Γ_{hom} can easily be achieved by slightly varying either the approximations or the averaging procedure. In fact, each of the theories explains specific aspects of the experimental data, but unfortunately, many of the theoretical assumptions have not been verified experimentally. Moreover, none of the models has withstood a critical test and is, therefore, not generally valid. A review of the various theoretical models and the derivation of different temperature dependences of Γ_{hom}, starting from the Redfield equations, can be found in Silbey's chapter in this book [238]. For other reviews and discussions, the reader may consult the literature [33, 246, 277, 307, 313].

Since 1983 most of the theories have tried to provide an understanding of the $T^{1.3}$-temperature dependence of the homogeneous linewidth found for many organic amorphous systems between 0.3 and 20 K [1, 24, 45, 49, 56, 86], and for one inorganic glassy system [289]. In this section we will only discuss a few of these models, in particular, those for which the physical implications can be compared to experimental results.

One of the first models that explained a $T^{1+\alpha}$ (with $0 \leqq \alpha \leqq 1$) dependence of Γ_{hom} at low temperatures was reported by *Jackson and Silbey* [245]. It is based on a combination of two dephasing processes, one resulting from coupling of the optical transition to TLS, giving a contribution to the linewidth Γ_{TLS}, and the other arising from coupling to librational modes of the guest in the host, leading to Γ_{lib}, such that $\Gamma_{hom} \propto \Gamma_{TLS} + \Gamma_{lib}$. For the TLS contribution, it was assumed that $\Gamma_{TLS} \propto T^{1.0}$ at low T (based on a dipole-dipole interaction between guest and TLS, and taking a constant TLS-density of states [307]). For the term Γ_{lib}, a phonon dephasing contribution was taken similar to that in crystals (see Section 4.1.2), but with a Gaussian distribution of librational or localized modes [245]. The latter yields $\Gamma_{lib} \propto T^4$ at very low T, and $\Gamma_{lib} T^{1.0}$ at high T. Thus, at very low T, the TLS-mechanism with a linewidth linear in T takes over, and it is expected that Γ_{hom} extrapolates to the fluorescence lifetime-limited value, $(2\pi T_1)^{-1}$, when $T \rightarrow 0$. Since at higher T librational modes can be thermally populated, both dephasing processes will play a role, as long as the coupling of the guest to the TLS is small and low enough librational frequencies are involved. In this regime it was estimated that $\Gamma_{hom} \propto T^{1.3}$ between ~ 2 K and 20 K [245]. At still higher temperatures ($T \gtrsim 20$ K), one would expected again a linear T-dependence of Γ_{hom}. The model does not predict, however, the conditions and temperatures for which a cross-over from $T^{1.3}$ to $T^{1.0}$ should occur (neither for low nor for high T).

Jackson and Silbey's model has been fitted to optical dephasing results on organic amorphous systems. From these fits, activation energies of about ~ 2 to 2.5 cm^{-1} have been estimated for various systems between 0.3 and 4.2 K [86, 88, 186]. The fitted values of E, however, increase when the measured T-range is extended to higher temperatures. For example, for resorufin in glycerol, E increases from 2.2 cm^{-1} to ~ 6 cm^{-1} (if a single activation energy is assumed; and from 2 to ~ 4 cm^{-1}, if a Gaussian distribution of E-values is used) when T is extended from 4.2 to 22 K [86]. These results, together with the fact that no evidence has been found so far for localized modes with such low energies from site-selection spectra in glassy systems, makes one wonder whether an interpretation of optical dephasing in terms of low-frequency librational modes is physically meaningful [86].

Lyo and Orbach [310] have suggested that phonons may be replaced by "fractons" in amorphous systems. "Fractons" [315] are short length-scale excitations on a "fractal" structure [316]. They become important above a critical frequency, $\omega_c \propto 1/L$, for which the system obeys self-similarity. L is a characteristic length beyond which the systems is Euclidian. These authors postulated an electrostatic multipolar interaction to account for the coupling between impurity and TLS. After averaging in the same way as previously done by Lyo [307], they found that $\Gamma_{hom} \propto T^{1+\mu+\bar{\bar{d}}-3\bar{\bar{d}}/s}$, where s depends on the type of coupling [310]. $\bar{\bar{d}}$ is the "fracton" or spectral dimensionality [315], which yields an anomalous density of phonon states, $N(\omega) \propto \omega^{\bar{\bar{d}}-1}$; μ is the exponent in the energy dependent density of TLS-states, $\rho(E) \propto E^\mu$, with $0 \leq \mu \leq 1$. If one assumed that polymers are topologically described by percolating networks [317], and if one accepts the conjecture of Alexander and Orbach that the latter are fractals with $\bar{\bar{d}} = 4/3$ [315], one obtains $\Gamma_{hom} \propto T^{1.3}$ under either of the following conditions: with a dipole-quadrupole interaction ($s = 4$) and a constant TLS-density of states ($\mu = 0$), or with a dipole-dipole interaction ($s = 3$) and $\mu = 0.3$.

One of the critical questions about the validity of this model is whether organic molecules incorporated in organic glasses and polymers [24], Nd^{3+} in fused silica [289], organic molecules embedded as guests in the bulk of amorphous silica [296], and organic molecules adsorbed on the surface of porous silica [297], which all show a $T^{1.3}$ dependence of Γ_{hom}, may be described by percolating networks. A second unclear point is the predicted cross-over from a $T^{1.3}$ to a $T^{1.0}$ dependence at low temperature [310]. Under what conditions should this transition precisely occur? On the other hand, it still remains to be proven whether a dipole-quadrupole or a dipole-dipole coupling dominates in the different amorphous systems studied. Finally, if "fracton" excitations were relevant at 0.1 K in fused silica, the "fractal length-scale" below which this glass obeys self-similarity would be very long ($>$ 2000 Å [310]). Some physical properties of silicate glass are probably in contradiction with this statement. Despite these problems, the "fracton" model [310] is elegant and appealing because of its simplicity.

Another theory, developed by *Hunklinger and Schmidt* [309], is based on "spectral diffusion" of the electronic transition due to elastic dipolar coupling to "flipping" TLS [281, 306]. As a consequence of this "flipping", the electronic transition is modulated, and the homogeneous linewidth is expected to broaden as a function of temperature and time. The time dependence should arise from the large distribution of TLS in the glass, which may have relaxation times over many orders of magnitude. By using an approach similar to that of ref. [306], and assuming a constant density of TLS-states, Hunklinger and Schmidt obtained an expression for the linewidth which depends linearly on T, and logarithmically on T and time. From the numerical evaluation of Γ_{hom}, an approximately T^α dependence of Γ_{hom} results, where α depends on the dominant relaxation mechanism: for a direct process, Γ_{hom} was found to be proportional to $T^{1.8}$, whereas for a Raman process, $\Gamma_{hom} \propto T^{1.3}$, at least between 0.4 and 5 K [309].

It should be remarked that in none of the hole-burning experiments on organic glassy systems for which $\Gamma_{hom} - \Gamma_0 \propto T^{1.3}$ between 0.3 and 20 K was found, a change of the holewidth was observed with time after burning, at a given temperature [56, 86, 214]. The holes in these experiments were probed by fluorescence excitation on a time scale of minutes to hours after burning (see Section 4.2.4). Furthermore, accumulated photon echo experiments on Pr^{3+} in silicate glass between 2 and 20 K on a time scale of µs yielded the same homogeneous widths as hole-burning experiments on a time scale of seconds to minutes [53] (see also Section 4.2.4). These results suggest that "spectral diffusion" processes, if they take place, would only occur at times shorter than µs in this system.

Two types of photon echo experiments on glasses have been interpreted in terms of theoretical models developed for this purpose. One of them tries to explain the two pulse ns-photon echo results obtained for Nd^{3+} in fused silica between 0.1 and 1 K [289a, b]. The model, which was worked out by *Huber et al.* [289c], is also based on "spectral diffusion of TLS" [281, 306]. It assumes an elastic dipole-dipole interaction between impurity and TLS, similarly as done in ref. [309], but with a density of TLS-states that varies with energy as $\rho(E) \propto E^\mu$, where $\mu = 0.3$ [318]. The calculation yields $\Gamma_{hom} \propto T^{1+\mu} = T^{1.3}$, in agreement with the experimental results. Unfortunately, the experiments have not been performed above 1 K, since a cross-over from $T^{1.3}$ to T^2 is expected, and should be measured. This argument is based on accumulated echo results for Nd^{3+} in ED-2 glass between 10 and 80 K [288], and FLN experiments on Nd^{3+} in phosphate glass between 80 and 220 K [287], which both yielded a T^2-dependence of Γ_{hom}.

The second model, developed to explain accumulated photon echo results on pentacene in PMMA was reported by *Molenkamp and Wiersma* [55]. It is based on optical Redfield theory [249, 250]. The approach is similar to that of ref. [307], but instead of postulating a dipole-quadrupole interaction between impurity and

TLS with a constant TLS-density of states as in ref. [307], an electrostatic dipole-dipole interaction in combination with a density of TLS-states varying as $\rho(E) \propto E^{0.3}$ [289c, 318] was assumed. This model also yields $\Gamma_{\text{hom}} - \Gamma_0 \propto T^{1.3}$.

Notice that in the two latter models, a $T^{1.3}$-dependence is only obtained by assuming a density of TLS-states that varies as E^μ, with $\mu = 0.3$. It is difficult, however, to justify this assumption, since there is neither a convincing proof that $\rho(E) = E^{0.3}$ is valid for every glass, nor a real physical need for this assumption for the interpretation of the specific heat experiments [318, 319]. Furthermore, in both models a dipole-dipole interaction had to be assumed in order to explain the experimentally found exponential echo decay. In principle, if a glass had a large distribution of TLS-relaxation times, one would neither expect an exponential decay of the echo signal nor a Lorentzian line shape, contrary to what has been observed so far.

Pietronero [320] has developed a model in which he assumed that TLS have hierarchical constraints [319]. As a consequence of this assumption, the available degrees of freedom for low temperature properties of glasses increase with T with a density $\rho(T) = \rho_0(1 + \rho_0 BT)$, where B is a constant and ρ_0 is the density of TLS in the limit $T \to 0$. With the further assumption of a dipole-dipole interaction [307], the linewidth becomes $\Gamma_{\text{hom}} \propto \rho T \simeq \rho_0 T + B\rho_0^2 T^2$. This T-dependence may explain the experimentally observed T^ν, with ν ranging from 1.3 for organic glasses to 2 for inorganic glasses. A critical test of this model could only be done if specific heat data were available for the organic glasses for which Γ_{hom} has been determined.

A $T^{1.3}$-dependence of Γ_{hom} can also be obtained with other approaches. For example, *Maynard* [312] has taken a dipole-quadrupole interaction between guest molecules and TLS, and assumed an elastic quadrupole associated with a librational mode of the guest [58, 60, 236, 237, 245]. With the further assumption of a density of thermally excited TLS equal to $\rho_0 kT$, he obtained $\Gamma_{\text{hom}} \propto (\rho_0 kT)^{4/3}$ which, for constant ρ_0, yields $\Gamma_{\text{hom}} \propto T^{1.3}$.

Osadko [313], on the other hand, used a non-perturbative theory, and suggested that the TLS should be included in the bath and treated in a similar way as the phonons. By assuming an interaction between the impurity and TLS that "flips" two TLS simultaneously, like in a Raman process, he could also obtain a $T^{1.3}$-dependence of Γ_{hom}.

Finally, *Reineker and Kassner* [277, 311, 314] have made assumptions on the specific form of the microscopic Hamiltonian, which includes the interactions between the optical transition and many TLS, and between these TLS and the phonons of the bath. By using a Mori projection operator method they derived the equation of motion for the correlation functions that describe the line shape. From a detailed numerical averaging procedure, they have shown that there are several ways to explain the $T^{1.3}$-dependence of the observed linewidth within this model [314], which depends on the details of the averaging procedure. It can also lead to changes in the T-dependence. In fact, the disadvantage of this model is that there is no analytical expression for the linewidth, but only a numerical approach [246].

All the theories for dephasing in glasses presented here are based on the TLS-model, which was developed to explain specific heat properties of glasses at low temperature [275]. The assumptions made to calculate the temperature dependence of Γ_{hom} in each of these models have not been tested, however, by an independent technique. Furthermore, we have seen above that the approximations and averaging methods may change the T-dependence of Γ_{hom}. The models, in fact, do not provide parameters from which Γ_{hom} or T can be calculated "a priori", in order to make quantitative predictions of the experimental results. It is therefore recommended that the fitting parameters obtained from optical linewidth experiments should be tested by other experimental methods, like specific heat or acoustic experiments. At this moment, it is difficult to decide which model is the correct one, if at all, and more work is needed.

4.2.6. *Optical Dephasing in Semi-Crystalline Polymers*

An experiment aimed at finding out whether any of the dephasing models for glasses discussed in the previous section really applies, has recently been reported [88]. Organic molecules in a semi-crystalline polyethylene (PE) host were chosen for this purpose. If librational or localized modes, as predicted by Jackson and Silbey [245], were important for optical dephasing in glasses, one would expect the temperature dependence of $\Gamma_{hom} - \Gamma_0$ in *semi-crystalline* systems to be steeper than in *amorphous* systems and, conceivably, to approach the exponential function observed for *crystalline* materials.

Dimethyl-s-tetrazine (DMST) diffused into previously molten, and subsequently solidified, *semi-crystalline* PE shows a much steeper dependence of $\Gamma_{hom} - \Gamma_0$ than $T^{1.3}$ between 0.3 and 4.2 K. The slope of the curve in a log-log plot of $\Gamma_{hom} - \Gamma_0$ versus T increases with the degree of crystallinity of the host [88]. This is illustrated in Fig. 70 for a quickly cooled PE sample, which yields a $T^{1.7}$ dependence between 1.2 and 4.2 K, and for a slowly cooled sample which shows a $T^{2.3}$ dependence in the same T-range. By decreasing the temperature, the T-dependence increases to $\sim T^4 - T^5$ between 0.7 and 1 K, approaching the behaviour of crystalline systems (see dashed line). Thus, the more ordered the direct vicinity of the excited guest molecule, the steeper the T-dependence in the low temperature regime. This argument is supported by the fact that DMST diffused into amorphous PMMA shows again a $T^{1.3}$ dependence, as previously found in other organic glassy systems [88]. These results suggest that HB measures the local guest-host interaction, that is, the degree of crystallinity in the direct environment of the guest molecule, in contrast to other techniques, which measure properties of the bulk of the host independent of the presence of the guest.

The experimental data obtained for amophous [86, 186] and semi-crystalline systems [88] were fitted with three of the theoretical models reported in the literature [237, 245, 320]. The purpose of this study was to estimate the relative influence of librational or localized modes, as compared to TLS modes, on the dephasing process. As regards the three theories examined, neither of them seems to offer a simple physical explanation of the $T^{1.3}$-dependence of $\Gamma_{hom} - \Gamma_0$. It appears, that low frequency localized modes [245] may contribute to the dephasing

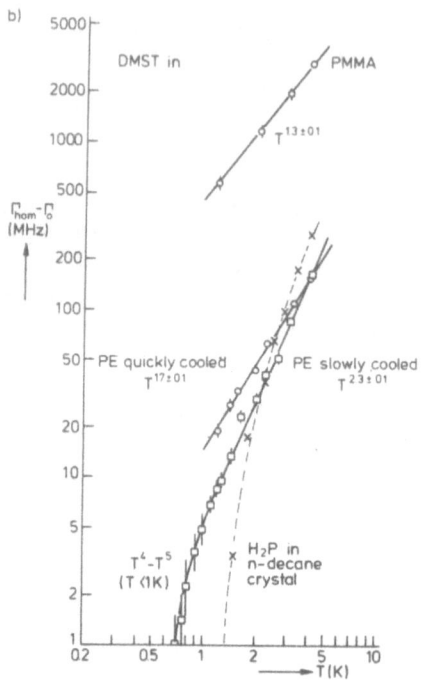

Fig. 70. Log-log plot of $\Gamma_{hom} - \Gamma_0$ vs T between 0.3 and 4.2 K. Top: Dimethyl-s-tetrazine (DMST) diffused and/or dissolved in amorphous PMMA. Notice that $\Gamma_{hom} - \Gamma_0 \propto T^{1.3 \pm 0.1}$. Bottom: DMST diffused into semi-crystalline polyethylene (PE). The quickly cooled sample follows $\Gamma_{hom} - \Gamma_0 \propto T^{1.7 \pm 0.1}$ between 1.2 and 4.2 K, whereas the slowly cooled sample follows $\Gamma_{hom} - \Gamma_0 \propto T^{2.3 \pm 0.1}$ for $T > 1$, and $\propto T^4 - T^5$ for $T < 1$ K [88]. Data of H_2P in an n-decane crystal are plotted for comparison [60].

in glasses but, unfortunately, there is no independent spectroscopic evidence for such librations yet, neither from site-selection not from hole-burning experiments [86, 88, 186]. From the results of ref. [88], it was concluded that TLS [274, 275], if present at all in semi-crystalline solids, are much more diluted than in amorphous materials. For a quantitative comparison between models and experiments, see refs. [86, 88, 186].

4.2.7. *Optical Dephasing in Mixed Organic-Inorganic Systems*

In order to shed more light on the excited state dynamics of amorphous materials, two classes of *mixed organic-inorganic* systems have recently been studied: *organic* glasses doped with Eu^{3+} [214a, b], and *organic* molecules as guests in *inorganic* silica glass [296, 297]. Two types of experiments were performed in the latter case: in the *bulk*, i.e. the organic guest molecules were built into the inorganic host before the amorphous silica glass was formed [296], and *adsorbed on the surface* of porous silica [296, 297].

We will first discuss the results obtained for amorphous silica-doped organic molecules. Samples of chlorin and oxazine-4 perchlorate were prepared by the sol-gel method and studied by hole-burning between 1.7 and 5.7 K [296]. The homogeneous linewidths, of the order of 500 MHz at 1.7 K, were similar to those observed in pure organic glassy systems. They also broadened as a function of temperature with a $T^{1.3}$-dependence. Preliminary experiments by the same group on these dye molecules *adsorbed* on porous silica at 1.7 K yielded upper limits for the widths, which were an order of magnitude larger than those in amorphous silica [296]. From the results, it was concluded that the geometric structure of the environment of the guest seems to be more important in the determination of the linewidth than the chemical composition of the amorphous host.

Recent hole-burning experiments on the $S_1 \leftarrow S_0$ 0—0 transition of resorufin *adsorbed* on porous silica (Vycor glass) between 0.3 and 4.2 K have shown that Γ_{hom} is, in fact, about one order of magnitude larger than for resorufin in polymers [297]. Moreover, Γ_{hom} extrapolates to the fluorescence lifetime-limited value of resorufin, $\Gamma_0 = (2\pi T_1)^{-1}$, but at 0.3 K is still far from this value. In order to reach Γ_0, temperatures down to at least 0.05 K would be necessary. Furthermore, it was found that also this mixed system follows $\Gamma_{\text{hom}} - \Gamma_0 \propto T^{1.3}$ [297]. This is shown in the log-log plot of $\Gamma_{\text{hom}} - \Gamma_0$ versus T of Fig. 71 for resorufin adsorbed on Vycor glass between 0.3 and 4.2 K, and for DMST on Vycor glass at 1.2 K. Notice

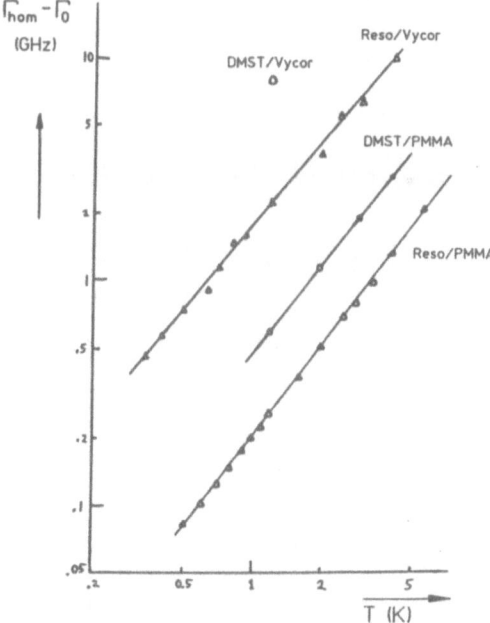

Fig. 71. Log ($\Gamma_{\text{hom}} - \Gamma_0$) versus log T for resorufin adsorbed on the surface of porous glass (Vycor) between 0.3 and 4.2 K, and for DMST adsorbed on Vycor at 1.2 K [297]. The values of $\Gamma_{\text{hom}} - \Gamma_0$ for both guests are about a factor of 10 larger than when incorporated in the bulk of PMMA. All samples follow $\Gamma_{\text{hom}} - \Gamma_0 \propto T^{1.3}$.

that the values of $\Gamma_{hom} - \Gamma_0$ for both guest molecules on porous silica are about a factor of 10 larger than in the bulk of PMMA, but follow the same $T^{1.3}$ law. Holes burnt in resorufin adsorbed on porous silica at 1.2 K do not broaden as a function of time after burning, when measured on a time scale of hours, and probed by fluorescence excitation [297]. Thus, no long-time "spectral-diffusion" processes seem to occur for organic molecules adsorbed on the surface of porous silica, as long as the molecules are in thermal contact with liquid helium. These results suggest that the substantially broader homogeneous linewidths are probably due to a higher degree of freedom of the guest, as compared to molecules incorporated in the *bulk* of the host [297].

Let us turn now to hole-burning experiments performed on Eu^{3+} as guest in *organic glasses* between 0.3 and 4.2 K. The results are in strong contrast to those found for either *pure organic* or *inorganic amorphous* systems. First, the inhomogeneous linewidths of Eu^{3+} in ethanol glass, water, ether, PMMA and polyvinyl-alcohol (PVA) are very small, ranging from a few GHz to ~ 20 cm^{-1} [214a, b], as compared to *pure* organic and inorganic glassy systems which have widths of a few hundred cm^{-1}. Fig. 72 shows excitation spectra of the $S_1 \leftarrow S_0$ 0—0 transition region of H$_2$P, and the $^5D_0 \leftarrow {}^7F_0$ region of Eu^{3+} both in ethanol glass at 1.2 K. Notice the large difference of almost two order of magnitude in the values of Γ_{hom}.

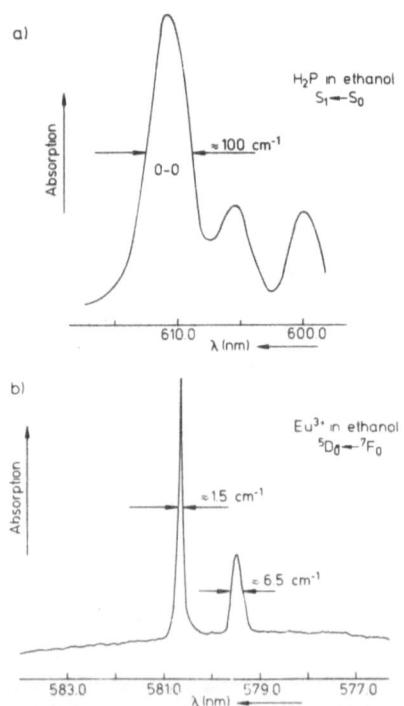

Fig. 72. Excitation spectra of (a) the $S_1 \leftarrow S_0$ 0—0 transition of H$_2$P, and (b) the $^5D_0 \leftarrow {}^7F_0$ transition of Eu^{3+} in ethanol glass at 1.2 K. Notice the large difference of almost two orders of magnitude between the inhomogeneous linewidths, Γ_{inh}, of the two spectra [214a, b].

Very narrow holes of a few MHz at liquid helium temperature could be burnt in the $^5D_0 \leftarrow {}^7F_0$ transition of Eu^{3+} in these glasses [214a, b]. The hole-burning mechanism here is due to population redistribution between the various nuclear quadrupole levels of the $^7F_0$ ground state of Eu^{3+} [53, 172a], whereas the lifetime of the holes, of the order of a few seconds to hours depending on the host, is limited by the spin-lattice relaxation between hyperfine levels [53]. In some hosts, like PMMA, the holewidths were limited by the laser jitter of ~ 2 MHz over the whole T-regime measured, whereas in others, residual linewidths, Γ_0, for $T \rightarrow 0$, were found of the order of $\Gamma_0 = \frac{1}{2}\Gamma_{hole} \simeq 2{-}10$ MHz [214b]. These widths are about three orders of magnitude larger than the value expected from the fluorescence decay time of the $^5D_0$ state of Eu^{3+}, and too large to be caused by coupling to fluctuating surrounding magnetic nuclei. In addition, distinct side-holes and anti-holes were observed symmetrically distributed at distances between ~ 20 and 300 MHz from the original hole. Such hole patterns are illustrated in Fig. 73 for Eu^{3+} in PVA and Eu^{3+} cryptate at 1.2 K. Multiple side-holes and anti-holes have been

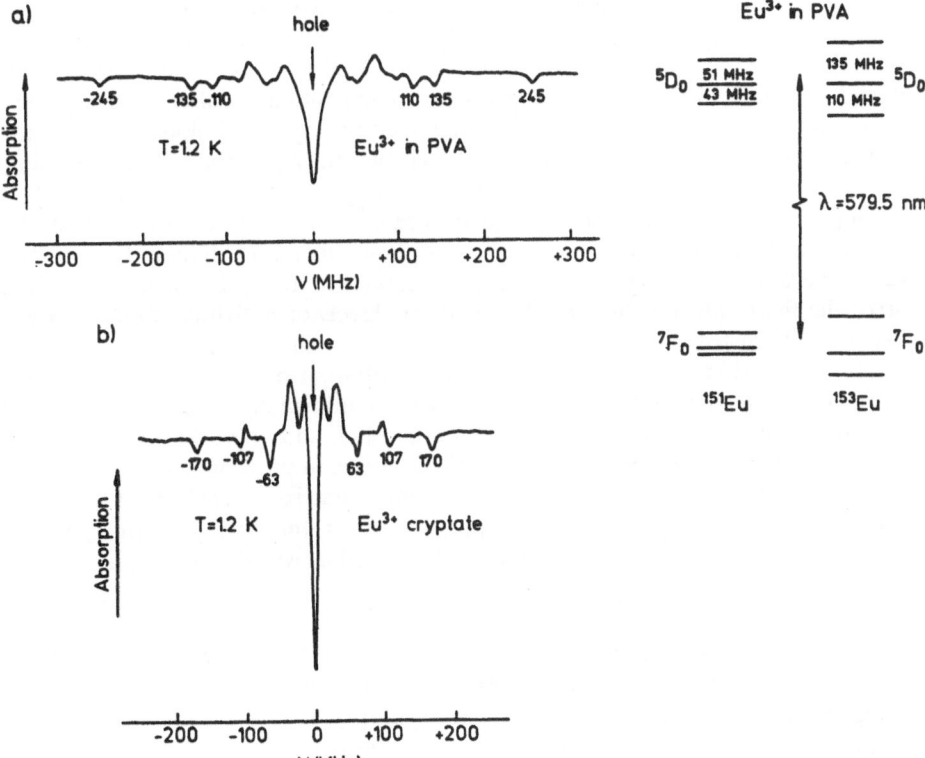

Fig. 73. Holes burnt in the $^5D_0 \leftarrow {}^7F_0$ transition of (a) Eu^{3+} in PVA at 579.5 nm, and (b) Eu^{3+} cryptate at 580.1 nm, both at 1.2 K. The symmetric pattern of side-holes and anti-holes corresponds to the nuclear quadrupole splittings in the ground ($^7F_0$) and excited ($^5D_0$) states of Eu^{3+}. The hyperfine splittings of the $^5D_0$-state were determined from the positions of the side-holes. Insert: Energy level scheme of the $^5D_0 \leftrightarrow {}^7F_0$ transition for the two isotopes of Eu^{3+} in PVA [214b].

reported for crystalline inorganic systems like EuP_5O_{15} [172a], $NaEu(WO_4)_2$ [321], and Eu^{3+} in $YAlO_3$ [322], $CaSO_4$ [323], and CaF_2 [323], but not, to our knowledge, in any type of glassy host.

Surprisingly, the temperature dependence of the homogeneous linewidth between 0.3 and 4.2 K was found to be exponential, instead of a $T^{1.3}$ power law, with activation energies between 1.4 and 5.6 cm^{-1} which depend on the coordination complex studied [214b]. These results indicate that the Eu^{3+} ion is surrounded in its first coordination sphere by *"crystalline-like"* host molecules which provide an ordered environment in the direct neighbourhood of the ion. These experiments prove once more that hole-burning is a very sensitive method to detect the structure of a material in the immediate vicinity of the excited guest molecule.

4.2.8. *External Field Effects on Amorphous Systems*

The observation of the effects of an external electric or magnetic field on the spectra of complex molecules embedded in solid hosts is generally limited by the large widths of the inhomogeneously broadened absorption bands, which amount to $\Gamma_{inh} \approx 1 - 10^2$ $cm^{-1} \approx 10^4 - 10^6$ MHz, depending on the nature of the host (crystalline or amorphous). As we have seen in Section 4.1.4, the frequency shift induced by the field is usually many orders of magnitude smaller than Γ_{inh}, and thus it cannot be detected by conventional spectroscopic techniques. Such small frequency shifts, however, can be made visible through optical hole-burning, because the hole represents a sharp frequency marker in the absorption band [42].

Complex molecules are not easily incorporated in crystalline hosts, but can be dissolved without difficulty in many amorphous materials like polymers and glasses. Because the molecules in this situation are randomly oriented in the matrix, the field-induced shifts of the electronic levels of individual molecules will be statistically distributed and give rise, not only to a shift or splitting of the hole, but also to an additional broadening and redistribution of its shape. By analyzing the profile of such a hole, it is possible to get information on the interaction between the guest-host system and the external field, and on various properties of the excited states of the guest molecules, like changes in the magnetic susceptibility (Zeeman effect), changes in the intrinsic and matrix-induced electric dipole moment (Stark effect), and physical parameters of the host like the polymer compressibility (pressure effect). In what follows, we will discuss these external field effects in more detail.

a. *Zeeman effect.* In Section 4.1.4a, we have presented the Zeeman effect on the organic molecules free-base porphin (H_2P) [42], free-base chlorin (H_2Ch) and its photoproduct (H_2Ch^*) [115, 116] in *single crystals* of *n*-alkanes. It was shown that by burning holes in zero field in the $S_1 \leftarrow S_0$ 0—0 transition at 4.2 K, Zeeman shifts of about 40 to 250 MHz/T^2 were measured in a magnetic field parallel to the out-of-plane molecular axis. From these shifts, values of $\chi^{z'z'}$, the change in magnetic susceptibility of the molecule on excitation to its lowest excited singlet state, and the matrix element $\Lambda = |\langle S_1 | L_z | S_2 \rangle|$ of the effective orbital angular momentum between the two lowest excited singlet states S_1 and S_2, could be obtained. With the magnetic field perpendicular to the molecular z-axis no Zeeman shift was observed.

Very few Zeeman experiments on *amorphous* systems have been reported so far. By means of site-selection spectroscopy, the $T_1 \leftarrow S_0$ 0—0 transition of coronene and 5-bromo-acenaphthene in ethanol and butyl-bromide at 4—6 K in magnetic fields up to $5\,T$ were investigated [324]. From the intensity distribution in the Zeeman quintet as a function of field strength, the zero-field radiative deactivation constants of the triplet spin sub-levels and the spin-lattice relaxation rates were determined.

In another study, very strong pulsed magnetic fields up to $40\,T$ with pulse durations of 2 ms were used in combination with hole-burning to investigate the Zeeman effect of the $S_1 \leftarrow S_0$ 0—0 transition of chlorin (H_2Ch) in a polystyrene film at 4.2 K [325a]. The results were preliminary, because the instrumental bandwidth (~ 1.5 cm^{-1}) was larger than the holewidths (~ 0.8 cm^{-1}). From the hole shift of about 1 cm^{-1} at $40\,T$, the value of the change in magnetic susceptibility of the molecules on excitation to the S_1 state was estimated to be approximately the same as previously found for H_2Ch in an n-octane single crystal [115]. Very similar experiments have been described in ref. [325b].

Zeeman shifts of ~ 100 MHz to a few GHz, and very accurate hole shapes in stationary fields of up to $7.5\,T$ have recently been reported for the $S_1 \leftarrow S_0$ 0—0 transition of free-base porphin (H_2P) as guest in *amorphous* polyethylene (PE) at 4.2 K [326]. Holes were burnt in zero field in two identical samples in two cryostats, of which one had a split-coil superconducting magnet and the other one was used as reference. In a similar way as done in ref. [42], a magnetic field was subsequently applied to the sample in the Zeeman cryostat, and the holes were probed simultaneously in both samples before and after applying the field. Hole shapes and frequency shifts could be accurately measured in this manner with respect to the hole burnt in the reference sample (See Fig. 74, left) [326]. The hole shifts were reversible with the magnetic field, and reorientation of the sample did not have any effect on the hole shape, which proved that the sample was isotropic.

Two approaches were used in ref. [326] to analyze the hole shape as a function of frequency and magnetic field for various holewidths. The first one is very simple, and gives an intuitive insight into the behaviour of randomly oriented molecules in a magnetic field. It yields a rather good value for the change in magnetic susceptibility, $\chi^{z'z'}$, although it is physically not quite correct. The second approach is more realistic, but also mathematically more cumbersome. It takes into account the anisotropy created in the molecular distribution by the polarized laser light in the burning and probing processes. The best fits to the experimental results for the two models yielded values of $\chi^{z'z'}$ equal to within 2%.

In order to calculate the hole shape, various assumptions were made [326]: first, a quadratic Zeeman shift dependence of the magnetic field, as previously observed for crystalline hosts [42] (see Section 4.1.4a). Further, an isotropic distribution of the excited molecules was assumed for the first approach. Taking into account that in the hole-burning process the laser excitation light is used to burn a hole and

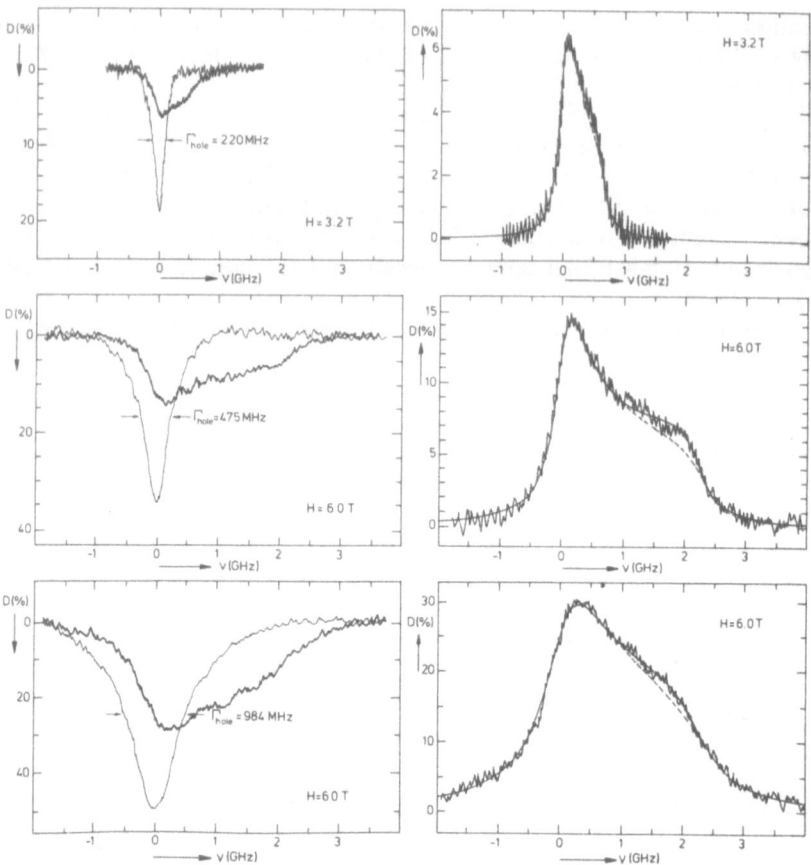

Fig. 74. Left: Effect of a magnetic field on holes burnt in zero field (thin curves) at 6145 Å in the $S_1 \leftarrow S_0$ 0–0 transition of free-base porphin (H₂P) in polyethylene at 4.2 K, for two different fields. Upper part: a field of $H = 3.2$ T is applied to a hole of 220 MHz width. Middle and lower parts: a field of $H = 6.0$ T is applied to holes with a width of 475 MHz and 984 MHz, respectively. Right: Fits of the calculated hole shapes (see text) to the experimental results for the same three holes as on the left. The best fit with eq. (12) (dashed lines) is obtained for $\frac{1}{2}\chi^{z'z'} = 61.9$ MHz/T^2, with eq. (14) (full lines) for $\frac{1}{2}\chi^{z'z'} = 60.6 \pm 1.4$ MHz/T^2 and $p = 0.78 \pm 0.02$ (from ref. [326]).

subsequently to probe it in a magnetic field, the relative number of guest molecules in zero field $N(\nu', t)\,\mathrm{d}\nu'$ in the frequency interval $(\nu', \nu' + \mathrm{d}\nu')$, is then given by [19, 27, 327, 328]

$$N(\nu', t)\,\mathrm{d}\nu' = N_0 \Phi \sigma \, \frac{It}{h\nu_b} \, g(\nu_b - \nu')\,\mathrm{d}\nu'. \tag{10}$$

Eq. (10) is valid for low burning fluences, It, in which I is the laser intensity and t is the burning time. N_0 is the total number of molecules, Φ the burning quantum yield, σ the total absorption cross section, $h\nu_b$ the energy of the absorbed photon, and $g(\nu_b - \nu')$ is a normalized the shape function of the $S_1 \leftarrow S_0$ transition, assumed to be Lorentzian. The polarization of the laser light was not taken into consideration in eq. (10). The hole shape function D, or hole depth, in a magnetic field is then given by the convolution of eq. (10) and the homogeneous absorption line shifted in the field: $\sigma g(\nu' + \Delta E' - \nu_p)$, with ν_p the probing frequency. Thus,

$$D(\nu) = \frac{1}{2} \int_0^\pi d\theta \sin \theta \int_{-\infty}^\infty N(\nu', t)\sigma g(\nu' + \Delta E' - \nu_p)\, d\nu'. \tag{11}$$

where the integration is over all frequencies $d\nu'$ and over the angular distribution function $dn(\theta)$. Since eq. (11) represents the convolution of two Lorentzian line shape functions, it yields another Lorentzian function $g(\nu + \Delta E')$ of width $\Gamma_{\text{hole}} = 2\Gamma_{\text{hom}}$, with $\nu = \nu_b - \nu_p$:

$$D(\nu) = \frac{1}{2} N_0 \Phi \sigma^2 \frac{It}{h\nu_b} \frac{1}{\sqrt{KH^2}} \frac{\Gamma_{\text{hole}}}{2\pi} \times$$

$$\times \int_\nu^{\nu + KH^2} \frac{1}{\sqrt{\nu' - \nu}} \frac{1}{(\nu'^2 + \Gamma_{\text{hole}}^2/4)}\, d\nu'. \tag{12}$$

In eq. (12), ν represents the frequency of the hole in zero field and $K = \frac{1}{2}\chi^{z'z'}$ is the only fitting parameter. This integral was numerically evaluated by Simpson's method. Fig. 75 shows a plot of the calculated hole depth D as a function of frequency ν, for a fixed value of the magnetic field, $H = 6.0T$, and for different zero-field holewidths. The areas under the curves have been normalized. Notice that all curves in Fig. 75 cross at a value of $\nu = 2.2$ GHz, which corresponds to the frequency shift observed at $H = 6.0T$ for H_2P oriented in a single crystal of n-octane with the molecular out-of-plane axis parallel to the field (in this case,

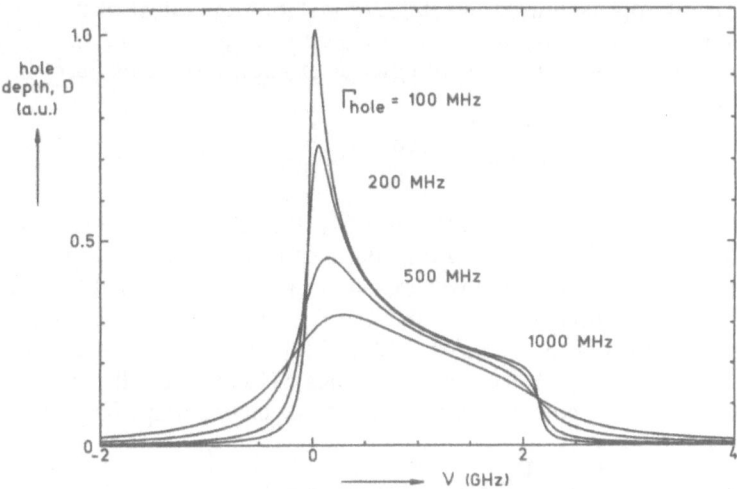

Fig. 75. Calculated magnetic field effect (see eq. (12)) on a hole burnt in zero field at $H = 6.0$ T, $\frac{1}{2}\chi^{z'z'} = 60$ MHz/T^2. All curves intersect at the cut-off value of the distribution function. The maximum intensity of the curve with $\Gamma_{\text{hole}} = 100$ MHz was arbitrarily taken equal to 1. All areas are normalized (from ref [326])

$\theta = 0$ and $\Delta E = KH^2$) [42]. A 3-dimensional plot of the calculated hole profile D as a function of the magnetic field strength H, for a fixed value of $\Gamma_{\text{hole}} = 500$ MHz, is reproduced in Fig. 76.

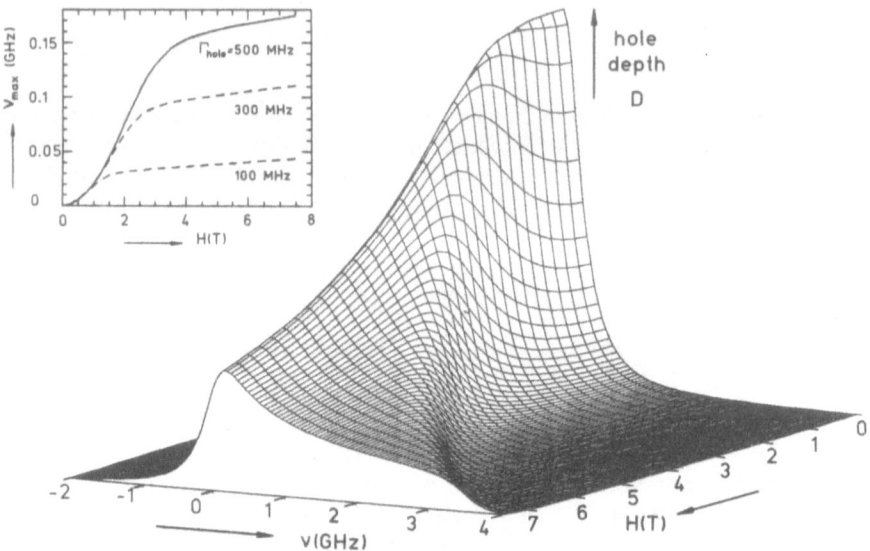

Fig. 76. Holes shape D as a function of frequency ν and magnetic field H for a zero-field hole of 500 MHz width (calculated with eq. (12)). Insert: Frequency shift of the maximum of the curve as a function of field for three different zero-field holewidths ($\Gamma_{\text{hole}} = 100$, 300 and 500 MHz) (from ref. [326]).

The second approach used in ref. [326] included the effect of the anisotropy of the excitation process, similar to the procedure used for the Stark effect in refs. [327, 328]. In this case, the relative number of guest molecules burnt in zero field is given by

$$N(\nu', t)\, d\nu' = N_0 \Phi \sigma\, \frac{I \cos^2 \alpha t}{h\nu_b}\, g(\nu_b - \nu')\, d\nu', \tag{13}$$

where all parameters are defined as in eq. (10), and α is the angle between the electric field vector $\mathbf{E}_0$ and $\boldsymbol{\mu}$, with $\boldsymbol{\mu}$ the electronic transition dipole moment of the molecule. Because the hole was probed in the magnetic field H with the laser light in the same direction as in the burning process, the contribution to the absorption coefficient is $\sigma \cos^2 \alpha$ [327, 328]. The hole shape function then becomes, after integration over all frequencies $d\nu'$ and over all possible molecular orientations, and after convolution of two Lorentzian line shape functions,

$$D(\nu) = C\, \frac{1}{\sqrt{KH^2}}\, \frac{\Gamma_{\text{hole}}}{2\pi} \int_{\nu}^{\nu + KH^2} \frac{1}{\sqrt{\nu' - \nu}} \left\{ A \left[1 + \left(\frac{\nu' - \nu}{KH^2} \right)^2 \right] + \right.$$
$$\left. + B \left(\frac{\nu' - \nu}{KH^2} \right) \right\} \frac{1}{(\nu'^2 + \Gamma_{\text{hole}}^2/4)}\, d\nu'. \tag{14}$$

Eq. (14) can be compared to eq. (12). The term in brackets represents the anisotropy created by the polarized laser light in the burning and probing processes. The constant C is given by

$$C = \frac{3}{128} \frac{N_0 \Phi I t \sigma^2}{h \nu_b}.$$

There are now two fitting parameters K and p, the latter being the fraction of laser light polarized perpendicular to the field H (with $A = -5p + 8$ and $B = 18p - 16$, see Appendix of ref. [326]).

The hole shape $D(\nu)$, calculated with both eq. (12) (dashed line) and eq. (14) (full line), is drawn in Fig. 74 (right) together with the experimental results. As seen from the figure, the line shapes obtained with both models are very similar, although the curve which includes the anisotropic distribution (eq. (14)) has a somewhat better fit. The values of $K = \frac{1}{2} \chi^{z'z'}$ obtained with both models are identical to within 2% ($\frac{1}{2} \chi^{z'z'} = 61.9$ vs 60.6 MHz/T^2, $p = 0.78 \pm 0.02$), because the value of the magnetic susceptibility is principally determined by the cut-off part of the curve, which justifies the use of the simple model with its intuitive insight. By following a similar calculation as in ref. [42], it was possible to determine the value of $\Lambda = |\langle S_1 | L_z | S_2 \rangle| = 5.64 \pm 0.06$, the matrix element of the effective orbital angular momentum operator L_z between the two lowest excited singlet states S_1 and S_2. This value is in good agreement with that previously found for the same molecule in a crystalline host. With this study it has been demonstrated that quantitative information on magneto-optical properties of excited states of complex biological molecules can be obtained from disordered materials with a resolution in the MHz regime.

b. *Stark effect.* The influence of an electric field on spectral holes burnt in *organic* molecules embedded in *crystalline* hosts at low temperature has only been studied in chlorin (H_2Ch) and its photoproduct (H_2Ch^*) in *n*-hexane and *n*-octane [114] (see Section 4.1.4b). From the linear increase of the hole splitting as a function of electric field, the magnitude and direction of the change in electric dipole moment on excitation to the first excited singlet state S_1 could be obtained. The results also yielded the orientation of the molecules in the *n*-alkane crystals.

In *amorphous* hosts, however, the hole not only splits in an electric field but also broadens. This effect is due to the statistical distribution of molecular orientations in the glass, as discussed for the case of a magnetic field. The first Stark effect experiment combined with hole-burning on an *organic glassy system* at low temperature was reported by Marchetti et al. for resorufin in PMMA [92]. A very small dipole moment change of 0.2 ± 0.05 D was inferred, which could not have been obtained by other methods. Since then experiments of this kind have been reported for various amorphous systems: chlorin in polystyrene (PS) [329a, b], PMMA [329b] and polyvinylbutyral (PVB) [62, 328, 329b, 330a, b, 331a], perylene [332], octaethylporphin [328, 330a, b] and Zn-tetrabenzoporphin [331a, b] in PVB, and octaethylchlorin [333] and pentacene [331b] in PMMA. All these amorphous systems exhibited a hole splitting which increased linearly with the applied electric field.

In general, one expects that the change in dipole moment of an individual molecule, when electronically excited, gives rise to an energy shift linear in field strength, whereas a quadratic shift is expected from polarization effects. Thus, the total energy shift to be measured by application of an electric field is [35, 329a, 334, 335]

$$\Delta \nu = -\frac{1}{h} \left(f \Delta \mu \cdot \mathbf{E} + \frac{1}{2} f^2 \mathbf{E} \cdot \Delta \hat{a} \cdot \mathbf{E} \right).$$ (15)

In eq. (15), $\mathbf{E}$ represents the external electric field given by V/d, where V is the voltage applied to the sample of thickness d. The factor $f = (\varepsilon + 2)/3$, with ε the dielectric constant, is the isotropic Lorentz internal field correction [336]. $\Delta \mu$ represents the change in dipole moment when going from the ground state to the excited state, and $\Delta \hat{a}$ the change in polarizability of the molecule on excitation.

The first calculations of the hole shape in the presence of an electric field for molecules in amorphous hosts have shown that the linear term in eq. (15) leads to a symmetric broadening and flattening of the hole, whereas the quadratic term in eq. (15) yields an asymmetrically displaced and broadened hole [329a]. The second term should only become important at very high fields. In these calculations, a statistical distribution of the field-induced shifts of the electronic levels was assumed, and the direction of polarization of the hole-burning and hole-probing light was taken into account (for details, see refs. [35, 329a]). Preliminary experiments on chlorin in polystyrene (PS) films at 4.2 K, with field strengths up to 6×10^5 V/cm, suggested a linear Stark effect, from which a value of $|\Delta \mu| = 0.4$ D was estimated [329a]. Very accurate hole-burning Stark effect measurements on the same molecule as guest in n-alkane single crystals, however, yielded a smaller value $|\Delta \mu| = 0.23 \pm 0.01$ [114], similar to that obtained in a PVB-film [62, 328].

A linear Stark effect has also been observed for perylene in PVB, although the perylene molecule has no intrinsic permanent electric dipole moment, neither in the S_0 nor in the S_1 state, thus $\Delta \mu = 0$ [332]. This effect has been attributed to matrix-induced dipole moments. The effective dipole moment change between the excited and the ground state induced by the host was assumed to be different for each individual molecule, and a Gaussian distribution function of the electric field-induced level shifts was taken [332]. By fitting the parameters of the model to the spectral hole measured, a huge effective matrix-induced electric dipole moment difference of $|\Delta \mu_{ind}| = 6.0$ D! was estimated, from which it was concluded that the environment has an important influence on the electric field-induced level shifts [332]. Although the latter may be true, the value of $\Delta \mu_{ind}$ seems unrealistically large.

Very detailed calculations of the influence of an applied electric field on spectral holes, together with systematic hole-burning Stark effect experiments have been reported for octaethylporphin (OEP) and chlorin in a PVB-matrix at 1.7 K by Meixner et al. [328]. The analysis of the hole shape was based on a model which involved the following parameters: the difference of the electric dipole moments in ground and excited states $\Delta \mu$, the corresponding transition moment

D, and the angle θ between $\Delta\mu$ and **D**. The guest molecules were assumed to be randomly oriented, and linearly polarized laser light was used as excitation source, which leads to an anisotropic effect in the hole shape [326, 327] (see also the discussion on Zeeman effect in amorphous systems in Section 4.2.8a). The hole shapes obtained in ref. [328] were shown to be dependent on the relative direction of the applied electric field and the polarization of the laser light. To describe the effect of the matrix-induced dipole moments, a statistical model was used based on spatial fluctuations of the local matrix fields. The results of the calculated line shapes, under different conditions, are reproduced in Fig. 77a, b (see ref. [328] for the equations used). Similar detailed calculations have later been reported in ref. [337].

The experimental hole shapes investigated in ref. [328] were recorded in transmission and by holographic detection [83]. For both amorphous systems studied, a linear Stark effect was observed, although the OEP molecules do not have an intrinsic dipole moment. As in the case of perylene in PVB [332], the results were interpreted in terms of matrix-induced dipole moments. Chlorin, by contrast, yielded not only a braodening of the hole in the field, but also a characteristic and pronounced dip. These results were explained in terms of a superposition of two effects, a change in molecular dipole moment on excitation, and the influence of the host dipole moments on the guest molecules. The latter

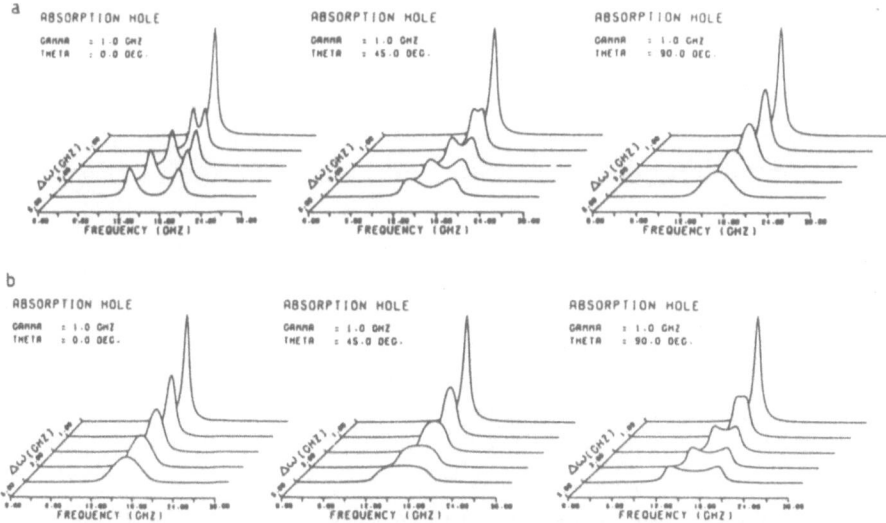

Fig. 77. Line shapes of absorption holes calculated with eq. (11) of ref. [328], for different values of the maximum frequency shift $\Delta\omega = |\Delta\mu_{eff}||E_s|\hbar$. A holewidth of 1 GHz was assumed at zero field. E_0 represents the polarization, and E_s the applied electric field. (a) $E_0 // E_s$. If $\theta = 0$ ($\Delta\mu // D$), the molecules will undergo a maximum frequency shift, and the signal is split into two lines displaced by $\pm\Delta\omega$ from the central position. If $\theta = \pi/2$, none of the excited molecules will show a Stark shift. (b) $E_0 \perp E_s$. For $\theta = 0$ a broadening is observed, whereas for $\theta = \pi/2$ a line shape with a dip results.

effect was found to be of the same magnitude for chlorin and OEP, $\Delta\mu_{ind}$ ≃ 0.06 − 0.10 D [328], a value which is much smaller than the 6 D reported for perylene in PVB [332]. From fits of the theoretical curves to the experimental results, a dipole moment difference of $\Delta\mu_{mol}$ = 0.28 ± 0.02 D was obtained for chlorin [328]. This value is consistent with those found in crystals of n-alkanes [114], and with the 0.26 D theoretically calculated [272]. Experimental hole shapes are shown in Fig. 78 for OEP and chlorin. Notice that for chlorin also a hole dip appears, in addition to broadening, at field strength above 6 KV/cm, whereas only a hole broadening is seen in OEP up to 30 KV/cm [328].

Hole shapes in electric fields have also been reported by Kador *et al.* [331a]. In their calculations the polarization of the laser light was not taken into account, which leads to line shapes that only flatten in the field, but do not show a dip. The model of ref. [331a] takes into consideration two cases: (a) the molecular dipole moment difference $\Delta\mu$ is the same for all molecules, and (b) a distribution of $\Delta\mu$ values is induced by the polar host. The first case was applied to polar, the second to unpolar dye molecules. The equations were fitted to hole-burning experiments performed on chlorin and Zn-tetrabenzo-porphin (Zn-TBP) in PVB at 2 K. The hole shapes obtained for the two molecules were different [331a], but similar to the line shapes found in ref. [328]. The curve of chlorin was fitted with $\Delta\mu$ = 0.21 ± 0.01 D, whereas for Zn-TBP a mean induced dipole moment difference of

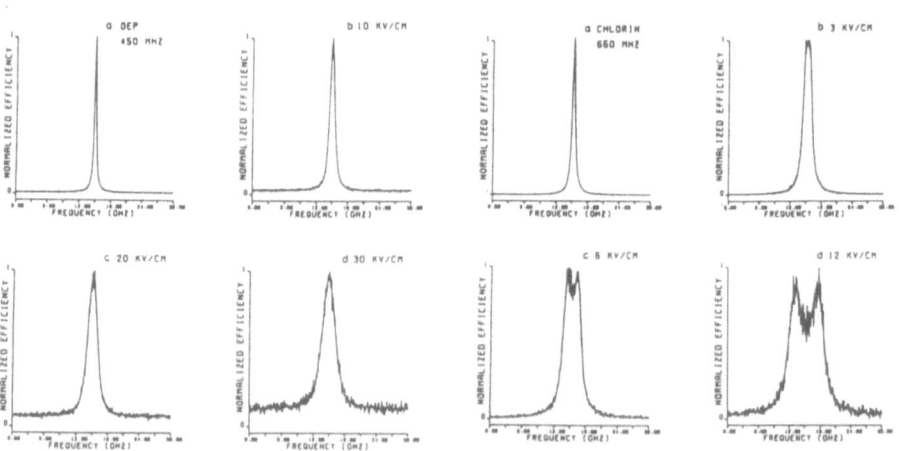

Fig. 78. Photochemical hole spectra recorded by holographic detection, $E_0 \perp E_s$. Left: OEP in PVB at 1.7 K. The hole was burnt at zero field. Electric fields of 10, 20 and 30 KV/cm were applied. The signal broadens due to a linear Stark effect. A value of $|\Delta\mu_{ind}|$ = 0.06 − 0.10 D was obtained from fits of theoretical curves to the experimental results (see text). Right: Chlorin in PVB at 1.7 K. Notice the hole broadening and the dip in the line shape on application of a field of 3, 6 and 12 KV/cm. The curves were fitted with $\theta = \pi/2$ and an additional matrix-induced distribution of dipole moments. Best fits were obtained with $\Delta\mu_{mol}$ = 0.28 ± 0.02 D and $\Delta\mu_{ind}$ = 0.10 D (from ref. [328]).

$\overline{\Delta\mu} = 0.17 \pm 0.01 \, D$ was estimated [331a]. The latter value is twice as large as the value obtained from the fit of a more refined calculation to the hole-burning results of a very similar molecule (OEP) in ref. [328].

In a recent hole-burning paper the combined effects of an electric field and pressure were discussed in terms of a common solvent shift model [331b]. For pentacene and Zn-TBP in PVB, a variation of the mean value $\overline{\Delta\mu}$ of the matrix-induced dipole moment across the inhomogeneous absorption band was observed, and interpreted in terms of gas-to-matrix shifts. Furthermore, a correlation was suggested between electron-phonon coupling strength and solvent shift [331b].

Finally, a few applications of hole-burning in electric fields should be mentioned. One of them is the determination of the homogenous linewidth with a laser at fixed frequency. This was possible by measuring the hole depth as a function of electric field at different temperatures, for chlorin and perylene in PVB [329b]. The results yielded a $T^{1.4 \pm 0.2}$ dependence of the holewidth in the electric field domain, in agreement with previous optical dephasing experiments on organic glassy systems in the frequency domain [24] (see also Section 4.2.2).

Another application is the use of modulated electric fields to study hole-burning parameters with high sensitivity. By measuring the second-harmonic amplitude of the Stark signal relative to the light intensity at zero field, and knowing the dipole moment change on excitation, very shallow hole depths of the order of 10^{-4} optical density units could be determined in the $S_1 \leftarrow S_0$ 0—0 transition of octaethylchlorin in PMMA [333].

The Stark effect has also been proposed in combination with hole-burning for optical memories. Its feasibility has been demonstrated by Wild et al. [330b], based on the idea that a hole can be used as a "bit" of information in the frequency domain [218]. With a laser at a fixed frequency, the data bits are recorded at selected electric field strengths and read out by sweeping the electric field. Holes were burnt into chlorin in PVB at 4.2 K, and data were obtained simultaneously as a function of frequency and applied voltage. Thus, writing and reading of optical information in the two dimensional frequency/voltage space is possible. This approach may lead to an increase in data storage capacity that would amount to 2500 K bits per spot at 1.5 K [330b].

c. *Pressure effect*. Spectral lines of organic molecular systems can also be influenced by external pressure. For example, PMMA matrices show pressure induced shifts of the order of 10^{-2} cm^{-1}/10^3 hPa. In order to observe such shifts with conventional spectroscopic techniques, however, large pressures of at least 10^6 hPa are needed. A more convenient method is to use spectral hole-burning. The first pressure experiments of this kind were reported by Richter et al. in 1984 [339]. By burning a hole in the $S_1 \leftarrow S_0$ 0—0 transition of free-base phthalocyanine (H$_2$Pc) as guest in PMMA and polyethylene (PE) at 1.5 K, it was possible to observe hole shifts and broadening at uniaxial pressures down to 100 hPa. The results were independent of the "pre-pressure" applied to the sample up to pressures of 3×10^4 hPa. A linear increase of the holewidth was observed at pressure differences Δp

≤ 500 hPa. Broadening and shift of the holes were reversible, and the effect was explained by assuming a statistical distribution of site energies in the glass. For Δp ≥ 500 hPa, a non-monotonic increase of the hole broadening versus Δp was observed, which was interpreted in terms of irreversible changes of TLS minimum energies of the molecular ground states [339]. The same group found later that this was an artifact which originated from the non-parallelity of the sample and led to irreversible strains [331a, 340].

Subsequent work on pressure-induced hole shifts and broadening was per-formed under hydrostatic pressure by the same group [331a, b, 340, 341]. The goal of these experiments was to get an insight into matrix parameters like compressibility, molecular interaction potentials and strain fields. It was stressed that the advantage of working with small pressures was to be close to the equilibrium positions of the involved molecular potentials, such that only lowest-order terms in the external perturbation needed to be considered [331a, b]. Holes were burnt into the $S_1 \leftarrow S_0$ 0—0 transition of H_2Pc in PMMA, PS and PE at 1.6 K. Pressures up to 20 MPa were applied, and liquid helium was used as hydrostatic pressure transmitting medium. Changes in line shape and width were reversible, and only pressure differences, instead of absolute pressures, were found to be important. An example of such a pressure induced hole shift and broadening is given in Fig. 79 (reproduced from ref. [331a]). It shows a frequency shift to the red of a hole burnt into the 0—0 transition of H_2Pc in PE under increasing

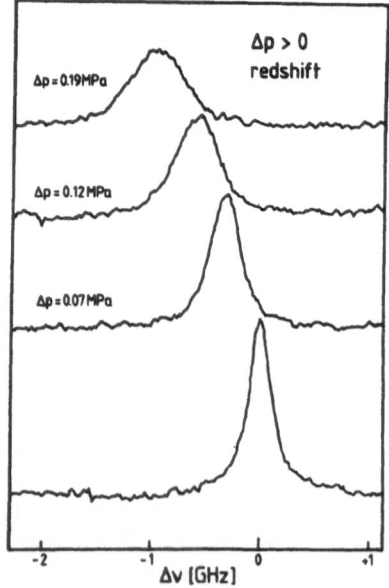

Fig. 79. Shift and broadening of a hole burnt into the $S_1 \leftarrow S_0$ 0—0 transition of phthalocyanine (H_2Pc) in PE at 1.6 K with increasing hydrostatic pressure (from refs. [331a, 340]).

hydrostatic pressure of a few hundred hPa. Blue shifts were observed with decreasing pressure. From plots of such shifts versus pressure (see Fig. 80), it was shown that the linear shift term is the dominant contribution at small pressure variations [331a]:

$$\Delta \bar{\nu} = \Delta \bar{\nu}_s (1 + 2\kappa \Delta p), \tag{17}$$

where $\Delta \bar{\nu}_s$ is the solvent shift and κ is the hydrostatic compressibility of the matrix. With eq. (17) it is possible to determine the compressibility of the matrix by measuring the pressure shifts $\Delta \bar{\nu} / \Delta p$ of the holes if the solvent shift of the host-guest system is known. Values for the compressibilities between 0.13 and 0.15 GPa^{-1} were obtained in this way from experiments on H_2Pc in PMMA, PE and PS [331a, 340] which agree within 20% with those obtained from mechanical measurements. A subsequent pressure induced hole-burning experiment on pentacene in PMMA yielded a value of $\kappa = 0.12 \ GPa^{-1}$ [331b], which is close to that previously found for the same sample of 0.15 $\pm$ 0.03 GPa^{-1} [331a]. The hole broadening, on the other hand, reflects the degree of order of the matrix around the guest molecule, which is also related to the inhomogeneous width.

The frequency shift and broadening of holes with variable pressure have been related in a quantitative manner to the solvent shift and the width of the inhomogeneous band, and the static, long-range attractive part of the guest-host interaction potential [341]. By measuring the dependence of hole shapes on temperature, in addition to pressure, the dynamical contributions of the interaction potential could be separated from the static contributions. It was concluded, that pressure dependent hole shapes are very sensitive probes to obtain accurate values of matrix compressibilities, which have shown to be independent of temperature between 4.2 and 77 K [341].

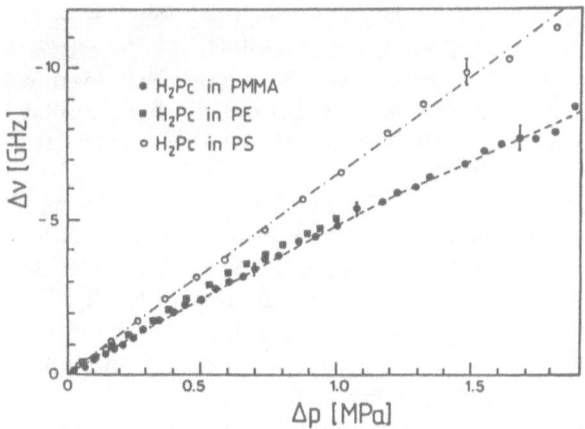

Fig. 80. Spectral shifts of holes as a function of increasing hydrostatic pressure. Notice that the linear shift term is the dominant contribution to the observed hole shift (from refs. [340, 341]).

5. Summary and Conclusions

Various aspects of spectral hole-burning have been discussed here. We have stressed its use as a tool for studying optical relaxation processes at low temperature, and have shown that hole-burning is a very easy and straightforward technique to determine the homogeneous linewidth, Γ_{hom}, of electronic transitions of impurity molecules diluted as guests in either crystalline or amorphous hosts. To get the correct value of Γ_{hom}, the holewidth has to be measured in a systematic way as a function laser power P and burning time t, and its value has to be extrapolated to $P = 0$ and $t = 0$.

Hole-burning experiments reported in the literature have been discussed in a chronological order with the purpose to present an insight into the historical development of this technique. The majority of the systems described here are organic molecules in organic matrices. They have been ordered according to their underlying hole-burning mechanism: intramolecular reactions that take place within the guest molecule, and intermolecular reactions occurring between the guest and the host. Within the first category, we have emphasized the experiments on porphyrin molecules and related biological compounds, because it is for the latter where the hole-burning method holds great promise for elucidating the dynamics of chromophore-protein interactions. Such interactions are difficult to study by other techniques due to the broad overlapping spectroscopic absorption bands of these complex biological systems.

Two systems have been chosen to illustrate intermolecular photochemical reactions: quinizarin (1,4-dihydroxyanthraquinone) in alcoholic glasses which undergoes a hydrogen-bonding reaction with the host, and the ionic dyes resorufin and cresyl violet in polar hosts, which in contrast to non-photochemical hole-burning systems, have a high hole-burning efficiency. We have further presented results on two-photon and photon-gated systems because of their potential for frequency domain optical storage. Their advantage is that the hole can be read without being erased. Finally, we have discussed non-photochemical hole-burning (NPHB), a process characteristic for glasses. We have seen that even photostable guest molecules, when incorporated into an amorphous host, exhibit this effect. The hole-burning efficiency of NPHB, however, is rather low ($\sim 10^{-5}$—10^{-6}) as compared to PHB ($\sim 10^{-2}$—10^{-3}).

The main application of hole-burning so far has been the study of excited state dynamics of molecules and ions as guests in solid matrices at low temperature. It has been shown that optical dephasing of the $S_1 \leftarrow S_0$ 0—0 transition of molecules in crystalline systems is due to phonon scattering to localized librational modes, which have lifetimes ranging from ~ 0.1 to 120 ps. These low-frequency modes of ~ 5—40 cm^{-1} appear as weak phonon side-bands in the optical spectra. Their frequencies depend on the tightness of the guest in the host lattice. At temperatures below 2 K, Γ_{hom} is determined only by the fluorescence lifetime T_1 of the guest molecule, $\Gamma_0 = (2\pi T_1)^{-1}$, whereas with increasing temperature, pure dephasing processes take place which broaden the hole and shift its frequency in an exponential manner, with an activation energy given by the frequency of the librational mode. Not only 0—0 transitions, but also vibronic relaxation times from higher vibrational S_1-states have been studied by hole-burning. Results for free-base porphin in various n-alkane crystalline hosts have shown that this type of relaxation is dominated by guest-host interactions: the tighter the fit of the guest molecule in the substitutional host cage, the faster the relaxation time. Similar optical relaxation and dephasing processes have been observed for other guest molecules in crystalline matrices by picosecond photon echoes.

For organic amorphous systems, however, the agreement between hole-burning and photon-echo results does not seem to be so clear yet. Many experiments by hole-burning have shown that Γ_{hom} extrapolates to $\Gamma_0 = (2\pi T_1)^{-1}$ when $T \to 0$, and that $\Gamma_{hom} - \Gamma_0$ follows a $T^{1.3}$ power law between 0.3 and, at least, 20 K. Recent accumulated picosecond photon echo experiments seem to agree with these results, but two-pulse photon echoes do not. More photon echo experiments are necessary in order to clarify the apparent contradictions. The problem is of actual interest, because the two techniques probe very different time scales of motions in glasses, of which very little is known so far. From this type of experiments we may learn more about the nature of "spectral diffusion" processes in amorphous materials at low temperature.

We have further shown that hole-burning experiments on organic molecules incorporated in semi-crystalline polymers and mixed organic-inorganic amorphous systems confirm the idea that the $T^{1.3}$ power law is related to the disorder surrounding the direct environmment of the guest. Many theories of optical dephasing in glasses, based on the TLS model, have been proposed to explain this $T^{1.3}$-dependence of the homogeneous linewidth, but none of them is thoroughly convincing, since they have not yet withstood a critical experimental test. In the calculations approximations are made which, depending on the averaging procedure, may lead to different power law dependences. Thus, most of the models do not predict a $T^{1.3}$ power law, but try to attain agreement of the calculations with the experimental results. In order for a theory to be satisfactory, it should also predict the quantitative behaviour of other properties of glasses, like specific heat.

A second, quite different application of hole-burning has also been presented: its use as a technique to measure very accurate frequency shifts. The effects of external fields can be studied in this way with a resolution in the MHz-regime, because the hole serves as a permanent sharp frequency marker in the spectrum. We have discussed the results of Zeeman and Stark effects on single crystals of n-alkanes doped with porphin molecules. From the energy shifts of the holes as a function of the field strength, parameters like the magnetic susceptibility, and the change in magnitude and direction of the electric dipole moment on excitation can be accurately determined. It has been demonstrated that these parameters can also be obtained in amorphous systems. In this case magnetic and electric fields applied to holes burnt in zero field yield hole shapes which are not only shifted, as in crystals, but also broadened. From an analysis of the line shapes, it was concluded that the parameters were consistent with those deduced for the same guest molecules in crystalline hosts. Thus, the technique is reliable, and can be used to study complex biological molecules. We have also seen that matrix compressibilities at low temperature can be obtained by measuring the effect of small external pressure differences on holes burnt in organic glassy systems. The values of the compressibilities agree well with those obtained by mechanical methods at much higher pressures.

Finally, it has been mentioned briefly that permanent hole-burning offers potential technological applications for data storage in the frequency domain. The presence of a hole at a particular frequency can be used to store one "bit" of information. With this technique it is possible, in principle, to increase the storage density by a factor of 10^3-10^4 compared to conventional optical recording. Problems encountered in the realization of such storage devices have been addressed, and references are given to detailed work in the subject.

In conclusion, we have tried to prove that hole-burning is a very simple spectroscopic technique of wide applicability. It has the advantage with respect to other, more sophisticated techniques, that it does not need samples of good optical quality, it can be applied to weak electronic transitions, and it is a well suited method for studying highly-resolved energy selected photochemical processes in the MHz-regime at low temperature.

6. Acknowledgements

I would like to thank many people who have collaborated in part of the work presented here. I am especially indebted to J. H. van der Waals for his continuous inspiration, and to R. M. Macfarlane for the enthusiasm with which he embarked on our joint MHz-resolution experiments at the IBM Research Laboratory in San José. The HB-results obtained in Leiden would not have been possible without the efforts of A. I. M. Dicker, H. P. H. Thijssen, H. Jansen, R. van den Berg, H. van der Laan, A. Visser, P. J. van der Zaag, and J. P. Galaup. I am also grateful to G. J. Small, J. Friedrich, D. Haarer, J. Deisenhofer, H. Michel, M. E. Michel-Beyerle, S. G. Boxer, R. M. Macfarlane, R. M. Shelby, W. E. Moerner, M. D. Levenson, W. Lenth, D. A. Wiersma, M. D. Fayer, P. Reineker and U. P. Wild for the reproduction of figures from their articles.

Finally, I acknowledge I. Zschokke and J. Fünfschilling and the co-authors of this book, J. Schmidt and R. Silbey, for their patience to wait for this contribution. I am particularly grateful to M. A. Kempe and L. van der Poel, for typing and correcting the manuscript many times, and always with a smile.

Many of the experiments described here would not have been possible without the financial help, in chronological order, of the 3. Physikalisches Institut of the University of Stuttgart (led by H. C. Wolf); the IBM Research Laboratory in San Jose, California; the University of Leiden; the Netherlands Foundation for Chemical Research (SON) and Physical Research (FOM) in collaboration with the Netherlands Organization for the Advancement of Pure Research (NWO); and the Max-Planck Institut für Festkörperphysik in Stuttgart (by invitation of K. Dransfeld and S. Hunklinger, during September 1982).

Note

The figures in this article that were taken from the literature have been reproduced with permission from the Journals mentioned in the references of the figure captions.

References

1. H. P. H. Thijssen, A. I. M. Dicker, and S. Völker, *Chem. Phys. Lett. 92* (1982), 7.
2. R. I. Personov, E. I. Al'shits, and L. A. Bykovskaya, *Opt. Commun.* **6** (1972), 169.
3. A. Szabo, *Phys. Rev. Lett.* **25** (1970), 924; *ibid.* **27** (1971), 323.
4. R. I. Personov, in: 'Spectroscopy and Excitation Dynamics of Condensed Molecular Systems' (eds. V. M. Agranovich and R. M. Hochstrasser, North Holland, Amsterdam, 1983), p. 555, and references therein.
5. N. A. Kurnit, I. D. Abella, and S. R. Hartmann, *Phys. Rev. Lett.* **13** (1964), 567.
6. R. G. Brewer, in: 'Frontiers in Laser Spectroscopy', Vol. 1 (eds. R. Balian, S. Haroche, and S. Lieberman, North Holland, Amsterdam, 1977), p. 342; *Physics Today* **30** (1977), 50.
7. R. L. Shoemaker, in: 'Laser and Coherence Spectroscopy' (ed. J. I. Steinfeld, Plenum Press, New York, 1978), p. 197.
8. T. J. Aartsma and D. A. Wiersma, *Phys. Rev. Lett.* **36** (1978), 1360.
9. W. H. Hesselink and D. A. Wiersma, in: 'Spectroscopy and Excitation Dynamics of Condensed Molecular Systems' (eds. V. M. Agranovich and R. M. Hochstrasser, North Holland, Amsterdam, 1983), p. 249, and references therein.
10. R. G. Brewer and R. L. Shoemaker, *Phys. Rev. Lett.* **27** (1971), 631; *Phys. Rev.* **A6** (1972), 2001.
11. A. Z. Genack, R. M. Macfarlane, and R. G. Brewer, *Phys. Rev. Lett.* **37** (1976), 1078.
12. R. G. De Voe and R. G. Brewer, *Phys. Rev. Lett.* **40** (1978), 862.
13. M. J. Burns, W. K. Liu, and A. H. Zewail, in: 'Spectroscopy and Excitation Dynamics of Condensed Molecular Systems' (eds. V. N. Agranovich and R. M. Hochstrasser, North Holland, Amsterdam, 1983), p. 301.
14. L. A. Riseberg, *Phys. Rev.* **A7** (1973), 671.
15. A. A. Gorokhovskii, R. K. Kaarli, and L. A. Rebane, *J. Exp. Theor. Phys. Lett.* **20** (1974), 216; *Opt. Commun.* **16** (1976), 282.
16. B. M. Kharlamov, R. I. Personov, and L. A. Bykovskaya, *Opt. Commun.* **12** (1974), 191.
17. H. de Vries and D. A. Wiersma, *Phys. Rev. Lett.* **36** (1976), 91.
18. S. Völker and J. H. van der Waals, *Mol. Phys.* **32** (1976), 1703.
19. S. Völker, R. M. Macfarlane, A. Z. Genack, H. P. Trommsdorff, and J. H. van der Waals, *J. Chem. Phys.* **67** (1977), 1759.
20. J. M. Hayes and G. J. Small, *Chem. Phys.* **27** (1978), 151.
21. 'Topics in Current Physics. Persistent Spectral Hole-Burning: Science and Applications', Vol. 44 (ed. W. E. Moerner, Springer Verlag, Berlin, 1988), and references therein.
22. W. E. Moerner, *J. Molec. Electr.* **1** (1985), 55.
23. S. Völker, in: 'Excited State Spectroscopy in Solids', XCVI Course of the Enrico Fermi Summer School of Physics, Varenna, Italy (ed. U. M. Grassano and N. Terzi, North Holland, Amsterdam, 1987), p. 363.
24. S. Völker, *J. Lumin.* **36** (1987), 251, and references therein.
25. L. A. Rebane, A. A. Gorokhovskii, and J. V. Kikas, *Appl. Phys.* **B29** (1982), 235, and references therein.
26. (a) G. J. Small, in: 'Spectroscopy and Excitation Dynamics of Condensed Molecular Systems' (eds. V. N. Agranovich and R. M. Hochstrasser, North Holland, Amsterdam, 1983), p. 515, and references therein.
     (b) R. Jankowiak and G. J. Small, *Sciences* **237** (1987), 618, and references therein.

27. J. Friedrich and D. Haarer, *Angew. Chemie, Int. Ed.* **23** (1984), 113.
28. H. P. H. Thijssen, R. van den Berg, and S. Völker, in: 'Photoreaktive Festkörper' (ed. H. Sixl, Wahl-Verlag, Karlsruhe, 1984), p. 763.
29. R. I. Personov and B. M. Kharlamov, *Laser Chem.* **6** (1986), 181.
30. R. M. Macfarlane and R. M. Shelby, *J. Lumin.* **36** (1987), 179.
31. R. M. Macfarlane and R. M. Shelby, *Cryst. Latt. Def. and Amorph. Mat.* **12** (1985), 417.
32. W. M. Yen, in: 'Optical Spectroscopy of Glasses' (ed. I. Zschokke, D. Reidel, Dordrecht, 1986), p. 23, and references therein.
33. 'Optical Linewidths in Glasses', special issue of *J. Lumin.* **36** (1987), pp. 179—329.
34. 'Optical Spectroscopy of Glasses', Series C: *Molecular Structures* (ed. I. Zschokke, D. Reidel, Dordrecht, 1986), and references therein.
35. M. Maier, *Appl. Phys.* **B41** (1986), 73, and references therein.
36. J. L. Skinner and D. Hsu, *Adv. Chem. Phys.* **65** (1986), 1, and references therein.
37. A. Szabo, *Phys. Rev.* **B11** (1975), 4512.
38. R. M. Shelby and R. M. Macfarlane, *Chem. Phys. Lett.* **64** (1979), 545.
39. A. I. M. Dicker, L. W., Johnson, S. Völker, and J. H. van der Waals, *Chem. Phys. Lett.* **100** (1983), 8.
40. H. de Vries and D. A. Wiersma, *Chem. Phys. Lett.* **51** (1977), 565.
41. S. Völker and R. M. Macfarlane, *IBM J. Res. Dev.* **23** (1979), 547, and references therein.
42. A. I. M. Dicker, M. Noort, S. Völker, and J. H. van der Waals, *Chem. Phys. Lett.* **73** (1980), 1.
43. H. P. H. Thijssen and S. Völker, *Chem. Phys. Lett.* **120** (1985), 496.
44. S. Völker and R. M. Macfarlane, *J. Chem. Phys.* **73** (1980), 4476.
45. H. P. H. Thijssen, R. van den Berg, and S. Völker, *Chem. Phys. Lett.* **97** (1983), 295.
46. R. M. Macfarlane and S. Völker, *Chem. Phys. Lett.* **69** (1980), 151.
47. F. Drissler, F. Graf, and D. Haarer, *J. Chem. Phys.* **72** (1980), 4996.
48. J. Friedrich, H. Wolfrum, and D. Haarer, *J. Chem. Phys.* **77** (1982), 2309.
49. H. P. H. Thijssen, R. van den Berg, and S. Völker, *Chem. Phys. Lett.* **120** (1985), 503.
50. D. M. Burland, F. Carmona, and J. Pacansky, *Chem. Phys. Lett.* **56** (1978), 221.
51. H. W. Lee, M. Gehrtz, E. E. Marinero, and W. E. Moerner, *Chem. Phys. Lett.* **118** (1985), 611.
52. J. M. Hayes, R. P. Stout, and G. J. Small, *J. Chem. Phys.* **71** (1981), 4266.
53. R. M. Macfarlane and R. M. Shelby, *Opt. Commun.* **45** (1983), 46.
54. R. Jankowiak and H. Bässler, *Chem. Phys. Lett.* **95** (1983), 310.
55. L. W. Molenkamp and D. A. Wiersma, *J. Chem. Phys.* **83** (1985), 1.
56. R. van den Berg and S. Völker, *Chem. Phys. Lett.* **127** (1986), 525.
57. R. M. Macfarlane and R. M. Shelby, in: 'Laser Spectroscopy VI' (eds. H. P. Weber and W. Lüty, Springer Verlag, Berlin, 1983), p. 113.
58. S. Völker, R. M. Macfarlane, and J. H. van der Waals, *Chem. Phys. Lett.* **58** (1978), 8.
59. T. P. Carter, B. L. Fearey, J. M. Hayes, and G. J. Small, *Chem. Phys. Lett.* **102** (1983), 272.
60. A. I. M. Dicker, J. Dobkowski, and S. Völker, *Chem. Phys. Lett.* **84** (1981), 415.
61. R. W. Olson, H. W. H. Lee, F. G. Patterson, M. D. Fayer, R. M. Shelby, D. P. Burum, and R. M. Macfarlane, *J. Chem. Phys.* **77** (1982), 2283.
62. F. A. Burkhalter, G. W. Suter, U. P. Wild, V. D. Samoilenko, N. V. Rasumova, and R. I. Personov, *Chem. Phys. Lett.* **94** (1983), 483.
63. C. A. Walsh and M. D. Fayer, *J. Lumin.* **34** (1985), 37; H. W. H. Lee, C. A. Walsh and M. D. Fayer, *J. Chem. Phys.* **82** (1985), 3948.
64. H. W. H. Lee, A. L. Huston, M. Gehrtz, and W. E. Moerner, *Chem. Phys. Lett.* **114** (1985), 491.
65. J. Friedrich, D. Haarer, and R. Silbey, *Chem. Phys. Lett.* **95** (1983), 119.
66. A. Gutiérrez, G. Castro, G. Schulte, and D. Haarer, in: 'Organic Molecular Aggregates' (eds. P. Reineker, H. Haken, and H. C. Wolf, Springer Verlag, Berlin, 1983), p. 206.
67. W. Breinl, J. Friedrich, and D. Haarer, *Phys. Rev.* **B34** (1986), 7271.

68. G. Schulte, W. Grond, D. Haarer, and R. Silbey, *J. Chem. Phys.* **88** (1988), 679.
69. E. Cuellar and G. Castro, *Chem. Phys.* **54** (1981), 217.
70. W. E. Moerner, F. M. Schellenberger, G. C. Bjorklund, P. Kaipa, and F. Lüty, *Phys. Rev.* **B32** (1985), 1270.
71. W. E. Moerner, T. P. Carter, and C. Bräuchle, *App. Phys. Lett.* **50** (1987), 430.
72. (a) H. P. H. Thijssen, thesis, University of Leiden (1985).
    (b) R. E. van den Berg, thesis, University of Leiden (1988).
73. J. Cachenaut, C. Man, P. Cerez, A. Brillet, F. Stoeckel, A. Jourdan, and F. Hartmann, *Rev. de Phys. Appliquée* **14** (1979), 685.
74. H. van der Laan and S. Völker, unpublished results.
75. H. P. H. Thijssen, R. van den Berg, S. Völker, J. H. van der Waals, and L. P. J. Husson, *Chem. Phys. Lett.* **111** (1984), 121; H. P. H. Thijssen, S. Völker, and J. H. van der Waals, in: 'Photoreaktive Festkörper' (ed. H. Sixl, Wahl-Verlag, Karlsruhe, 1984), p. 749.
76. W. E. Moerner, W. Lenth and G. C. Bjorklund, in: 'Topics in Current Physics. Persistent Spectral Hole-Burning: Science and Applications', Vol. 44 (ed. W. E. Moerner, Springer Verlag, Berlin, 1988), Chapter 7, and references therein.
77. G. C. Bjorklund, *Opt. Lett.* **5** (1980), 15; W. Lenth, C. Ortiz, and G. C. Bjorklund, *Opt. Lett.* **6** (1981), 351.
78. G. C. Bjorklund, M. D. Levenson, W. Lenth, and C. Ortiz, *App. Phys.* **B32** (1983), 145.
79. P. Pokrowsky, W. E. Moerner, F. Chu, and G. C. Bjorklund, *Opt. Lett.* **8** (1983), 280; M. Gehrtz, G. C. Bjorklund, and E. A. Whittaker, *J. Opt. Soc. Am.* **B2** (1985), 1510.
80. (a) A. L. Huston and W. E. Moerner, *J. Opt. Soc. Am.* **B1** (1984), 349.
    (b) W. E. Moerner and A. L. Huston, *App. Phys. Lett.* **48** (1986), 1181.
81. M. D. Levenson, R. M. Macfarlane, and R. M. Shelby, *Phys. Rev.* **B22** (1980), 4915.
82. B. Dick, *Chem. Phys. Lett.* **143** (1988), 186.
83. A. Renn, A. J. Meixner, U. P. Wild, and F. A. Burkhalter, *Chem. Phys.* **93** (1985), 157.
84. P. Saari, P. Kaarli, and A. Rebane, *J. Opt. Soc. Am.* **B3** (1986), 527, and references therein.
85. A. I. M. Dicker, thesis, University of Leiden (1982).
86. R. van den Berg, A. Visser, and S. Völker, *Chem. Phys. Lett.* **144** (1988), 105.
87. H. P. H. Thijssen, R. van den Berg, and S. Völker, *Chem. Phys. Lett.* **103** (1983), 23.
88. H. P. H. Thijssen and S. Völker, *J. Chem. Phys.* **85** (1986), 785.
89. J. H. F. van Kooten, thesis, University of Leiden (1984), p. 50.
90. J. Friedrich and D. Haarer, in: 'Optical Spectroscopy of Glasses' (ed. I. Zschokke, D. Reidel, Dordrecht, 1986), p. 149.
91. A. C. Anderson, in: 'Amorphous Solids. Low Temperature Properties' (ed. W. A. Phillips, Springer Verlag, Berlin, 1981), p. 65.
92. A. P. Marchetti, M. Scozzafava, and R. H. Young, *Chem. Phys. Lett.* **51** (1977), 424.
93. A. F. Childs and A. H. Francis, *J. Chem. Phys.* **89** (1985), 466.
94. W. Breinl, J. Friedrich, and D. Haarer, *J. Chem. Phys.* **80** (1984), 3496.
95. R. T. Brundage and W. M. Yen, *Phys. Rev.* **B33** (1986), 4436.
96. M. Berg, C. A. Walsh, L. R. Narasimhan, K. A. Littau, and M. D. Fayer, *Chem. Phys. Lett.* **139** (1987), 66; *J. Chem. Phys.* **88** (1988), 1564.
97. R. Bonnett, in: 'The Porphyrins', Vol. 1 (ed. D. Dolphin, Academic Press, New York, 1978), Chapter 1.
98. A. A. Gorokhovskii and L. A. Rebane, *Opt. Commun.* **20** (1977), 144.
99. A. A. Gorokhovskii and L. A. Rebane, *Sov. Phys. Solid State* **19** (1977), 1996.
100. K. Mauring and R. Avarmaa, *Chem. Phys. Lett.* **81** (1981), 446.
101. H. Jansen and S. Völker, unpublished results.
102. A. A. Gorokhovskii, V. Korrovits, V. Palm, and M. Trummal, *Chem. Phys. Lett.* **125** (1986), 355.
103. C. B. Storm and Y. Teklu, *J. Am. Chem. Soc.* **94** (1972), 1745; C. B. Storm, Y. Teklu, and E. A. Sokoloski, *Ann. N. Y. Acad. Sci.* **206** (1973), 631.

104. R. J. Abraham, G. E. Hawkes, and K. M. Smith, *Tetrahedron Lett.* **16** (1974), 1483.
105. K. N. Solov'ev, I. E. Zalesski, V. N. Kotlo, and S. F. Shkirman, *J. Exp. Theor. Phys. Lett.* **17** (1973), 332.
106. I. Y. Chan, W. G. van Dorp, T. J. Schaafsma, and J. H. van der Waals, *Mol. Phys.* **22** (1971), 741.
107. G. Jansen, M. Noort, G. W. Canters, and J. H. van der Waals, *Mol. Phys.* **35** (1978), 283.
108. W. G. van Dorp, M. Soma, J. A. Kooter, and J. H. van der Waals, *Mol. Phys.* **28** (1974), 1551.
109. W. G. van Dorp, W. H. Shoemaker, M. Soma, and J. H. van der Waals, *Mol. Phys.* **30** (1975), 1701.
110. G. Jansen, M. Noort, N. van Dijk, and J. H. van der Waals, *Mol. Phys.* **39** (1980), 865.
111. G. W. Canters, J. van Egmond, T. J. Schaafsma, and J. H. van der Waals, *Mol. Phys.* **24** (1972), 1203.
112. T. R. Koehler, *Mol. Cryst. Liquid Cryst.* **50** (1979), 93.
113. S. Völker and R. M. Macfarlane, *Mol. Cryst. Liquid Cryst.* **50** (1979), 213.
114. A. I. M. Dicker, L. W. Johnson, M. Noort, and J. H. van der Waals, *Chem. Phys. Lett.* **94** (1983), 14.
115. A. I. M. Dicker, M. Noort, H. P. H. Thijssen, S. Völker, and J. H. van der Waals, *Chem. Phys. Lett.* **78** (1981), 212.
116. A. I. M. Dicker, J. Dobkowski, M. Noort, S. Völker, and J. H. van der Waals, *Chem. Phys. Lett.* **88** (1982), 135.
117. H. Scheer and H. Inhoffen, in: 'The Porphyrins', Vol. 2 (ed. D. Dolphin, Academic Press, New York, 1978), p. 64; C. Weiss, in: 'The Porphyrins', Vol. 3, p. 211.
118. H. P. H. Thijssen and S. Völker, *Chem. Phys. Lett.* **82** (1981), 478.
119. H. Jansen and J. Lugtenburg, unpublished results.
120. R. J. Platenkamp, H. J. den Blanken, and A. J. Hoff, *Chem. Phys. Lett.* **76** (1980), 35.
121. K. K. Rebane and R. A. Avarmaa, *J. Photochem.* **17** (1981), 311.
122. A. R. Gutiérrez, J. Friedrich, D. Haarer, and H. Wolfrum, *IBM J. Res. Dev.* **26** (1982), 198.
123. (a) L. Kador, G. Schulte, and D. Haarer, *J. Chem. Phys.* **90** (1986), 1264.
(b) M. Romagnoli, W. E. Moerner, F. M. Schellenberger, M. D. Levenson, and G. C. Bjorklund, *J. Opt. Soc. Am.* **B1** (1984), 341.
124. E. I. Zenkevich, A. M. Schulga, A. V. Chernook, and G. P. Gurinovich, *Chem. Phys. Lett.* **109** (1984), 306; E. I. Zenkevich, A. M. Schulga, I. V. Filatov, A. V. Chernook, and G. P. Gurinovich, *Chem. Phys. Lett.* **120** (1985), 63.
125. (a) R. A. Avarmaa and K. K. Rebane, *Spectrochim. Acta* **41A** (1985), 1365.
(b) R. A. Avarmaa and K. K. Rebane, *Stud. Biophys.* **48** (1975), 209.
(c) J. Fünfschilling and D. F. Williams, *Photochem. Photobiol.* **22** (1975) 151; *ibid.* **26** (1977), 109.
126. R. A. Avarmaa and K. H. Mauring, *Zh. Prikl. Spektrosk.* **28** (1978), 658.
127. J. Hála, I. Pelant, L. Parma, and K. Vacek, *J. Lumin* **24/25** (1981), 803; *ibid.* **26** (1981), 117.
128. L. A. Bykovskaya, F. F. Litvin, R. I. Personov, and Yu. V. Romanovskii, *Biofizika* **25** (1980), 13.
129. K. K. Rebane and R. A. Avarmaa, *Chem. Phys.* **68** (1982), 191.
130. G. Jansen, thesis, University of Leiden (1977).
131. G. W. Canters and J. H. van der Waals, in: 'The Porphyrins', Vol. 3 (ed. D. Dolphin, Academic Press, New York, 1978), p. 531, and references therein.
132. R. A. Avarmaa, K. Mauring, and A. Suisalu, *Chem. Phys. Lett.* **77** (1981), 88.
133. R. A. Avarmaa, *Chem. Phys. Lett.* **46** (1977), 279; *Mol. Phys.* **37** (1979), 441; R. A. Avarmaa and T. J. Schaafsma, *Chem. Phys. Lett.* **71** (1980), 339.
134. T. P. Carter and G. J. Small, *Chem. Phys. Lett.* **120** (1985), 178.
135. R. P. H. Kooijman, T. J. Schaafsma, and J. F. Kleibeuker, *Photochem. Photobiol.* **26** (1977), 235.
136. R. J. Platenkamp, *Mol. Phys.* **45** (1982), 113.
137. P. J. Angiolillo, J. S. Leigh, and J. M. Vanderkooi, *Photochem. Photobiol.* **36** (1982), 133.
138. H. Koloczek, J. Fidy, and J. M. Vanderkooi, *J. Chem. Phys.* **87** (1987), 4388.

139. R. A. Avarmaa, I. Renge, and K. Mauring, *FEBS Lett.* **167** (1984), 186.
140. I. Renge, K. Mauring, and R. Avarmaa, *Biochim. Biophys. Acta* **766** (1984), 501.
141. S. G. Boxer, D. S. Gottfried, D. J. Lockhart, and T. R. Middendorf, *J. Chem. Phys.* **86** (1987), 2439.
142. T. Tanako, *J. Mol. Biol.* **110** (1977), 537, 569.
143. J. C. Hanson and B. P. Schoenborn, *J. Mol. Biol.* **153** (1981), 117.
144. I. S. Yang and A. C. Anderson, *Phys. Rev.* **B35** (1987), 9305.
145. G. P. Singh, H. J. Schink, H. v. Lohneysen, F. Parak, and S. Hunklinger, *Z. Phys.* **B55** (1984), 23.
146. I. S. Yang and A. C. Anderson, *Phys. Rev.* **B34** (1986), 2942.
147. J. Friedrich, H. Scheer, B. Zickendraht-Wendelstadt, and D. Haarer, *J. Am. Chem. Soc.* **103** (1981), 1030.
148. J. Friedrich, H. Scheer, B. Zickendraht-Wendelstadt, and D. Haarer, *J. Chem. Phys.* **74** (1981), 2260.
149. W. Köhler, J. Friedrich, R. Fischer, and H. Scheer, *Chem. Phys. Lett.* **143** (1988), 169.
150. G. Feher and M. Y. Okamura, in: 'The Photosynthetic Bacteria' (eds. R. K. Clayton and W. R. Sistrom, Plenum Press, New York, 1978), p. 840.
151. J. Deisenhofer, O. Epp, K. Miki, R. Huber, and H. Michel, *J. Mol. Biol.* **180** (1984), 385.
152. D. Holten, C. Hogenson, M. W. Windsor, C. C. Schenk, W. W. Parson, A. Migus, R. L. Fork, and C. V. Shank, *Biochim. Biophys. Acta* **592** (1980), 461.
153. C. Kirmaier, D. Holten, and W. W. Parson, *FEBS Letters* **185** (1985), 76.
154. S. R. Meech, A. J. Hoff, and D. A. Wiersma, *Chem. Phys. Lett.* **121** (1985), 287; *ibid., Proc. Natl. Acad. Sci. USA* **83** (1986), 9464.
155. S. G. Boxer, D. J. Lockart, and T. R. Middendorf, *Chem. Phys. Lett.* **123** (1986), 476; *ibid., FEBS Lett.* **200** (1986), 237.
156. S. Völker and R. M. Macfarlane, *Chem. Phys. Lett.* **61** (1979), 421.
157. (a) D. J. Lockhart and S. G. Boxer, *Biochemistry* **26** (1987), 664.
      (b) D. J. Lockhart and S. G. Boxer, *Chem. Phys. Lett.* **144** (1988), 243.
158. J. M. Hayes and G. J. Small, *J. Chem. Phys.* **90** (1986), 4928.
159. J. K. Gillie, B. L. Fearey, J. M. Hayes, G. J. Small, and J. H. Golbeck, *Chem. Phys. Lett.* **134** (1987), 316.
160. V. G. Maslov, A. S. Chunaev, and V. V. Tugarinov, *Mol. Biol.* **15** (1981), 788; V. G. Maslov and A. S. Chunaev, *Mol. Biol.* **16** (1982), 479.
161. N. W. Woodbury, M. Becker, D. Middendorf, and W. W. Parson, *Biochemistry* **24** (1985), 7516; J. L. Martin, J. Breton, A. J. Hoff, A. Migus, and A. Antonetti, *Proc. Nat. Acad. Sci. USA* **83** (1986), 957.
162. (a) J. K. Gillie, J. M. Hayes, G. J. Small, and J. H. Golbeck, *J. Chem. Phys.* **91** (1987), 5524.
      (b) J. M. Hayes, J. K. Gillie, D. Tang, and G. J. Small, *Biochim. et Biophys. Acta* **932** (1988), 287.
163. V. A. Shuvalov, A. V. Klevanik, A. V. Kryukov, and B. Ke, *FEBS Lett.* **107** (1979), 313.
164. K. J. Vink, S. de Boer, J. J. Plijter, A. J. Hoff, and D. A. Wiersma, *Chem. Phys. Lett.* **142** (1987), 433.
165. (a) R. T. Harley and R. M. Macfarlane, *J. Phys. C. Lett.* **16** (1983), L 1121.
      (b) W. S. Brocklesby, R. T. Harley, and A. S. Plaut, *Phys. Rev.* **36** (1987), 7941.
166. R. M. Macfarlane and R. M. Shelby, *Phys. Rev. Lett.* **42** (1979), 788.
167. B. Dick and B. Nickel, *Chem. Phys.* **110** (1986), 131.
168. L. E. Erickson, *Phys. Rev.* **B16** (1977), 4731.
169. R. M. Shelby and R. M. Macfarlane, *Opt. Commun.* **27** (1978), 399.
170. R. M. Macfarlane, R. M. Shelby, and D. P. Burum, *Opt. Lett.* **6** (1981), 593.
171. R. M. Macfarlane, D. P. Burum, and R. M. Shelby, *Phys. Rev.* **B29** (1984), 2390.
172. (a) R. M. Macfarlane, R. M. Shelby, A. Z. Genack, and D. A. Weitz, *Opt. Lett.* **5** (1980), 462.
      (b) R. M. Macfarlane and R. M. Shelby, *Opt. Lett.* **6** (1981), 96.
173. D. P. Burum, R. M. Shelby, and R. M. Macfarlane, *Phys. Rev.* **B25** (1982), 3009.
174. R. M. Shelby, A. C. Tropper, R. T. Harley, and R. M. Macfarlane, *Opt. Lett.* **8** (1983), 304.

175. R. M. Macfarlane and J. C. Vial, *Phys. Rev.* **B36** (1987), 3511.
176. A. T. Gradyushko and M. P. Tsvirko, *Opt. Spectr.* **31** (1971), 291.
177. G. Jansen and M. Noort, *Spectrochim. Acta* **32A** (1976), 747.
178. R. J. Platenkamp and M. Noort, *Mol. Phys.* **45** (1982), 97.
179. A. R. Gutiérrez, *Chem. Phys. Lett.* **74** (1980), 293.
180. J. Friedrich, J. D. Swalen, and D. Haarer, *J. Chem. Phys.* **73** (1980), 705.
181. J. Friedrich and D. Haarer, *Chem. Phys. Lett.* **74** (1980), 503.
182. J. Friedrich and D. Haarer, *J. Chem. Phys.* **79** (1983), 1612.
183. F. Graf, H. K. Hong, A. Nazzal, and D. Haarer, *Chem. Phys. Lett.* **59** (1978), 217.
184. (a) Y. Iino, T. Tani, M. Sakuda, H. Nakahara, and K. Fukuda, *Chem. Phys. Lett.* **140** (1987), 76.
      (b) T. Tani, A. Itani, Y. Iino, and M. Sakuda, *J. Chem. Phys.* **88** (1988), 1272.
      (c) M. Yoshimura, M. Maedea, and T. Nakayama, *Chem. Phys. Lett.* **143** (1988), 342.
185. B. L. Feary, F. P. Carter, and G. J. Small, *J. Chem. Phys.* **87** (1983), 3590.
186. R. van den Berg and S. Völker, *Chem. Phys. Lett.* **137** (1987), 201.
187. R. van den Berg and S. Völker, *Chem. Phys.* **128** (1988), 257 (special issue, eds. M. D. Fayer and J. L. Skinner).
188. E. M. Kosower, *J. Am. Chem. Soc.* **80** (1958), 3253, 3261, 3267.
189. B. M. Kharlamov, L. A. Bykovskaya, and R. I. Personov, *Chem. Phys. Lett.* **50** (1977), 407.
190. B. L. Fearey, T. P. Carter, and G. J. Small, *Chem. Phys.* **101** (1986), 279.
191. C. A. Walsh, M. Berg, L. R. Narasimhan, and M. D. Fayer, *Chem. Phys. Lett.* **130** (1986), 6; *J. Chem. Phys.* **86** (1987), 77.
192. R. M. Hochstrasser and D. S. King, *J. Am. Chem. Soc.* **97** (1975), 4760; B. Dellinger, D. S. King, R. M. Hochstrasser, and A. B. Smith, *J. Am. Chem. Soc.* **99** (1977), 3197.
193. D. M. Burland and D. Haarer, *IBM J. Res. Dev.* **23** (1979), 534, and references therein.
194. J. Meiler and J. Friedrich, *Chem. Phys. Lett.* **134** (1987), 263.
195. A. Kiermeier, K. Dietrich, E. Riedle, and H. J. Neusser, *J. Chem. Phys.* **85** (1986), 6983.
196. W. E. Moerner and M. D. Levenson, *J. Opt. Soc. Am.* **B2** (1985), 915.
197. W. Lenth and W. E. Moerner, *Opt. Commun.* **58** (1986), 249, and references therein.
198. A. Winnacker, R. M. Shelby, and R. M. Macfarlane, *Opt. Lett.* **10** (1985), 350.
199. R. M. Macfarlane, W. S. Brocklesby, P. D. Bloch, and R. T. Harley, *Opt. Commun.* **58** (1986), 25.
200. R. M. Macfarlane and J. C. Vial, *Phys. Rev.* **B34** (1986), 1.
201. A. J. Silversmith, W. Lenth, and R. M. Macfarlane, private communication.
202. M. Iannone, G. W. Scott, D. Brinza, and D. R. Coulter, *J. Chem. Phys.* **85** (1986), 4863.
203. T. P. Carter, C. Bräuchle, V. Y. Lee, M. Manavi, and W. E. Moerner, *Opt. Lett.* **12** (1987), 370; *J. Phys. Chem.* **91** (1987), 3998.
204. J. M. Clemens, R. M. Hochstrasser, and H. P. Trommsdorff, *J. Chem. Phys.* **80** (1984), 1744.
205. M. Dubs, H. H. Günthard, *Chem. Phys. Lett.* **64** (1979), 105; *J. Mol. Struct.* **60** (1980), 311.
206. M. Dubs, L. Ermanni, and H. H. Günthard, *J. Mol. Spectrosc.* **91** (1982), 458.
207. W. E. Moerner, A. R. Chraplyvy, A. J. Sievers, and R. H. Silsbee, *Phys. Rev.* **B28** (1983), 7244; *Phys. Rev.* **B29** (1984), 4791, and references therein.
208. R. C. Spitzer, W. P. Ambrose, and A. J. Sievers, *Opt. Lett.* **11** (1986), 428.
209. B. M. Kharlamov, R. I. Personov, and L. A. Bykovskaya, *Opt. Spectros.* **39** (1975), 7013.
210. (a) B. L. Fearey, R. P. Stout, J. M. Hayes, and G. J. Small, *J. Chem. Phys.* **78** (1983), 7013.
      (b) J. M. Hayes and G. J. Small, *Chem Phys.* **27** (1978), 151.
211. (a) R. Jankowiak and H. Bässler, *Chem. Phys. Lett.* **95** (1983), 124.
      (b) R. Jankowiak and H. Bässler, *Chem. Phys. Lett.* **101** (1983), 274.
212. R. Jankowiak, H. Bässler, and B. Silbey, *Chem. Phys. Lett.* **125** (1986), 139.
213. A. Elschner and H. Bässler, *Chem. Phys.* **112** (1987), 285.
214. (a) R. van den Berg and S. Völker, *J. Lumin.* **38** (1987), 25.
      (b) R. van den Berg and S. Völker, *Chem. Phys. Lett.* **150** (1988), 491.
215. R. M. Macfarlane and R. M. Shelby, *Opt. Commun.* **42** (1982), 346.
216. R. M. Macfarlane and R. S. Meltzer, *Opt. Commun.* **52** (1985), 320.

217. R. M. Macfarlane, R. M. Shelby, and A. Winnacker, *Phys. Rev.* **B33** (1986), 4207.
218. G. Castro, D. Haarer, R. M. Macfarlane, and H. P. Trommsdorff, U. S. Patent 4101976, July 18 (1978); D. Haarer, R. V. Pole and S. Völker, U. S. Patent 4103346, July 25 (1978); R. M. Macfarlane, U. T. Müller-Westerhoff and S. Völker, *IBM Tech. Discl. Bull.* **22** (1980), 3352.
219. H. de Vries and D. A. Wiersma, *J. Chem. Phys.* **72** (1980), 1851.
220. T. J. Aartsma and D. A. Wiersma, *Chem. Phys. Lett.* **42** (1976), 520.
221. T. J. Aartsma and D. A. Wiersma, *Chem. Phys. Lett.* **54** (1978), 415.
222. D. E. Cooper, R. W. Olson, and M. D. Fayer, *J. Chem. Phys.* **72** (1980), 2332.
223. J. B. W. Morsink, T. J. Aartsma, and D. A. Wiersma, *Chem. Phys. Lett.* **49** (1977), 34.
224. K. Duppen, L. W. Molenkamp, J. B. W. Morsink, D. A. Wiersma, and H. P. Trommsdorff, *Chem. Phys. Lett.* **84** (1981), 421.
225. L. W. Molenkamp, D. P. Weitekamp, and D. A. Wiersma, *Chem. Phys. Lett.* **99** (1983), 382.
226. P. Brout and V. M. Visscher, *Phys. Rev. Lett.* **9** (1962), 54.
227. D. Hsu and J. L. Skinner, *J. Chem. Phys.* **87** (1987), 54, and references therein.
228. W. H. Hesselink and D. A. Wiersma, *J. Chem. Phys.* **73** (1980), 648.
229. L. W. Molenkamp and D. A. Wiersma, *J. Chem. Phys.* **80** (1984), 3054.
230. (a) P. J. F. Verbeek, A. I. M. Dicker, and J. Schmidt, *Chem. Phys. Lett.* **56** (1978), 585.
     (b) P. J. F. Verbeek, thesis, University of Leiden (1979); J. Schmidt, chapter on 'Spin-Lattice Relaxation', in this book, p. 3.
     (c) J. Schmidt, chapter on 'Spin-Lattice Relaxation', in this book, p. 3.
231. T. R. Koehler, *J. Chem. Phys.* **72** (1980), 3389.
232. S. Völker and R. M. Macfarlane, unpublished results.
233. R. Kubo and T. Tomita, *J. Phys. Soc. Japan* **9** (1954), 888.
234. P. W. Anderson, *J. Phys. Soc. Japan* **9** (1954), 316.
235. H. M. McConnell, *J. Chem. Phys.* **28** (1958), 430.
236. C. B. Harris, *J. Chem. Phys.* **67** (1977), 5607.
237. R. M. Shelby, C. B. Harris, and P. A. Cornelius, *J. Chem. Phys.* **70** (1979), 34.
238. R. Silbey, chapter on 'Relaxation Theory', in this book, and references therein.
239. J. L. Skinner and D. Hsu, *J. Phys. Chem.* **90** (1986), 4931, and references therein.
240. K. E. Jones and A. H. Zewail in: 'Advances in Laser Chemistry' (ed. A. H. Zewail, Springer Verlag, Berlin, 1978), and references therein.
241. I. S. Osad'ko, in: 'Spectroscopy and Excitation Dynamics of Condenses Molecular Systems' (eds. V. M. Agranovich and R. M. Hochrasser, North Holand, Amsterdam 1983), p. 437, and references therein.
242. C. A. van 't Hof and J. Schmidt, *Chem. Phys. Lett.* **36** (1975), 457; *Chem. Phys. Lett.* **42** (1976), 73.
243. P. J. F. Verbeek, C. A. van 't Hof, and J. Schmidt, *Chem. Phys. Lett.* **51** (1977), 292.
244. J. Schmidt, chapter on 'Spin-Lattice Relaxation', in this book, and references therein.
245. B. Jackson and R. Silbey, *Chem. Phys. Lett.* **99** (1983), 331.
246. R. Silbey and K. Kassner, *J. Lumin.* **36** (1987), 283, and references therein.
247. D. E. McCumber and M. D. Sturge, *J. Appl. Phys.* **34** (1963), 1682.
248. B. di Bartolo, 'Optical Interactions in Solids' (Wiley, New York, 1968), and references therein.
249. Ph. de Bree and D. A. Wiersma, *J. Chem. Phys.* **70** (1979), 790.
250. A. G. Redfield, *IBM J. Res. Dev.* **1** (1957), 19; *Adv. Magn. Res.* **1** (1965), 1.
251. D. Hsu and J. L. Skinner, *J. Chem. Phys.* **83** (1985), 2107.
252. K. K. Rebane and P. M. Saari, *J. Lumin.* **16** (1978), 223.
253. R. M. Hochrasser and C. A. Nyi, *J. Chem. Phys.* **70** (1979), 1112.
254. T. Tamm and P. Saari, *Chem. Phys.* **40** (1979), 311.
255. A. Amirav, U. Even, and J. Jortner, *Opt. Commun.* **32** (1980), 266; *Chem. Phys. Lett.* **71** (1980), 12.
256. P. L. De Cola, R. M. Hochrasser, and H. P. Trommsdorff, *Chem. Phys. Lett.* **72** (1980), 1.
257. W. H. Hesselink and D. A. Wiersma, *J. Chem. Phys.* **74** (1981), 886.
258. A. I. M. Dicker and S. Völker, *Chem. Phys. Lett.* **87** (1982), 481.
259. A. A. Gorokhovskii and J. Kikas, *Opt. Commun.* **21** (1977), 272.

260. S. Völker and R. M. Macfarlane, *J. Lumin.* **18/19** (1979), 213.
261. R. M. Macfarlane and R. M. Shelby, *Opt. Lett.* **9** (1984), 533.
262. R. T. Harley and R. M. Macfarlane, *J. Phys.* **C16** (1983), 1507.
263. R. T. Harley and R. M. Macfarlane, *J. Phys. C Lett.* **16** (1983), L 395.
264. W. E. Moerner, P. Pokrowsky, F. M. Schellenberger, and G. C. Bjorklund, *Phys. Rev.* **B33** (1986), 5702.
265. G. W. Canters, G. Jansen, M. Noort, and J. H. van der Waals, *J. Phys. Chem.* **80** (1976), 2253.
266. G. W. Canters, M. Noort, G. Jansen, and J. H. van der Waals, Proc. Eur. Congr. Mol. Spectrosc. 12th. (Elsevier, Amsterdam 1976), p. 445.
267. N. van Dijk, M. Noort, S. Völker, G. W. Canters, and J. H. van der Waals, *Chem. Phys. Lett.* **71** (1980), 415.
268. W. G. van Dorp, G. W. Canters, and J. H. van der Waals, *Chem. Phys. Lett.* **35** (1975), 450.
269. N. van Dijk, thesis, University of Leiden (1981), Chapter 4; W. A. J. A. van der Poel, University of Leiden (1984), Chapter 2.
270. M. Gouterman, in: 'The Porphyrins', Vol. 3 (ed. D. Dolphin, Academic Press, New York, 1978), Chapter 1, and references therein.
271. A. J. McHugh, M. Gouterman, and C. Weiss, *Theoret. Chim. Acta* **24** (1972), 346.
272. J. D. Petke, G. M. Maggiora, L. L. Shipman, and R. E. Christoffersen, *J. Mol. Spectry.* **73** (1978), 311; C. Weiss, H. Kobayashi and M. Gouterman, *J. Mol. Spectry.* **16** (1965), 415.
273. A. E. Bell, *Proc. Soc. Infor. Disp.* **24** (1983), 17, and references therein.
273. R. C. Zeller and R. O. Pohl, *Phys. Rev.* **B4** (1971), 2029.
274. For a review on glasses, see: 'Amorphous Solids. Low Temperature Properties' (ed. W. A. Phillips, Springer Verlag, Berlin, 1981).
275. W. A. Phillips, *J. Low Temp. Phys.* **7** (1972), 351; P. W. Anderson, B. I. Halperin, and C. M. Varma, *Phil. Mag.* **25** (1972), 1.
276. S. Hunklinger and W. Arnold, in: 'Physical Acoustics', Vol. XII (ed. W. P. Mason and R. N. Thurston, Academic Press, New York, 1976), p. 155.
277. P. Reineker and K. Kassner, in: 'Optical Spectroscopy of Glasses' (ed. I. Zschokke, D. Reidel, Dordrecht, 1986), p. 65.
278. P. M. Selzer, D. L. Huber, D. S. Hamilton, W. M. Yen, and M. J. Weber, *Phys. Rev. Lett.* **36** (1976), 813.
279. P. Avouris, A. Campion, and M. A. El-Sayed, *J. Chem. Phys.* **67** (1977), 3397.
280. J. R. Morgan and M. A. El-Sayed, *Chem. Phys. Lett.* **84** (1981), 213.
281. T. L. Reinecke, *Solid State Comm.* **32** (1979), 1103.
282. S. K. Lyo and R. Orbach, *Phys. Rev.* **B22** (1980), 4223.
283. P. Reineker and H. Morawitz, *Chem. Phys. Lett.* **86** (1982), 359; H. Morawitz and P. Reineker, *Solid State Comm.* **42** (1982), 609.
284. G. S. Dixon, R. C. Powell, and Xu Gang, *Phys. Rev.* **B33** (1986), 2713.
285. S. Alexander and R. Orbach, *J. de Physique* **43** (1982), L-625.
286. J. Hegarty and W. M. Yen, *Phys. Rev. Lett.* **43** (1979), 1126.
287. J. M. Pellegrino, W. M. Yen, and M. J. Weber, *J. Appl. Phys.* **51** (1980), 6332.
288. R. M. Shelby, *Opt. Lett.* **8** (1983), 88.
289. (a) Hegarty, M. M. Broer, B. Golding, J. R. Simpson, and J. B. MacChesney, *Phys. Rev. Lett.* **51** (1983), 2033.
(b) M. M. Broer, B. Golding, W. H. Haemmerle, J. R. Simpson, and D. L. Huber, *Phys. Rev.* **B33** (1986), 4160.
(c) D. L. Huber, M. M. Broer, and B. Golding, *Phys. Rev. Lett.* **52** (1984), 2281; *ibid.*, *Phys. Rev.* **B33** (1986), 7297.
290. F. J. Bergin, J. F. Donegan, T. J. Glynn, and G. F. Imbusch, *J. Lum.* **34** (1986), 307.
291. J. Fünfschilling and I. Zschokke-Gränacher, *Chem. Phys. Lett.* **110** (1984), 315.
292. (a) P. J. Flory, *Science* **188** (1975), 1268.
(b) R. Zallen, 'The Physics of Amorphous Solids' (ed. J. Wiley, New York, 1983), Chapter 3.

293. H. P. H. Thijssen, S. Völker, M. Schmidt, and H. Port, *Chem. Phys. Lett.* **94** (1983), 537.
294. A. A. Gorokhovskii, Y. V. Kikas, V. V. Palm, and L. A. Rebane, *Sov. Phys. Sol. State* **23** (1981), 602.
295. W. Breinl and J. Friedrich, *Chem. Phys. Lett.* **145** (1988), 107.
296. R. Locher, A. Renn, and U. P. Wild, *Chem. Phys. Lett.* **138** (1987), 405.
297. H. van der Laan and S. Völker, to be published.
298. (a) S. de Silvestri, A. M. Weiner, J. G. Fujimoto, and E. P. Ippen, *Chem. Phys. Lett.* **112** (1984), 195.
     (b) H. Nakatsuka, M. Fujiwara, and R. Kuroda, *J. de Physique* **C7** (1985), 511.
     (c) C. H. Brito Cruz, R. L. Fork, W. H. Knox, and C. V. Shank, *Chem. Phys. Lett.* **132** (1986), 341.
299. O. Haida, H. Suga, and S. Seki, *J. Chem. Therm.* **9** (1977), 1113.
300. R. van den Berg and S. Völker, unpublished results.
301. (a) W. Breinl, J. Friedrich, and D. Haarer, *Chem. Phys. Lett.* **106** (1984), 487.
     (b) W. Breinl, J. Friedrich, and D. Haarer, *J. Chem. Phys.* **81** (1984), 3915.
302. J. Friedrich and D. Haarer, *J. de Physique* **C7** (1985), 357.
303. J. Zimmermann, *Cryogenics*, January 1984, 27.
304. P. J. van der Zaag, J. P. Galaup, and S. Völker, to be published.
305. J. R. Klauder and P. W. Anderson, *Phys. Rev.* **125** (1962), 912.
306. J. L. Black and B. I. Halperin, *Phys. Rev.* **B16** (1977), 2879.
307. (a) S. K. Lyo, *Phys. Rev. Lett.* **48** (1982), 688; S. K. Lyo, in: 'Electronic Processes in Organic Molecular Aggregates' (eds. P. Reineker, H. Haken and H. C. Wolf, Springer Verlag, Berlin, 1983), p. 215.
     (b) S. K. Lyo, in: 'Optical Spectroscopy of Glasses' (ed. I. Zschokke, D. Reidel, Dordrecht, 1986), p. 1.
308. D. L. Huber, *J. Non-Cryst. Solids* **51** (1982), 241.
309. S. Hunklinger and M. Schmidt, *Z. Phys.* **B54** (1984), 93.
310. S. K. Lyo and R. Orbach, *Phys. Rev.* **B29** (1984), 2300.
311. P. Reineker, H. Morawitz, and K. Kassner, *Phys. Rev.* **B29** (1984), 4546.
312. R. Maynard, *J. de Physique* **C7** (1985), 325.
313. I. S. Osad'ko, *J. Exp. Theor. Phys.* **39** (1984), 426; *Chem. Phys. Lett.* **115** (1985), 411; *J. Exp. Theor. Phys. Lett.* **90** (1986), 1453.
314. K. Kassner and P. Reineker, *Chem. Phys.* **106** (1986), 345, 371; K. Kassner and P. Reineker, *Phys. Rev.* **B35** (1987), 828.
315. S. Alexander and R. Orbach, *J. Physique* **43** (1982), L-625.
316. B. B. Mandelbrot, 'The Fractal Geometry of Nature' (ed. W. H. Freeman and Co., San Francisco, 1982).
317. P. G. de Gennes, 'Scaling Concepts in Polymer Physics' (ed. Cornell University, Ithaca, 1979), p. 138.
318. J. C. Lasjaunias, A. Ravex, M. Vandorpe, and S. Hunklinger, *Solid State Comm.* **17** (1975), 1045.
319. C. M. Varma, R. Dynes, and J. Banavar, *J. Phys.* **C15** (1982), L 1221.
320. L. Pietronero, unpublished results.
321. N. B. Manson, *Optics Comm.* **44** (1982), 32.
322. R. M. Shelby and R. M. Macfarlane, *Phys. Rev. Lett.* **47** (1981), 1172.
323. A. J. Silversmith and N. B. Manson, *J. Lumin.* **31/32** (1984), 848.
324. B. M. Kharlamov, E. I. Al'shitz, and R. I. Personov, *Opt. Comm.* **44** (1983), 149; *ibid., Sov. Phys. JETP* **60** (1984), 428.
325. (a) N. I. Ulitskii, B. M. Kharlamov, A. M. Pyndyk, and R. I. Personov, *Opt. Spectrosc.* **59** (1985), 560.
     (b) N. I. Ulitskii, B. M. Kharlamov, and R. I. Personov, *Chem. Phys.* **122** (1988), 1.
326. R. van den Berg, H. van der Laan, and S. Völker, *Chem. Phys. Lett.* **142** (1987), 535.
327. W. Köhler, W. Breinl, and J. Friedrich, *J. Phys. Chem.* **89** (1985), 2473.

328. A. J. Meixner, A. Renn, and U. P. Wild, *J. Phys. Chem.* **90** (1986), 6777.
329. (a) V. D. Samoilenko, N. V. Razumova, and R. I. Personov, *Opt. Spectrosc.* **52** (1982), 346.
     (b) V. I. Ivanov, R. I. Personov, and N. V. Rasumova, *Opt. Spectrosc.* **58** (1985), 6.
330. (a) A. J. Meixner, A. Renn, S. E. Bucher, and U. P. Wild, XIth Molecular Crystal Symposium, Lugano (1985), p. 198.
     (b) U. P. Wild, S. E. Bucher, and F. A. Burkhalter, *Appl. Opt.* **24** (1985), 1526.
331. (a) L. Kador, R. Personov, W. Richter, Th. Sesselmann, and D. Haarer, *Polymer J.* **19** (1987), 61.
     (b) Th. Sesselmann, L. Kador, W. Richter, and D. Haarer, *Europhys. Lett.* **5** (1988), 351.
332. V. Bogner, P. Schätz, R. Seel, and M. Maier, *Chem. Phys. Lett.* **102** (1983), 267.
333. O. N. Korotaev, N. M. Surin, A. I. Yurchenko, V. I. Glyadkovsky, and E. I. Donskoi, *Chem. Phys. Lett.* **110** (1984), 533.
334. H. Labhart, *Adv. Chem. Phys.* **13** (1967), 179.
335. W. Liptay, in 'Excited States', Vol. 1 (Academic Press, New York, 1974), p. 129.
336. R. P. Feynman, R. B. Leighton, and M. Sands, 'The Feynman Lectures on Physics', Vol. 2 (Addison-Wesley, Reading, 1966), Chapter 32.
337. P. Schätz and M. Maier, *J. Chem. Phys.* **87** (1987), 809.
338. H. W. Offen, in 'Organic Molecular Photophysics', Vol. 1 (ed. J. B. Birks, London, New York, 1973), p. 103.
339. W. Richter, G. Schulte, and D. Haarer, *Opt. Commun.* **51** (1984), 412.
340. Th. Sesselmann, W. Richter, and D. Haarer, *Europhys. Lett.* **2** (1986), 947.
341. Th. Sesselmann, W. Richter, D. Haarer, and H. Morawitz, *Phys. Rev.* **B36** (1987), 7601.

RELAXATION THEORY APPLIED TO SCATTERING OF EXCITATIONS AND OPTICAL TRANSITIONS IN CRYSTALS AND SOLIDS

ROBERT SILBEY

Department of Chemistry, Massachusetts Institute of Technology, Cambridge, MA 02139, U.S.A.

1. Introduction

In this chapter, we will discuss the application of standard quantum relaxation theory to the problems of excitation transport and scattering in molecular solids and dephasing of optical transitions in solids and glasses, all at low temperatures. In the next section, we will review the standard theory based on the weak coupling approximation (Redfield theory), and apply a number of simple microscopic models to discuss the dynamics and temperature dependence of various processes. In the third section we apply these ideas to the study of the scattering of delocalized excitations in low-temperature solids, including the effects of impurities to low order. We compare our theoretical results to the experimental results of Schmidt (this volume). In the fourth section, we discuss the homogeneous optical line shapes of localized excitations in crystals and glasses, presenting a number of the dynamical models used to discuss this phenomenon. The many models used to describe the dephasing of the optical transition are compared and it is shown how to derive the results of these models from the standard Redfield theory. Unfortunately, because many of the microscopic parameters are unknown, it is not yet possible to decide which is the correct model. This is discussed in detail in Section 4.

2. Theory

2.1. THE REDFIELD EQUATIONS

The quantum theory of relaxation [1] is based on the idea that the quantum system, S, under study in the laboratory is a small part of the entire system which consists of S and a large reservoir, R. For example, S might be a molecule embedded in a crystal lattice, and R would be all the degrees of freedom (or eigenstates) of the lattice, e.g. the phonons and vibrations. In another case, S might be the spin variables of the molecule, and R *all* the other variables in the problem. It is even possible to think of the electronic states of a large molecule as comprising S and the vibrational and rotational variables of the molecule as R. The other components of the theory of relaxation are that S and R are *weakly* interacting, and that the properties of S vary slowly in time compared to the relaxation times of the

243

J. Fünfschilling (ed.), Relaxation Processes in Molecular Excited States, 243–276.

reservoir, R. This mismatch of time scales allows us to formulate the equations of motion of the system in a straightforward manner, and allows the Redfield equations [2] to be derived.

The total Hamiltonian of the combined system is

$$H = H_S + H_R + V \equiv H^{(0)} + V \tag{2.1}$$

where H_S is the system part, H_R the reservoir and V the interaction between them. This way of writing the Hamiltonian implies that we know H_S, H_R and V. In order to make the weak coupling approximation, V must be small in some sense. If it is not, all is not lost. In some cases, it is possible to find a new definition of the system which includes a large part of V, so that the remainder is small. For example, in order to describe a molecule embedded in a solid, can we use the gas phase electronic wave functions or states in such a description? Or should we use wave functions perturbed by the other molecules in the solid for the description of H_S? The latter will be preferable in most cases, if it is possible. In what follows, we will assume that a good choice of $H^{(0)}$ has been made and derive the Redfield equations based on that.

The basic operator which is used to describe the time evolution of the entire system is the density matrix, ρ, which satisfies the equation

$$\dot{\rho} = -i[H, \rho] \equiv -iL\rho \tag{2.2}$$

where L is the operation defined by the commutator with H. The average value of any operator of S, e.g. A, is

$$\langle A(t) \rangle = \mathrm{Tr}\{\rho(t)A\} = \mathrm{Tr}_S\{A \cdot \mathrm{Tr}_R\rho(t)\} = \mathrm{Tr}_S\{A\sigma(t)\} \tag{2.3}$$

The initial trace is over the states of the combined system, which can be written as the trace over the states of S and R separately since the total trace is invariant to changing the representation. The reduced density matrix $\sigma(t)$ is an operator in S space only since the trace has been taken over the states of R. It is $\sigma(t)$ we wish to find, and which is approximated by the solution to the Redfield equation. Using the eigenstates of $H_S\{|n\rangle\}$ as a basis we can write

$$\langle A(t) \rangle = \sum_{m,n} A_{nm}\sigma_{mn}(t) \tag{2.4}$$

Thus the relaxation of $\langle A(t) \rangle$ is governed by that of the appropriate matrix elements of $\sigma(t)$. Note that the diagonal matrix elements $\sigma_{nn}(t)$ are populations of the states $|n\rangle$, and the off diagonal elements σ_{mn} are coherences between states $|m\rangle$ and $|n\rangle$.

There are a variety of routes to find the equation of motion of $\sigma(t)$ [1, 2, 3], and the results often look different because of the use of different projection operators, etc. Since the exact equation of motion is useless for most purposes, one must use an approximate form. The standard approximations are those mentioned above: weak coupling (i.e. second order perturbation theory) and separation of time scales (i.e. the reservoir R relaxes to its equilibrium state sufficiently quickly). Sometimes one must go beyond these approximations, or extract information in another manner [4], but we will not concern ourselves with these cases. After making these approximations, we arrive at the Redfield equations for $\sigma_{mn}(t)$:

$$\dot{\sigma}_{mn}(t) = -i\omega_{mn}\,\sigma_{mn}(t) - \sum_{mnpq} T_{mn,\,pq}\,\sigma_{pq}(t) \tag{2.5}$$

with $\omega_{mn} = \varepsilon_m - \varepsilon_n$, the difference in the eigenvalues of H_S in the eigenstates $|m\rangle$ and $|n\rangle$, and the relaxation matrix $T_{mn,\,pq}$ independent of time. We have also assumed in this derivation that at $t = 0$ the reservoir could be characterized by an equilibrium density matrix (e.g. at temperature T), independent of S, so that $\rho(0) = \sigma(0)\rho_R^{eq}$. Finally, we have neglected the frequency shifts of the system S states due to V, to get

$$T_{mn,\,pq} = \delta_{mp}\sum_r t_{nrrq}(\omega_{qr}) + \delta_{nq}\sum_r t_{mrrp}(\omega_{pr}) -$$
$$- t_{pmnq}(\omega_{qn}) - t_{qnmp}(\omega_{pm}) \tag{2.6}$$

with

$$t_{pmnq}(\omega) = \frac{1}{2}\int_{-\infty}^{+\infty} d\tau\, e^{i\omega\tau}\langle V_{pm}(\tau)V_{nq}(0)\rangle_R \tag{2.7}$$

Here, $V_{mp} \equiv \langle m|\,V\,|p\rangle$ is an operator in R,

$$V_{mp}(\tau) = e^{iH_R\tau}\,V_{mp}\,e^{-iH_R\tau} \tag{2.8}$$

and the brackets with the subscript R represent an average over the equilibrium density matrix of R:

$$\langle B\rangle_R \equiv \mathrm{Tr}_R\left\{\frac{e^{-\beta H_R}}{z}\,B\right\} = \sum_\alpha e^{-\beta E_\alpha}\,B_{\alpha\alpha}\,\bigg/\,\sum_\alpha e^{-\beta E_\alpha} \tag{2.9}$$

with $|\alpha\rangle$ an eigenstate of H_R and $\beta = (k_B T)^{-1}$. Derivations of all these can be found in the literature [1, 3].

It is useful to examine this rather complex equation, which allows a tremendous range of behavior, in a few simple cases. First, let us look at dynamics of the population variables, $\sigma_{nn}(t)$. We note

$$\dot{\sigma}_{nn}(t) = -\sum_p T_{nn,\,pq}\,\sigma_{pp}(t) - \sum_{p,\,q}{}' T_{nn,\,pq}\,\sigma_{pq}(t) \tag{2.10}$$

with the prime indicating $p \neq q$. Since $\sigma_{pq}(t)$ has a natural frequency of ω_{pq}, we can hope that it decays quickly and that the last set of terms does not play a major role, so we can neglect it. Then we see that the populations obey a master equation [5], with the transition rates given by ($n \neq p$)

$$-T_{nn,\,pp} = 2t_{pnnp}(\omega_{pn})$$

$$= \int_{-\infty}^{+\infty} d\tau\; e^{i\omega_{pn}\tau}\, \langle V_{pn}(\tau) V_{np}(0) \rangle \tag{2.11}$$

$$= 2\pi \sum_{a,\,\gamma} \frac{e^{-\beta E_a}}{z} |\langle pa| V |n\gamma\rangle|^2\; \delta(\varepsilon_p + \varepsilon_a - \varepsilon_n - \varepsilon_\gamma)$$

$$T_{nn,\,nn} = -\sum_{p\,\neq\,n} T_{pp,\,nn}.$$

This is the Golden Rule expression [6] for the rate of going from $|p\rangle$ to $|n\rangle$ starting from an equilibrium distribution of reservoir states. Thus the rate of decay of population variables (T_1^{-1}) is governed by the Golden Rule rates.

Now, consider the dynamics of an off diagonal element, σ_{mn}. Then,

$$\dot{\sigma}_{mn} = [-i\omega_{mn} - T_{mn,\,mn}]\sigma_{mn} - {\sum_{p,\,q}}'' T_{mn,\,pq}\, \sigma_{pq} \tag{2.12}$$

(with the double prime indicating that the single term $m = p$, $n = q$ is omitted in the sum). Look at $T_{mn,\,mn}$

$$\begin{aligned}
T_{mn,\,mn} &= \sum_r t_{nr,\,rn}(\omega_{nr}) + \sum_r t_{mr,\,rm}(\omega_{mr}) - \\
&\quad - t_{mm,\,nn}(0) - t_{nn,\,mm}(0) \\[4pt]
&= \sum_{r\,\neq\,n} t_{nr,\,rn}(\omega_{nr}) + \sum_{r\,\neq\,m} t_{mr,\,rm}(\omega_{mr}) \\
&\quad + t_{nn,\,nn}(0) + t_{mm,\,mm}(0) - t_{nn,\,mm}(0) - t_{mm,\,nn}(0)
\end{aligned} \tag{2.13}$$

so that

$$\begin{aligned}
T_{mn,\,mn} &= \frac{1}{2} \sum_{r\,\neq\,n} T_{nn,\,rr} + \frac{1}{2} \sum_{r\,\neq\,n} T_{mm,\,rr} + \\[4pt]
&\quad + \frac{1}{2} \int_{-\infty}^{+\infty} \langle\{V_{mm}(\tau) - V_{nn}(\tau)\}\{V_{mm}(0) - V_{nn}(0)\}\rangle
\end{aligned}$$

Thus the decay of $\sigma_{mn}(T_2^{-1})$ is governed (if we can neglect coupling to all the other σ_{pq}) by the sum of one half the decay rate of the population of m and one half the population decay rate of n, and another term, the pure dephasing rate. All of this is well known; we just indicate that this equation contains all the usual terms.

Often, we speak about these terms in a slightly different manner. By focussing on the states of S and the matrix elements of the perturbation V in those states, V_{mn}, we can think about diagonal energy fluctuations, V_{nn}, and off diagonal fluctuations V_{nm}, $n \neq m$. The diagonal terms represent instantaneous changes in the *energy* of the state $|m\rangle$ due to the interaction with the reservoir and the off diagonal terms represent fluctuations in the interaction matrix elements between states $|n\rangle$ and $|m\rangle$ due to the reservoir. This leads to various stochastic theories [7] which treat the reservoir as a source of fluctuations and postulate time dependences for $V_{nm}(t)$ and $V_{nn}(t)$ to be used in the formulas for T. These are powerful approaches, but have to be treated with care especially at low temperature [8] where the reservoir may be more complicated than can be easily accommodated by these methods. In this chapter, we will treat simple microsopic models for H_R and V, so that direct calculations can be made.

After all these approximations have been made, the equation for the density matrix elements is still formidable. One can arrange the elements σ_{nm} as a row vector, σ, and the equation then becomes (in an obvious notation)

$$\dot{\sigma}(t) = -[i\omega + T]\sigma(t) \tag{2.15}$$

so that, formally at least,

$$\sigma(t) = \{\exp-(i\omega + T)t\}\,\sigma(0) \tag{2.16}$$

Thus the eigenvalues of the complex matrix, $i\omega - T$, are the oscillatory frequencies and decay rates for the density matrix elements, and the usual methods for finding eigenvalues and vectors can be used to study relaxation.

2.2. SIMPLE MODELS

Assume that the reservoir is a collection of harmonic oscillators, such as the acoustic phonons of a solid or a set of optical modes (such as librations) with a narrow frequency range. Assume that the system under study has only two levels, $|1\rangle$ and $|2\rangle$. Then we can write

$$H_S = \varepsilon_1|1\rangle\langle 1| + \varepsilon_2|2\rangle\langle 2|$$

$$H_R = \sum_\lambda \frac{1}{2}\{p_\lambda^2 + \omega_\lambda^2 q_\lambda^2\} = \sum_\lambda \omega_\lambda\left(a_\lambda^+ a_\lambda + \frac{1}{2}\right) \tag{2.17}$$

with p_λ and q_λ the momentum and coordinate of mode λ and a_λ the usual boson annihilation operator [6]. We take the interaction, V, to be

$$V = V_{11}|1\rangle\langle 1| + V_{22}|2\rangle\langle 2| + V_{12}|1\rangle\langle 2| + V_{21}|2\rangle\langle 1| \tag{2.18}$$

with V_{nm} operators in the phonon system which we will specify depending on our model. If we are concerned with the optical absorption between states $|1\rangle$ and $|2\rangle$, we must consider the density matrix elements $\sigma_{12}(t)$ and $\sigma_{21}(t)$. At the simplest level, we know that the decay of $\sigma_{12}(t)$ is governed by (see above) $T_{12,12}$:

$$T_{12,12} = \tfrac{1}{2}(T_{22,11} + T_{11,22}) +$$

$$+ \frac{1}{2} \int_{-\infty}^{+\infty} d\tau \langle \{V_{22}(\tau) - V_{11}(\tau)\} \{V_{22}(0) - V_{11}(0)\} \rangle \qquad (2.19)$$

Let us calculate each of these terms for a specific model for V_{nm}. The simplest model is to take V_{nm} to be a combination of linear and quadratic phonon coordinates ($q_\lambda = (2\omega_\lambda)^{-1/2}(a_\lambda + a_\lambda^+)$):

$$V_{nm} = \sum_\lambda g_\lambda^{nm}(a_\lambda + a_\lambda^+) + \sum_{\substack{\lambda,\lambda' \\ \lambda \neq \lambda'}} f_{\lambda\lambda'}^{nm}(a_\lambda + a_\lambda^+)(a_{\lambda'} + a_{\lambda'}^+) \qquad (2.20)$$

First we must calculate $V_{nm}(t)$; however this is straightforward for bosons, since $a_\lambda(t) = e^{-i\omega_\lambda t} a_\lambda(0)$ and $a_\lambda^+(t) = e^{+i\omega_\lambda t} a_\lambda^+(0)$. Now we must calculate the relevant averages; this is also simple for bosons, since

$$\langle a_\lambda a_\lambda^+ \rangle = (\bar{n}_\lambda + 1) = (1 - e^{-\beta\omega_\lambda})^{-1};$$

$$\langle a_\lambda^+ a_\lambda \rangle = \bar{n}_\lambda = e^{-\beta\omega_\lambda}(1 - e^{-\beta\omega_\lambda})^{-1} \qquad (2.21)$$

Thus

$$\langle V_{nm}(\tau) V_{pq}(0) \rangle = \sum_\lambda g_\lambda^{nm} g_\lambda^{pq} \phi_\lambda(\tau) + 2 \sum_{\lambda,\lambda'} f_{\lambda\lambda'}^{nm} f_{\lambda\lambda'}^{pq} \phi_\lambda(\tau) \phi_{\lambda'}(\tau) \qquad (2.22)$$

with

$$\phi_\lambda(\tau) = (\bar{n}_\lambda + 1)e^{-i\omega_\lambda\tau} + \bar{n}_\lambda e^{i\omega_\lambda\tau} \qquad (2.23)$$

Now we can calculate the relevant terms. For example,

$$T_{22,11} = -\int_{-\infty}^{+\infty} d\tau\, e^{i\omega_{12}\tau} \langle V_{21}(\tau) V(0) \rangle$$

$$= -2\pi \sum_{\lambda,\lambda'} |g_\lambda^{21}|^2 \{(\bar{n}_\lambda + 1)\delta(\omega_{12} - \omega_\lambda)\} + \bar{n}_\lambda \delta(\omega_{12} + \omega_\lambda) -$$

$$- 4\pi \sum_{\lambda,\lambda'} |f_{\lambda\lambda'}^{21}|^2 \{(n_\lambda + 1)(n_{\lambda'} + 1)\delta(\omega_{12} - \omega_\lambda - \omega_{\lambda'})\} + \qquad (2.24)$$

$$+ \bar{n}_\lambda(\bar{n}_{\lambda'} + 1)\delta(\omega_{12} + \omega_\lambda - \omega_{\lambda'}) +$$

$$+ (\bar{n}_\lambda + 1)\bar{n}_{\lambda'}\, \delta(\omega_{12} - \omega_\lambda + \omega_{\lambda'}) +$$

$$+ \bar{n}_\lambda \bar{n}_{\lambda'}\, \delta(\omega_{12} + \omega_\lambda + \omega_{\lambda'}).$$

For concreteness, take $\omega_{12} > 0$, then the first sum only contains terms in $(\bar{n}_\lambda + 1)$, and these represent one phonon emission terms. That is, the system can go from $|1\rangle$ to $|2\rangle$ by emitting a phonon of frequency $\omega_\lambda = \omega_{12}$, and this has the usual spontaneous and stimulated parts. The second sum has two types of allowed processes: the first term is a two phonon emission process where $\omega_{12} = \omega_\lambda + \omega_{\lambda'}$, the next two terms are (Raman) two phonon process where one phonon is emitted and one absorbed so that, e.g., $\omega_{12} = \omega_\lambda - \omega_{\lambda'}$. Whether these processes are allowed depends on the phonon dispersion or bandwidth compared to ω_{12}. Note that detailed balance holds, that is

$$\frac{T_{11,22}}{T_{22,11}} = e^{-\beta\omega_{12}} \tag{2.25}$$

which can be checked by using the relation $\bar{n}_\lambda/(\bar{n}_\lambda + 1) = e^{-\beta\omega_\lambda}$ and the energy conserving delta functions.

To calculate the pure dephasing rate, we find

$$\Gamma^{PD} \equiv \int_{-\infty}^{+\infty} d\tau \, \langle \{V_{22}(\tau) - V_{11}(\tau)\} \{V_{22}(0) - V_{11}(0)\} \rangle \tag{2.26}$$

$$= 8\pi \sum_{\lambda,\lambda} |f_{\lambda\lambda'}^{22} - f_{\lambda\lambda'}^{11}|^2 \, \bar{n}_\lambda(\bar{n}_\lambda + 1)\delta(\omega_\lambda - \omega_{\lambda'}).$$

Since there are no phonon states at $\omega_\lambda = 0$, all the other terms give no contribution. Since $\bar{n}_\lambda \to 0$ as $T \to 0$, we see that the pure dephasing rate goes to zero as $T \to 0$. Of course, all of this is well known [9] and our point is didactic. One last point can be made: if g_λ^{nm} and $f_{\lambda\lambda'}^{nm}$ depend on the frequencies ω_λ only, then the sums over λ can be made into integrals over the phonon density of states $\rho(\omega)$, by using

$$N^{-1} \sum_\lambda F(\omega_\lambda) = \int d\omega \, F(\omega) \, N^{-1} \sum_\lambda \delta(\omega - \omega_\lambda) \equiv \int_0^{\omega_D} d\omega F(\omega)\rho(\omega) \tag{2.27}$$

where ω_D is the cutoff frequency (Debye frequency for acoustic modes).

The reader will have noticed that nothing has been said about the coupling constants g_λ and $f_{\lambda\lambda'}$. These depend on phonon frequency and the system, S, in a way which may be found, *in principle*, from electronic structure calculations. Very little has been done on this problem, so we usually take these to be adjustable parameters. However, in the case of acoustic phonons, this frequency dependence can be found for low frequencies (sound waves) from classical theories, and it is found [10] that $g_\lambda \sim \omega_\lambda^{1/2}$ and $f_{\lambda\lambda'} - \omega_\lambda^{1/2} \omega_{\lambda'}^{1/2}$.

Thus the pure dephasing rate for acoustic phonon scattering is given in the above example by

$$\Gamma^{PD} = c \int_0^{\omega_D} d\omega \, \omega^2 \, [\bar{n}(\omega) + 1]\rho^2(\omega) \tag{2.28}$$

where c is a collection of constants. At high $T(T \gg \theta_D)$, $\Gamma_{PD} \sim T^2$ and at low T $(T \ll \theta_D) \Gamma_{PD} \sim T^7$, since $\rho(\omega) \sim \omega^2$ at low ω.

For systems with more states, the analysis can still be done; however more complex behavior is to be expected. In the next section we discuss a few examples of these.

3. Relaxation and Scattering of Excitations at Low Temperature

3.1. DIMER TRIPLET STATES

The simplest system to study excitation transfer and scattering is the naphthalene dimer triplet state [11]. Although it is simple, it exhibits a range of complex behavior and is almost a textbook case of the phenomena we are discussing.

Two types of dimers have been studied: those made of translationally equivalent (AA) pairs of naphthalene embedded in a host crystal and those made of translationally inequivalent pairs (AB). We will focus on the latter. The three triplet sublevels of the lowest triplet state of naphthalene were first studied more than 25 years ago [12], and the zero field splittings are well known $(D - 0.1 \text{ cm}^{-1}$ $E \sim -0.01 \text{ cm}^{-1})$. The excitation exchange matrix element between inequivalent molecules, J, is $\sim 1.2 \text{ cm}^{-1}$, so to lowest order the states of the system can be taken to be linear combinations of the localized excited states of the two molecules $|A\rangle$ and $|B\rangle$:

$$|\pm\rangle = \frac{1}{\sqrt{2}} \{|A\rangle \pm |B\rangle\}; E_\pm = \pm J. \tag{3.1}$$

Using the known fine structure tensor for each molecule, we can find the eigenvalues of the six states of this system (2 electronic levels with 3 spin states each). Because of the matrix element J, the splittings and principal axes of the two levels are not equal, and this gives rise to the interesting effects seen in the experiments [11]. These six states of the system are eigenstates of the Hamiltonian only in the absence of interaction of the system with the phonons (i.e. the thermal bath). The effect of the phonons will be (in the limit of weak coupling) to scatter an excitation from one state of another. These scattering processes may be absorption or emission of one phonon, or two phonons (e.g. a Raman process), or many phonons. Any scattering process which takes the excitation from one state to another will lead to dephasing of the ESR transitions and decay of the electron spin echo signal. In order to understand these effects, we must find the (reduced) density matrix of the system, σ, which governs the dynamics of the triplet states in the presence of scattering processes.

To describe these experiments, we write the reduced density matrix equations for the 6 state system:

$$\dot{\sigma}_{nm} = -i\omega_{nm}\sigma_{nm} - \sum_{k,\ell} T_{nm;\,k\ell}\,\sigma_{k\ell} \tag{3.2}$$

In order to solve this for the 36 matrix elements σ_{nm}, we could assume some microscopic scattering process which then yields the $T_{nmk\ell}$ from our analysis above, and then diagonalize the 36×36 matrix. On the other hand, we can try to isolate the important terms for the experimental variables and treat only a subset of terms. To do this we remember that the eigenvalues of the complex matrix $+i\omega + \mathbf{T}$ are the frequencies and relaxation rates (widths) of the problem; thus, a perturbation theory argument suggests that if we are interested in a frequency near ω_{34}, for example, then we need only consider those elements σ_{nm} whose zeroth order frequencies are close to ω_{34}. A better analysis would also include all those σ_{nm} strongly coupled to σ_{34}. This procedure allows us to make a great deal of progress with little computation, and retains a simple picture.

In the present case, we are interested in a certain ESR transition in the lower of the two triplet levels whose behavior is governed in the absence of scattering by, say, σ_{12}. In the upper triplet level, there is an ESR line with almost the same frequency and governed by σ_{45}. Scattering of the excitation between the two levels will occur due to the phonon-excitation interactions in V. We assume that the phonon-excitation interaction is independent of spin variables and since the spin axes of the upper level are rotated slightly with respect to those of the lower level, some spin lattice relaxation can occur in the following way [11]. An excitation in the lower level can be scattered to the higher level by absorption of a phonon. The excitation remains in the upper level for a short time and then emits a phonon and returns to the lower level. Since the ESR frequency in the upper state is different from that in the lower state, some dephasing has occurred; since the spin axes are also slightly rotated with respect to one another, some spin relaxation has also taken place.

Isolating the relevant reduced density matrix elements, we have for the lower triplet variables

$$\dot{\sigma}_{12} = -i\omega_{12}\,\sigma_{12} - \Gamma\sigma_{12} + \Gamma'\sigma_{45}$$
$$\dot{\sigma}_{11} - \dot{\sigma}_{22} = -\Gamma(\sigma_{11} - \sigma_{22}) + \Gamma'(\sigma_{44} - \sigma_{55}) \tag{3.3}$$
$$\sigma_{21} = \sigma_{12}^*$$

where Γ is the scattering rate from the lower to the upper level and Γ' the rate from upper to lower (so that $\Gamma' \geqslant \Gamma$). This comes directly from the formula for $T_{12,12}$ in the last section, where we have neglected pure dephasing entirely. The upper triplet variables can be written in their natural axis system or in the axis system of the lower level. The latter makes the relaxation terms easier to visualize, so that

$$\dot{\sigma}_{45} = -i\omega_{45}\cos 2\theta\,\sigma_{45} + i\omega_{45}\sin 2\theta(\sigma_{44} - \sigma_{55}) - \Gamma'\sigma_{45} + \Gamma\sigma_{12}$$
$$\dot{\sigma}_{44} - \dot{\sigma}_{55} = -i\omega_{45}\sin 2\theta(\sigma_{45} - \sigma_{54}) - \Gamma'(\sigma_{44} - \sigma_{55}) + \Gamma(\sigma_{11} - \sigma_{22}). \tag{3.4}$$

Using $\sigma_{nm} = \sigma_{mn}^*$ and converting to the variables $r_1 = (\sigma_{12} + \sigma_{21})/2$, $R_1 = (\sigma_{45} + \sigma_{54})/2$, $r_2 = (\sigma_{12} - \sigma_{21})/2i$, $R_2 = (\sigma_{45} - \sigma_{54})/2i$, $r_3 = (\sigma_{11} - \sigma_{22})$, $R_3 = (\sigma_{44} - \sigma_{55})$ yields the usual modified Bloch equations [13]. We can extract some relevant information immediately by first considering the case $\theta = 0$ (i.e. unrotated axes).

Then the eigenvalues of these equations are found by diagonalizing three 2×2 matrices; these are given by

$$\lambda_{\pm} = -i\left(\frac{\omega_{12} + \omega_{45}}{2}\right) - \left(\frac{\Gamma + \Gamma'}{2}\right) \pm \left[\Gamma\Gamma' + \frac{1}{4}(-i\omega_{12} + i\omega_{45} - \Gamma + \Gamma')^2\right]^{1/2} \lambda_{\pm}^{*}$$

$$\lambda_0 = 0, \lambda_0' = \Gamma + \Gamma'. \tag{3.5}$$

We interpret these as the positions and widths $(1/2T_2)$ of the two EPR lines $(\lambda_{\pm})$ and $(\lambda_{\pm}^{*})$ and the longitudinal relaxation (T_1^{-1}) rates (λ_0, λ_0'). Note that in the absence of phonon absorption, $\Gamma = 0$ and so $\lambda_- = -i\omega_{45} - \Gamma'$ and $\lambda_+ = -i\omega_{12}$ so that the EPR line of the lower triplet is a delta function while that of the upper triplet is broadened by phonon emission processes. If the two EPR lines have identical frequencies $(\omega_{12} = \omega_{45})$, then one line is not broadened and the other is broadened by $\Gamma + \Gamma'$. Note also that if $\Gamma = \Gamma'$ (i.e. equal scattering up and down), the $\lambda_{\pm}$ are just those found by simple exchange theory [14]. As the present example makes clear, $\Gamma/\Gamma' \cong e^{-2\beta|J|}$ since the upward process requires phonons of energy $2|J|$ while the downward process can occur via a spontaneous emission process. In the usual experimental case that the difference in frequencies $\delta = \omega_{12} - \omega_{45}$ obeys the inequality

$$\frac{\delta}{\Gamma} \ll 1$$

we find for the narrow line, $(\Delta = 2|J|)$

$$\lambda_+ = -i\left[\omega_{12} - \frac{\delta e^{-\beta\Delta}}{(1 + e^{-\beta\Delta})}\right] - \frac{(\delta^2/\Gamma')e^{-\beta\Delta}}{(1 + e^{\beta\Delta})^3} \tag{3.6}$$

so that the width of this line increases with increasing temperature in an activated manner, in agreement with the earlier theories and experiments [11, 14, 15]. Thus, in the fast exchange limit $\delta_2/\Gamma'^2 \ll 1$, the line position will vary as $\delta\, e^{-\beta\Delta}(1 + e^{-\beta\Delta})^{-1}$, while the width will vary as $(\delta^2/\Gamma')e^{-\beta\Delta}/(1 + e^{-\beta\Delta})^3$ so that from a temperature study one can find δ and Γ' [11].

   From our simple models of phonon-scattering in the last section, we find that Γ' also has a temperature dependence. In fact from eq. (2.24), for a 1 phonon process,

$$\Gamma' = 2\pi\, g^2(\Delta)\rho(\Delta)\,[\bar{n}(\Delta) + 1] \tag{3.7}$$

using the same notation as above. This predicts that the EPR width will vary with temperature as

$$Re\,\lambda_+ \propto \frac{e^{-\beta\Delta}(1 - e^{-\beta\Delta})}{(1 + e^{-\beta\Delta})^3} \tag{3.8}$$

so that the width first increases and then descreases as the temperature is raised: a typical exchange broadening and then narrowing. A more complete analysis of this temperature behavior for this case can be found in Reineker *et al.* [15]. The agreement with experiment [11] is qualitatively good, but not perfect.

If we consider two phonon effects in Γ', we find from eq. (2.26)

$$\Gamma' = 4\pi \int_0^\Delta d\omega \, \rho(\omega)\rho(\Delta - \omega)\, [\bar{n}(\omega) + 1]\, [\bar{n}(\Delta - \omega) + 1] f^2(\omega, \Delta - \omega)$$

$$(3.9)$$

$$+ 8\pi \int_0^{\omega_D} d\omega \, \rho(\omega)\rho(\Delta + \omega)\bar{n}(\omega)\, [\bar{n}(\Delta + \omega) + 1] f^2(\omega, \Delta + \omega).$$

The first term is the spontaneous emission of two phonons (note that ω must be less than Δ in this case so that this term is small but nonzero). The second term is the Raman process where a phonon of ω is absorbed and a phonon of $\omega + \Delta$ is emitted while the system makes a transition from the upper level to the lower. From our analysis of similar terms in Γ_{PD} we can see that for $T > \Delta$, we expect this term to behave as T^7, so that as the Raman terms become more important, Γ' becomes large and faster exchange takes place, leading to a narrow line at the average frequency. In this limit, it is more correct to think of a localized description of the excitations (i.e. localized on A or B) and being scattered by phonons from one site to the other.

If we now consider the case $\theta \neq 0$, so that the spin axes are not parallel in the two dimer states, we must diagonalize a 6×6 matrix to find the eigenvalues representing T_2^{-1} and T_1^{-1}. Perturbation calculations have been done by Verbeek and Schmidt [11], Levinsky and Brenner [11], Vollmann [11], Dietz *et al.* [11] and Silbey [15a]. The result *at low temperature* is that the slow decay of the population variable is given by

$$(1/T_1) = \Gamma \sin^2 2\theta \, \frac{(\omega_{45}/\Gamma')^2}{1 + (\omega_{45}/\Gamma')^2},$$

$$(3.10)$$

while the transverse relaxation rate $(1/T_2)$ and the frequency of the lower transition (ε) are given by the rather complex expressions:

$$\frac{1}{T_2} = \Gamma \left\{ 1 - \frac{\left(\dfrac{1 + \cos 2\theta}{2}\right)^2}{1 + \delta^2/\Gamma'^2} - \frac{\left(\dfrac{1 - \cos 2\theta}{2}\right)^2}{1 + (\omega_{12} + \omega_{45})^2/\Gamma'^2} - \frac{\dfrac{1}{2}(\sin 2\theta)^2}{1 + \omega_{12}^2/\Gamma'^2} \right\}$$

$$\varepsilon = \omega_{12} - \frac{\delta \, e^{-\beta\Delta} \left(\dfrac{1 + \cos 2\theta}{2}\right)^2}{1 + \delta^2/\Gamma'^2} +$$

$$+ \frac{(\omega_{12} + \omega_{45})e^{-\beta\Delta} \left(\dfrac{1 - \cos 2\theta}{2}\right)^2}{1 + (\omega_{12} + \omega_{45})^2/\Gamma'^2} + \frac{1}{2} \frac{\omega_{12} \, e^{-\beta\Delta} \sin^2 2\theta}{1 + \omega_{12}^2/\Gamma'^2}$$

showing the strong dependence on the angle of rotation of the spin axes, and the scattering rates Γ and Γ'.

Unfortunately, at higher temperatures the solution to the equations can only be found numerically; however the physical model of the phonon-scattering of the excitation between the two states leading to an accumulation of phase errors and population relaxation seems to be borne out in the experiments of Schmidt and co-workers [11]. Problems remain however: although the theory predicts a turnover in the linewidth (T_2^{-1}) as a function of rising temperature, and this is seen in the experiment, the quantitative prediction of theory is not in agreement with experiment. In fact the theory predicts that for one phonon processes, the equivalent dimer should show the exchange narrowing at a lower temperature than the inequivalent dimer, in disagreement with experiment. In order to bring the theory into agreement with experiment, we can postulate that two different relaxation processes play a role, but that they have very different coupling strengths to the two types of dimers. That is, we suppose that *both* a linear term in phonon coordinate and a quadratic term in these coordinate are present in the interaction and contribute in *different* proportions for the two different dimers. We can then fit the experiment, but we are unable at the present time to explain why the coupling should be so different for the two.

3.2. ONE-DIMENSIONAL TRIPLET EXCITONS

In tetrachlorobenzene crystals, the linear stacks of molecules give rise to excited triplet states which are beautiful examples of one-dimensional exciton states [16]. These have been studied by a number of workers over the last few years, notably by Schmidt and co-workers [17].

The latter group has studied the population of the various exciton states as a function of time after initial preparation in the optically accessible state at the top of the exciton band. In order to describe this system using the Redfield theory, considering only the states in the band, would require an unreasonably large number of density matrix elements. Thus to make the system manageable and still be able to examine the population dynamics, we restrict our attention to the population variables, σ_{kk}. By neglecting off diagonal elements (coherences) we find that σ_{kk} obeys the master equation

$$\dot{\sigma}_{kk} = \sum_{k'} W_{kk'}\, \sigma_{k'k'} - \left(\sum_{k' \neq k} W_{k'k} \right) \sigma_{kk}. \tag{3.11}$$

In order to make progress at this level we must calculate the $W_{kk'}$.

The electronic Hamiltonian for the linear array of N identical and equivalent molecules with the energy transfer matrix between nearest neighbors, J can be written [18]

$$H = \varepsilon_0 \sum_n |n\rangle \langle n| + J \sum_n \{|n\rangle \langle n+1| + |n+1\rangle \langle n|\}$$

$$\tag{3.12}$$

$$= \sum_k \varepsilon(k) |k\rangle \langle k|$$

where $|n\rangle$ represents an excitation on site n of energy ε and

$$\varepsilon(k) = \varepsilon_0 + 2J \cos k$$

$$\tag{3.13}$$

$$|k\rangle = N^{-1/2} \sum_n e^{ikn} |n\rangle$$

Here, k is the quasimomentum along the chain direction. If the chain can be taken to be very long, k is a conserved quantity, so when we introduce phonon-scattering terms the change in k due to the change in state of the excitation must be compensated by an opposite change in k of the phonons. The phonons are not one dimensional, so we must consider only their quasimomentum along the chain $q_{\parallel}$ being conserved. For one phonon processes, we can write the total Hamiltonian as

$$H = \sum_k \varepsilon(k) |k\rangle \langle k| + \sum_q \omega_q \left(a_q a_q^+ + \frac{1}{2} \right) +$$

$$+ \sum_{k, k', q} G_{kk'}(q) |k'\rangle \langle k| (a_q + a_{-q}^+)$$

$$\tag{3.14}$$

where the a_q^+ and a_q are the creation and annihilation operators for the phonons. Consider a one phonon-scattering process from $|k\rangle$ to $\langle k'|$. Two conditions must be met: $k' = k + q_{\parallel}$ and energy conservation $\varepsilon(k') = \varepsilon(k) + \omega_q$. For TCB, these two conditions cannot be met [16], because the exciton bandwidth is so small ($\sim 1.3 \ cm^{-1}$) compared to the phonon bandwidth. That is, if we match the momenta we cannot match the energy. It gets even worse for multiphonon emission or absorption processes; only if we consider a mixed (i.e. Raman) process, is it possible to satisfy the conditions. This means that at very low temperatures when the density of phonons is low, there should be no phonon-exciton scattering. [A corollary of this is that the highest state in the exciton band which happens to be at $k = 0$ has no width due to phonon emission at $T = 0$. This is in contrast with most other systems, where such emission processes dominate the decay of the higher states in the band at $T = 0$.] Since the only allowed processes will be Raman processes, we predict that the scattering rate of excitons from one state to another should depend on T^n with $n = 7$. The experiments of van Strien and Schmidt [17] show that this is incorrect. The data clearly show a weak temperature dependence in this rate (see the chapter by Schmidt in this volume for details) characteristic of one phonon processes. In order to explain this observation, we must give up the condition of conservation of quasimomentum (we cannot of course give up conservation of energy). In order to do this, we can introduce some form of *weak* impurity effects. Three come to mind immediately: (1) an impurity (or misaligned or rotated molecule) in a nearby stack of TCB molecules can induce a nontranslationally invariant phonon-exciton interaction, e.g.

$$V_1 = \sum_q G_{nq} | n \rangle \langle n | (a_q + a_{-q}^+) = \sum_{k, q} \frac{G_{nq} e^{i(k - k')n}}{N^{1/2}} | k \rangle \langle k' | (a_q + a_{-q}^+) \quad (3.15)$$

where n is the site nearest to the impurity (more nearby sites can be included without problem); (2) the finite size of the stack, N, due to impurities at each end implies that k is not strictly conserved because the phonon wave vector or quasimomentum is defined over a large length than the exciton wave vector. This mechanism will thus decrease in importance as the length of the stack increases; and (3) an impurity in the middle of the chain will change the wave functions (i.e. mix the k states) so that conservation of quasimomentum also breaks down. This mechanism depends on the strength of the impurity potential in the stack. We [19] have considered all these mechanisms and using the relaxation theory described in Section 2, calculated the scattering rates from state k to k'. All three mechanisms give rise to a one phonon process, and all depend linearly on the impurity concentration, but they have distinct dependences on the initial and final exciton states.

In order to mimic the physical system, we took an anisotropic but approximate phonon dispersion relation, $\omega(q) = \{q_\parallel^2 c_\parallel^2 + q_\perp^2 c_\perp^2\}^{1/2}$ with two sound speeds, $c_\parallel$ and $c_\perp$. The important feature of the calculation was that since the exciton-phonon coupling constant G_{nq} is proportional to $\omega_q^{1/2}$ and the density of phonon states is proportional to ω_q^2 at small ω, the scattering rate is proportional to ω_q^3, but since ω_q must equal the difference in exciton energies, the rate is proportional to $|(E_k - E_{k'})^3|$. Thus larger energy jumps are favored by these mechanisms. The results of the calculation [19] are complicated; however, we were able to show, for very narrow exciton bands that mechanism 1 should be favored over mechanisms 2 and 3 by at least the square of the ratio of the phonon bandwidth to the exciton bandwidth. Detailed simulations of the time dependence of the populations using the rates found from these mechanisms in the master equation showed that mechanism 1 gave the best agreement with experiment [20] at least in the very low T regime.

As the temperature is raised, we expect two phonon (Raman) processes to become important and eventually dominate. Since such processes can occur without impurities, i.e. in the translationally invariant system, we [21] have calculated the rates for this mechanism both with and without impurities. The perturbation term responsible for the phonon exciton scattering is (see Eq. 2.20))

$$V = \sum_{k, k', q, q'} \eta(q, q') | k' \rangle \langle k | (a_q + a_{-q}^+) (a_{q'} + a_{-q'}^+). \quad (3.16)$$

In the deformation potential approximation, η^2 is proportional to $\omega_q \omega_{q'}$. In the translationally invariant system, only processes in which the wave vector along the stack axis is conserved. Although this does not forbid Raman processes, it does have an effect on the temperature dependence of the scattering rate. To see this, note that a typical process will be the scattering of the exciton from k and k' while a phonon of frequency ω_q is absorbed and one of frequency $\omega_{q'}$ is emitted, so

$$\varepsilon(k) + \omega_q = \varepsilon(k') + \omega_{q'}$$
$$k + q_1 = k' + q_1'$$
(3.17)

Let us assume that $\varepsilon(k') - \varepsilon(k) < 0$, so that $\omega_{q'} > \omega_q$. Now the phonon frequency as given above implies that $\omega_{q'} > c_1|q_1'|$ and $\omega_q > c_1|q_1|$ so that

$$\omega_{q'} > c_1|k - k' + q_1|$$
$$\omega_q > c_1|q_1|$$
(3.18)

Thus

$$\omega_{q'} + \omega_q > c_1|k - k'|$$
$$\omega_{q'} - \omega_q = \varepsilon(k) - \varepsilon(k') = \Delta\varepsilon > 0$$
(3.19)

and thus both phonons must have frequencies greater than $\omega_0 = \frac{1}{2}\{c_1|k - k'| - \Delta\varepsilon\} > 0$. The last inequality is true for very narrow exciton bands as in TCB. In wide band cases this fails so that there is no lower bound on the phonon frequency for allowed processes. Since the conservation of wave vector leads to this lower bound on ω, the impurity induced Raman processes do not have this restriction.

Jackson and Silbey have considered these two mechanisms and find at low T ($T < \theta_{\text{Debye}}$) (again for $\varepsilon(k') < \varepsilon(k)$) for the leading term,

$$W_{k'k}(\text{pure}) - T^6 \exp(-\beta\omega_0)$$
$$W_{k'k}(\text{impurity}) \sim c\, T^7$$
(3.20)

where c is the impurity concentration. Note that the wave vector restriction in the pure case lowers the temperature dependence. These authors also discussed the perturbation results arising from taking the linear exciton-phonon coupling to second order and found that the leading term here was $W_{k'k} \sim T^5$. Crude estimates of all the coefficients suggest that the pure crystal Raman term should dominate so that $W_{k'k} \sim e^{-\omega_0/k_B T}\, T^6$. Since ω_0 depends on $|k - k'|$, this predicts that small energy changes will be favored for this mechanism and that it should have a strong T dependence. These are in strong contrast to the low T results (where impurity induced one phonon processes dominate). The experiments indicate that scattering increases above $T \sim 2K$ with a rate proportional to $\sim T^{6.4}$, indicating reasonably good qualitative agreement between theory and experiment.

This simple model for the one-dimensional exciton system can be used to discuss other aspects of the experiments [17] and qualitative agreement is found. However, some puzzles remain to be understood; for example, since the natural abundance of C^{13} is -0.1%, one in sixteen TCB molecules will have a C^{13} which might play the role of disruptive impurities (that is, limit the effective chain length for the triplet excitons). If they do play this role, then the chain lengths will be on average smaller than previously thought. Although all the scattering processes we have discussed will remain valid, their relative importance may change. (This may lead to smaller energy scattering events playing a more important role, for example.)

3.3. CONCLUSION

The application of the Redfield formalism to the scattering of excitations in condensed phases at low temperatures, as exemplified by the two cases we have studied, leads to good agreement between experiment and theory. This suggests that, at least for excited triplet states at low temperature, the basic assumptions of weak coupling and a separation of time scales holds. The interesting test of the formalism will come when the type of experiments mentioned here (i.e. the ability to study the populations of excited states in time) are done on systems whose relaxation rates are in the picosecond regime. For at these temperatures, the phonons may not relax in such times; hence the necessary separation of time scales will not obtain and memory effects will play a role.

4. Optical Hole-Burning in Glasses and Crystals

In recent years, there has been a great deal of interest in determining the homogeneous linewidth of an optical transition in both crystals and glasses. The central problem is to uncover the homogeneous line in an inhomogeneously broadened transition. In a crystal the inhomogeneous broadening is relatively small compared to that in a glass; however at low temperatures this broadening can be large compared to the homogeneous width even in a crystal. Various methods have been suggested to uncover the homogeneous line [22—28], one of which is hole-burning [22, 23a, 25—28] (others are photon echo spectroscopy [23b] and fluorescence line narrowing [24]). Hole-burning comes in two flavors: photochemical (PHB) and nonphotochemical (NPHB); we will not be concerned with experimental techniques, etc., but focus only on the homogeneous width of an optical transition in a crystal or a glass.

4.1. PHOTOCHEMICAL HOLE-BURNING IN CRYSTALS

In a beautiful set of experiments, Völker and co-workers [25] studied PHB of free base porphin in substitutional sites in alkane crystals. Upon excitation, the two hydrogens in the inner part of this molecule can rearrange to form a physically distinct (although chemically identical) molecule which absorbs at a different frequency. Thus the optical absorption of the remaining molecules of the original type shows a "hole" of Lorentzian shape, assumed to be the convolution of the homogeneous line with itself [25]; therefore, the width of this hold is twice the homogeneous width of the transition at that temperature. In these systems, the width increases with temperature with an activation energy equal to the frequency of a known libration (or optical mode) and can be understood then by librational relaxation and dephasing. To be more explicit, we assume that the system, S, consists of a four level system: ground electronic state with and without a librational quantum:

$$H_S = \sum_{n=0, 1} \{(E^e + n\Omega^e)|en\rangle\langle en| + (E^g + n\Omega^g)|gn\rangle\langle gn|\} \tag{4.1}$$

Here, $|en\rangle$ and $|gn\rangle$ are the excited and ground state electronic states with n quanta of librational energy, and we have assumed that the frequency of the libration is not the same in the upper and lower electronic states. The reservoir consists of the phonons:

$$H_R = \sum_{\lambda} \omega_{\lambda} \{a_{\lambda}^{+} a_{\lambda} + 1/2\} \tag{4.2}$$

and the interaction is assumed to be librational relaxation:

$$V = \sum_{\lambda} (a_{\lambda} + a_{\lambda}^{+}) \{f_{\lambda}^e [|e1\rangle\langle e0| + |e0\rangle\langle e1|] \\ + f_{\lambda}^g [|g1\rangle\langle g0| + |g0\rangle\langle g1|]\} \tag{4.3}$$

so that one phonon processes can scatter a libration from $n = 1$ to $n = 0$ and vice versa. For generality we have assumed that the coupling constant is different in the ground and excited states.

Since this is a 4 level model system, there are 16 density matrix elements to consider. However for optical absorption we can restrict our attention to those matrix elements between the relevant states, i.e. those involved in the transition near $E^e - E^g$: $\sigma_{e0, g0}$ and $\sigma_{e1, g1}$. For simplicity, we label the states in the following way

$$|g0\rangle \equiv |a\rangle \quad |g1| \equiv |b\rangle \\ |e0| \equiv |c\rangle \quad |e1\rangle \equiv |d\rangle \tag{4.4}$$

Then the Redfield equations are

$$\dot{\sigma}_{ca} = -[i\omega_{ca} + T_{ca, ca}]\sigma_{ca} - T_{ca, db}\sigma_{db} \\ \dot{\sigma}_{db} = -[i\omega_{db} + T_{db, db}]T_{db} - T_{db, ca}\sigma_{ca} \tag{4.5}$$

Here $\omega_{ca} = E^e - E^g$ and $\omega_{db} = E^e - E^g + \Omega^e - \Omega^g$. From the general theory of Section 2, we have for our assumed form of V:

$$T_{ca,\,ca} = \frac{1}{2}\int_{-\infty}^{+\infty} d\tau \langle V_{ab}(\tau) V_{ba}(0)\rangle e^{-i\Omega^g\tau} + \frac{1}{2}\int_{-\infty}^{+\infty} d\tau \langle V_{cd}(\tau) V_{dc}(0)\rangle e^{-i\Omega^e\tau}$$

$$T_{db,\,db} = \frac{1}{2}\int_{-\infty}^{+\infty} d\tau \langle V_{ba}(\tau) V_{ab}(0)\rangle e^{+i\Omega^g\tau} + \frac{1}{2}\int_{-\infty}^{+\infty} d\tau \langle V_{dc}(\tau) V_{cd}(0)\rangle e^{+i\Omega^e\tau}$$

$$\tag{4.6}$$

$$T_{ca,\,db} = -\frac{1}{2}\int_{-\infty}^{+\infty} d\tau \langle V_{dc}(\tau) V_{ab}(0)\rangle e^{i\Omega^g\tau} - \frac{1}{2}\int_{-\infty}^{+\infty} d\tau \langle V_{ba}(\tau) V_{cd}(0)\rangle e^{i\Omega^e\tau}$$

$$T_{db,\,ca} = -\frac{1}{2}\int_{-\infty}^{+\infty} d\tau \langle V_{cd}(\tau) V_{ba}(0)\rangle e^{-i\Omega^g\tau} - \frac{1}{2}\int_{-\infty}^{+\infty} d\tau \langle V_{ab}(\tau) V_{dc}(0)\rangle e^{-i\Omega^e\tau}.$$

Using the above form of V, we find

$$T_{ca,\,ca} = \pi \sum_\lambda \{ |f_\lambda^{(g)}|^2\, \bar{n}_\lambda\, \delta(\omega_\lambda - \Omega^g) + |f_\lambda^{(e)}|^2\, \bar{n}_\lambda\, \delta(\omega_\lambda - \Omega^e) \}$$

$$T_{db,\,db} = \pi \sum_\lambda \{ |f_\lambda^{(g)}|^2\, (\bar{n}_\lambda + 1)\, \delta(\omega_\lambda - \Omega^g) + |f_\lambda^{(e)}|^2\, (\bar{n}_\lambda + 1)\, \delta(\omega_\lambda - \Omega^e) \}$$

$$\tag{4.7}$$

$$T_{ca,\,db} = -\pi \sum_\lambda \{ f_\lambda^{(e)} f_\lambda^{(g)} (\bar{n}_\lambda + 1)\, \delta(\omega_\lambda - \Omega^g) + f_\lambda^e f_\lambda^g (\bar{n}_\lambda + 1)\, \delta(\omega_\lambda - \Omega^e) \}$$

$$T_{db,\,ca} = -\pi \sum_\lambda \{ f_\lambda^{(e)} f_\lambda^{(g)} (\bar{n}_\lambda) \, [\delta(\omega_\lambda - \Omega^g) + \delta(\omega_\lambda - \Omega^e)] \}.$$

As usual, $T_{ca,\,ca}$ is the sum of one half the decay rates of level a and level c (the zero quanta libration state) while $T_{db,\,db}$ is one half the sum of the decay rates of levels b and d (the one quantum of libration states). Detailed balance holds for these decay rates, as can be seen from the form. If we assume further that $f_\lambda^{(e)} \approx f_\lambda^{(g)}$, and $\Omega^e \approx \Omega^g$, then

$$T_{ca,\,db} = -T_{db,\,db} = -\Gamma'$$
$$T_{db,\,ca} = -T_{ca,\,ca} = -\Gamma$$

$$\tag{4.8}$$

where Γ' is the decay rate of the one quantum libration by phonon emission and Γ the decay rate of the zero quantum libration by phonon absorption.

Before we go further, we should point out that we have left out some interactions. One is trivial: the interaction of the excited electronic states, $|c\rangle$. and $|d\rangle$ with the radiation field gives rise to the radiative lifetime of these states, which adds a term Γ_0 to the rate of decay of σ_{ca} and σ_{db}. Other interactions which we have left out are for example, two phonon processes leading to pure dephasing of these transitions of phonon-scattering to higher librational states. These processes can have different effects on σ_{ca} and σ_{db}. We can lump all of these into phenomenological additional decay rates γ and γ' so that the final equations are

$$\dot{\sigma}_{ca} = -[i\omega_{ca} + \Gamma + \gamma]\,\sigma_{ca} + \Gamma'\sigma_{db}$$

$$\dot{\sigma}_{db} = -[i\omega_{db} + \Gamma' + \gamma']\,\sigma_{db} + \Gamma\sigma_{ca}. \tag{4.9}$$

Note that these resemble the equations for the relevant matrix elements of the naphthalene dimer states discussed above. As before, the eigenvalues of this two-dimensional problem gives the homogeneous optical line position and width:

$$\lambda_{\pm} = i\left(\frac{\omega_{ca} + \omega_{db}}{2}\right) + \left(\frac{\Gamma + \Gamma' + \gamma + \gamma'}{2}\right) \pm$$

$$\pm \left\{\left[-i\left(\frac{\omega_{ca} - \omega_{db}}{2}\right) + \left(\frac{\Gamma + \gamma - \Gamma' - \gamma'}{2}\right)\right]^2 + \Gamma\Gamma'\right\}^{1/2}. \tag{4.10}$$

Once again, this resembles a typical exchange problem except that $\Gamma \cong e^{-\beta\Omega}\Gamma'$ where Ω is the librational frequency. Note that $(\omega_{ca} - \omega_{db}) = \Omega^g - \Omega^e \equiv -\delta$, and that this model can be described physically as an exchange between the 0—0 and the 1—1 lines in the librational progression.

In the limit that $\gamma \approx \gamma'$ this reduces to the same set of equations as in the naphthalene dimer. Thus for fast exchange $(\delta/\Gamma' \ll 1)$ the frequency of the narrow line is

$$\text{Im } \lambda_+ \cong \omega_{ca} + \frac{\delta\Gamma}{\Gamma + \Gamma'} = \omega_{ca} + \frac{\delta e^{-\beta\Omega}}{1 + e^{-\beta\Omega}} \tag{4.11}$$

and the width

$$\text{Re } \lambda_+ \cong \gamma + \frac{\delta^2}{(\Gamma + \Gamma')^3} = \gamma + \frac{\delta^2}{\Gamma'}\frac{e^{-\beta\Omega}}{(1 + e^{-\beta\Omega})^3} \tag{4.12}$$

Thus at low temperatures, activated shifts and widths should be seen. In the opposite limit that $\delta/\Gamma' \gg 1$ (slow exchange), then the two lines are separated by more than their widths and the 0—0 line has a position and width given by

$$\text{Im } \lambda_+ = \omega_{ca} + \Gamma\Gamma'/\delta$$

$$\text{Re } \lambda_+ \cong \Gamma + \gamma + \frac{\Gamma\Gamma'(\Gamma' - \Gamma)}{\delta^2} \cong \gamma + e^{-\beta\Omega}\Gamma' \tag{4.13}$$

Thus in this case the width is governed by the population decay rates of the states. Since Γ is just what one would calculate for the Golden Rule rate of decay of $|a\rangle$ and $|c\rangle$ due to phonon-scattering to the higher librational states without asking about the return rates, this is sometimes called the uncorrelated phonon-scattering rate [23b, c].

By comparing these model calculations to experiment, Völker and co-workers

[25] were able to show that the fit was excellent and the activation energies indeed compared well to the observed librational modes. Thus it is now well established that for this system and others [23], librational relaxation is the dominant dephasing process contributing to the homogeneous line width.

4.2. HOLE-BURNING IN GLASSES

When an optically absorbing molecule or ion is put into a glass, the inhomogeneous broadening of an optical transition is usually quite large (several hundred cm^{-1} is not uncommon). In such systems both PHB (as described above for free base porphin) and NPHB can occur. Nonphotochemical hole-burning (see G. J. Small et al. [22] for references) occurs presumably because while the molecule is in its excited state, the site environment around it can change (causing the optical transition frequency to shift). Upon returning to the ground state, the net effect is that there are less molecules absorbing at the original frequency, so a hole is formed. There are many questions remaining to be answered before a description of this process is at hand. We refer the reader to the book [28] edited by Moerner for recent work on a number of problems in the general area. In the present article, we will only discuss the temperature dependence of hole widths and try to understand this puzzling phenomenon. As we will see, there are many models which can give rise to the observed dependence and at the present time, no clear way of choosing among them is available.

When Völker and co-workers [27] studied hole-burning in glasses as they had studied it in crystals, they found that the width of the hole no longer had an activated temperature dependence, but increased with T as T^{α} with $\alpha \simeq 1.3$ for $0.3\,K < T < 20\,K$. At this time, these workers have studied a large number of different organic amorphous systems and it seems that this behavior is very common for these glasses. A recent study of an ion in glassy silica [29] also gave the result T^{α}, $\alpha = 1.3$ for the dephasing rate (by a photon echo decay measurement). Why should the temperature dependence be so different in glasses and crystals? The answer lies in the low frequency modes found in glasses, the so-called tunneling systems or two level systems (TLS). These were postulated by Anderson et al. [30] and by Phillips [30] to explain the anomalous specific heat and thermal conductivity of glasses. Briefly, the specific heat of a glass at extremely low temperature ($T < 1\,K$) shows an almost linear dependence which these authors derived based on the conjecture that in glasses, there were many tunneling states (from nearby energy minima found by moving atoms a small distance) which were (nearly) degenerate. If such two level systems exist then they will contribute to the specific heat, since

$$C = \frac{dU}{dT} = \frac{d}{dT} \Sigma \omega \bar{n}_{TLS}(\omega) = \frac{d}{dT} \int_0^{\omega_c} d\omega \; \omega \bar{n}_{TLS}(\omega) \, \rho_{TLS}(\omega) \quad (4.14)$$

where ω is the excitation energy of the TLS, $\bar{n}_{TLS}(\omega)$ is the equilibrium population of the upper state in the TLS, $\rho(\omega)$ is the density of the TLS frequencies (splittings) and ω_c is an upper cutoff for the TLS frequency. We have assumed that the lowest TLS frequency is 0. Since $\bar{n}_{TLS}(\omega)$ is a function of $\beta\omega(\beta = 1/kT)$, we have, at low T

$$C \propto \frac{d}{dT} T^2 \int_0^{\beta\omega_c} dx\, x\bar{n}_{TLS}(x)\, \rho_{TLS}(k_B Tx) \sim T^{1+\gamma} \qquad (4.15)$$

where we have assumed $\rho(\omega) \sim \omega^\gamma$ at small ω. From their conjectures about the tunneling systems, Anderson et al. [30] found that γ was about zero, so that there was a constant density of states at zero frequency. This model was also able to explain the anomalous thermal conductivity of glasses at low termperatures, although some inconsistencies remained [31]. Recent studies of the specific heat of glassy silica at very low T have found that $C \propto T^{1.3}$ which suggests that $\gamma = 0.3$ [32]; however, explanations [33] based on the time dependence of the specific heat at these very low T can explain the data with $\gamma = 0$, and other models have appeared [34] which suggest that perhaps the number density of TLS is not fixed, but can also change slightly with temperature leading to an *apparent* $\alpha = 0.3$. At the present time, the Anderson–Phillips model seems to work well for all the thermodynamic properties, but there may be some problems remaining, including the exact form of the density of states of the TLS at low frequency.

The situation in regard to explaining the holewidths is no better. There are a large number of models to explain the temperature dependence, most of which assume a constant density of TLS states (i.e. $\gamma = 0$). These models differ in the nature of the coupling of the TLS to the molecule (or ion) which absorbs the photon, or the manner in which the averaging over TLS parameters is done, or include more than one process, or evaluate the linewidth in different approximations. The initial model of Lyo and Orbach [35] for FLN in inorganic glasses predicted the width due to TLS-phonon interactions would vary as T^2 at low T, although subsequent analysis by Morawitz and Reineker [36] suggested that it would vary as T^2 only at extremely low T and change over to T at moderately low T. Lyo [37] then suggested a new model which predicted a linear T dependence at low T. Small and co-workers [38] in an early and general model of the same TLS-phonon interactions found the holewidth varied as $T^{1+\varepsilon}$, $0 < \varepsilon < 1$ in the temperature range of interest. Lyo and Orbach [40], suggesting the importance of fracton modes, found T or $T^{4/3}$ depending on whether a dipolar or quadrupolar interaction was considered, and Maynard [41] suggested an elastic quadrupolar interaction was responsible for $T^{4/3}$ result. Jackson et al. [39] suggested that librations were still playing a role in optical transitions in molecular systems and this when *added* to the low T result of Lyo gave an apparent $T^{1.3}$. Finally, Osadko [42] has recently suggested that the TLS themselves be treated as excitations similar to phonons in a thermal reservoir, and, using a two-excitation interaction term finds a T dependence consistent with $T^{1.3}$. Most of these models assume $\gamma = 0$, as in the original TLS models; however other authors assume $\gamma = 0.3$ so that the density of TLS states does not follow the Anderson et al. and Phillips form [30]. From this assumption, the $T^{1.3}$ follows. Clearly the definitive explanation for this phenomenon is not at hand (or if it is, we cannot distinguish it from the others as yet).

In the remainder of this section, we will discuss the effects of the phonon-TLS interactions on the homogeneous linewidth of an optical transition, using the Redfield formalism outlined above. We will derive the formula for an optical center interacting with one TLS and show how this agrees with the results of all the above authors in particular limits. By averaging over the distribution of TLS, their final results can be recovered. Our analysis shows that many of the theoretical predictions are based on the same basic physical model, with either different assumptions of the relative importance of the terms, or different approximations in the final part of the analysis.

We will now derive the general formula for the linewidth of an optical center interacting with a single TLS. In the ground electronic state of the optical center there are then two states (the two states of the TLS) which we label $|a\rangle$ and $|b\rangle$, $|a\rangle$ being the lower state. In the excited electronic state of the optical center the two states are labeled $|c\rangle$ and $|d\rangle$, $|c\rangle$ being lower in energy. We assume that the optical transition is largely $|a\rangle \rightarrow |c\rangle$ and $|b\rangle \rightarrow |d\rangle$, so that we are dealing with a two transition problem. We can easily include the other two transitions $|a\rangle \rightarrow |d\rangle$ and $|b\rangle \rightarrow |c\rangle$, at the expense of complicating the final formulas so much that it is difficult to see the physical mechanisms. We therefore neglect these weaker transitions.

We assume that interaction with the phonons causes transitions between $|d\rangle$ and $|c\rangle$ and between $|b\rangle$ and $|a\rangle$. The interaction is assumed to be a one phonon process so that

$$V = V_{ba}(|b\rangle\langle a| + |a\rangle\langle b|) + V_{cd}(|c\rangle\langle d| + |d\rangle\langle c|) \tag{4.16}$$

where V_{ba} and V_{cd} are one *phonon* operators (to be specified below). That is, we assume that one phonon absorption processes cause transitions from $|a\rangle$ to $|b\rangle$ and $|c\rangle$ to $|d\rangle$ and one phonon emission processes cause the reverse transitions. We could also add

$$V'(|d\rangle\langle d| - |c\rangle\langle c|)$$

where V' is a phonon operator. This sort of term makes the energy difference $E_d - E_c$ fluctuate and leads to *pure* dephasing terms in the width; if we assume that V' is a one phonon (creation or annihilation) operator, this term yields zero contribution to the width. If it is a two phonon term (i.e. Raman type process), it will yield (see above) a T^7 behavior at low T. We therefore neglect these terms.

The Redfield equations for the relevant density matrix operators become

$$\dot{\sigma}_{ca} = -[i\omega_{ca} + T_{ca, ca} + \gamma_{rad}]\,\sigma_{ca} - T_{ca, db}\,\sigma_{cb}$$

$$\dot{\sigma}_{db} = -[i\omega_{db} + T_{db, db} + \gamma_{rad}]\,\sigma_{db} - T_{db, ca}\,\sigma_{ca} \tag{4.17}$$

where we have, as usual, neglected the coupling to density matrix elements oscillating at very different frequencies. Also, we have included the radiative width, γ_{rad}, in the equation. Using the form V above, we have

$$T_{db, db} = \tfrac{1}{2} (W_{d \rightarrow c} + W_{b \rightarrow a})$$

$$T_{ca, ca} = \tfrac{1}{2} (W_{c \rightarrow d} + W_{a \rightarrow b})$$

$$T_{db, ca} = \frac{1}{2} \int_{-\infty}^{+\infty} d\tau\, e^{i\omega_{ab}\tau} \langle V_{cd}(\tau) V_{ba}(0) \rangle -$$

$$- \frac{1}{2} \int_{-\infty}^{+\infty} d\tau\, e^{i\omega_{cd}\tau} \langle V_{ab}(\tau) V_{cd}(0) \rangle$$

$$(4.18)$$

$$T_{ca, db} = - \frac{1}{2} \int_{-\infty}^{+\infty} d\tau\, e^{i\omega_{ba}\tau} \langle V_{cd}(\tau) V_{ba}(0) \rangle$$

$$- \frac{1}{2} \int_{-\infty}^{+\infty} d\tau\, e^{i\omega_{dc}\tau} \langle V_{ab}(\tau) V_{cd}(0) \rangle$$

where the Golden Rule rates are written as in Section 2,

$$W_{c \rightarrow d} = \int_{-\infty}^{+\infty} d\tau\, e^{i\omega_{cd}\tau} \langle V_{cd}(\tau) V_{cd}(0) \rangle$$

$$W_{a \rightarrow b} = \int_{-\infty}^{+\infty} d\tau\, e^{i\omega_{ab}\tau} \langle V_{ab}(\tau) V_{ab}(0) \rangle$$

$$(4.19)$$

$$W_{d \rightarrow c} = e^{+\beta\omega_{dc}} W_{c \rightarrow d}$$

$$W_{b \rightarrow c} = e^{-\beta\omega_{dc}} W_{a \rightarrow \beta}$$

In general the eigenvalues of the 2 × 2 matrix for the Redfield equations are

$$\lambda_{\pm} = -i \left(\frac{\omega_{ca} + \omega_{db}}{2} \right) - \left(\frac{T_{ca, ca} + T_{db, db}}{2} \right) \pm$$

$$\pm \left\{ \left[-i \left(\frac{\omega_{ca} + \omega_{db}}{2} \right) - \left(\frac{T_{ca, ca} + T_{db, db}}{2} \right) + T_{ca, db} T_{db, ca} \right] \right\}^{1/2}. \quad (4.20)$$

This of course allows fast and slow exchange, and all intermediate possibilities.

Now we must model our TLS so that we have a physical picture for all the mathematical terms. The model we take is that the TLS has a Hamiltonian in the basis of localized states (localized in either the left-hand well or the right-hand well of the double well)

$$H_{TLS}^{(0)} = \frac{1}{2} \begin{pmatrix} -\Delta & K \\ K & \Delta \end{pmatrix}$$

$$(4.21)$$

with eigenvalues $\pm \frac{1}{2}(\Delta^2 + K^2)^{1/2}$. Here, Δ is the energy splitting between the two localized states (left and right) and K is the tunneling matrix element between them. When the optical center is near the double well, they interact and both Δ and K are changed. These changes depend on the electronic state of the optical center so that in the ith state (i is ground or excited), the TLS Hamiltonian is now

$$H_{\text{TLS}}^{(i)} = \frac{1}{2} \begin{pmatrix} -\Delta_i & K_i \\ K_i & \Delta_i \end{pmatrix} \tag{4.22}$$

with eigenvalues $\pm \frac{1}{2}(\Delta_i^2 + K_i^2)^{1/2}$. The eigenstates of this Hamiltonian can be written as

$$\begin{pmatrix} \cos\theta_i \\ \sin\theta_i \end{pmatrix} \quad \text{and} \quad \begin{pmatrix} -\sin\theta_i \\ \cos\theta_i \end{pmatrix} \tag{4.23}$$

with $\tan 2\theta_i = K_i/\Delta_i$ or $\sin 2\theta_i = K_i/(\Delta_i^2 + K_i^2)^{1/2}$. Thus the energies of our four states are

$$\begin{aligned}
E_a &= \varepsilon_g - \tfrac{1}{2}(\Delta_g^2 + K_g^2)^{1/2} \\
E_b &= \varepsilon_g + \tfrac{1}{2}(\Delta_g^2 + K_g^2)^{1/2} \\
E_c &= \varepsilon_e - \tfrac{1}{2}(\Delta_e^2 + K_e^2)^{1/2} \\
E_d &= \varepsilon_e + \tfrac{1}{2}(\Delta_e^2 + K_e^2)^{1/2}
\end{aligned} \tag{4.24}$$

so

$$\begin{aligned}
\omega_{ca} &= (\varepsilon_e - \varepsilon_g) - \tfrac{1}{2}(\Delta_e^2 + K_e^2)^{1/2} + \tfrac{1}{2}(\Delta_g^2 + K_g^2)^{1/2} \\
\omega_{db} &= (\varepsilon_e - \varepsilon_g) + \tfrac{1}{2}(\Delta_e^2 + K_e^2)^{1/2} - \tfrac{1}{2}(\Delta_g^2 + K_g^2)^{1/2}.
\end{aligned} \tag{4.25}$$

The usual assumption is that $K_e = K_g$, $\Delta_g = \Delta - \delta$ and $\Delta_e = \Delta + \delta$, so that

$$\frac{\omega_{ca} + \omega_{db}}{2} = (\varepsilon_e - \varepsilon_g) \tag{4.26}$$

$$\begin{aligned}
\omega_{ca} - \omega_{db} &= [(\Delta - \delta)^2 + K^2]^{1/2} - [(\Delta + \delta)^2 + K^2]^{1/2} \\
&\cong \frac{2\delta\Delta}{(\Delta^2 + K^2)^{1/2}} + \theta(\delta^3).
\end{aligned}$$

Note that $2\delta = \Delta_e - \Delta_g$, so that δ is a measure of the difference between the *splitting of the TLS when the optical center* is excited and when it is in the ground state. This difference is caused by the different interactions of the TLS with the optical center in those two states and is assumed to be dependent on the distance of the TLS from the center, i.e. $\delta \propto 1/r^s$.

In addition, the phonon-TLS interaction is usually taken to be diagonal in the *localized* TLS states, $|L\rangle$ and $|R\rangle$

$$V_{\text{TLS-phonon}} = \sum_k g_k(b_k + b_k^\dagger)[|L\rangle\langle L| - |R\rangle\langle R|] \tag{4.28}$$

These represent changes (fluctuations) in the localized energies of the TLS due to phonons. Since in the ground electronic state

$$[|L\rangle\langle L| - |R\rangle\langle R|]|g\rangle\langle g| = \cos 2\theta_g[|a\rangle\langle a| - |b\rangle\langle b|] - $$
$$- \sin 2\theta_g[|a\rangle\langle b| + |b\rangle\langle a|] \qquad (4.29)$$

and in the excited state

$$[|L\rangle\langle L| - |R\rangle\langle R|]|e\rangle\langle e| = \cos 2\theta_e[|c\rangle\langle c| - |d\rangle\langle d|] - $$
$$- \sin 2\theta_e[|c\rangle\langle d| + |d\rangle\langle c|] \qquad (4.30)$$

Then

$$V = \sum_k g_k(b_k + b_k^\dagger) \{\cos 2\theta_g[|a\rangle\langle a| - |b\rangle\langle b|]$$
$$-\sin 2\theta_g[|a\rangle\langle b| + |b\rangle\langle a|]$$
$$+\cos 2\theta_e[|c\rangle\langle c| - |d\rangle\langle d|]$$
$$-\sin 2\theta_e[|c\rangle\langle d| + |d\rangle\langle c|]. \qquad (4.31)$$

As we said above, the diagonal terms lead to pure dephasing which varies as T^7 at low T (and T^2 at high T); we therefore neglect them at low T, and find

$$W_{d \to c} = 2\pi \sum_k |g_k|^2 (\bar{n}_k + 1) \sin^2 2\theta_e \, \delta(\omega_k - \omega_{dc})$$

$$W_{b \to a} = 2\pi \sum_k |g_k|^2 (\bar{n}_k + 1) \sin^2 2\theta_g \, \delta(\omega_k - \omega_{ba})$$
$$\qquad (4.32)$$

$$T_{ca, \, db} = -\pi(\sin 2\theta_e)(\sin 2\theta_g) \sum_k |g_k^2| \, (\bar{n}_k + 1) \, [\delta(\omega_k - \omega_{dc}) + \delta(\omega_k - \omega_{ba})]$$

$$T_{db, \, ca} = -\pi(\sin 2\theta_e)(\sin 2\theta_g) \sum_k |g_k^2| \, (\bar{n}_k) \, [\delta(\omega_k - \omega_{dc}) + \delta(\omega_k - \omega_{ba})].$$

Note that we can write, using (4.32) and the definition of $\sin 2\theta_i$,

$$T_{ca, \, db} = -\frac{1}{2}\left[\frac{\omega_{dc}}{\omega_{ba}} W_{d \to c} + \frac{\omega_{ba}}{\omega_{dc}} W_{b \to a}\right]$$
$$\qquad (4.33)$$
$$T_{db, \, ca} = -\frac{1}{2}\left[\frac{\omega_{dc}}{\omega_{ba}} W_{c \to d} + \frac{\omega_{ba}}{\omega_{dc}} W_{a \to b}\right]$$

where

$$\omega_{dc} = (\Delta_e^2 + K^2)^{1/2} \approx (\Delta^2 + K^2)^{1/2} + \frac{\delta\Delta}{(\Delta^2 + K^2)^{1/2}}$$

and

$$\omega_{ba} = (\Delta_g^2 + K^2)^{1/2} \approx (\Delta^2 + K^2)^{1/2} - \frac{\delta\Delta}{(\Delta^2 + K^2)^{1/2}}.$$

Note that in principle $W_{d \to c}$ does not equal $W_{b \to a}$, but in fact differs from it because of the term δ. Many authors assume that $W_{d \to c} = W_{b \to a} \equiv R_1$, the downward scattering rate) and take δ into account *only* in the imaginary terms appearing in Eq. (4.20). Other authors *have* taken the δ dependence of the rates $W_{i \to j}$ into account, and find different functional forms for the widths.

Before examining the approximations and models for the widths, the formula for the line shape should be written down. In order to find the line shape $(I(\omega))$, one must find the Fourier transform of the transition moment correlation function $C(t)$;

$$I(\omega) = \text{Re} \int_0^\infty dt \, e^{-i\omega t} C(t) \tag{4.34}$$

where

$$C(t) = (1, 1) \cdot \exp[-(i\boldsymbol{\omega} + \mathbf{T} + \gamma_{\text{rad}} \mathbf{I})t] \cdot \begin{pmatrix} P_a \\ P_b \end{pmatrix}. \tag{4.35}$$

Here, we have written the frequency (ω) and relaxation $(\mathbf{T})$ terms in Eq. (4.17) as two-dimensional matrices, and have inserted the initial (equilibrium) populations of levels $|a\rangle$ and $|b\rangle$. The final form for the line shape is found by taking into account the many TLS' which interact with the optical center by assuming their effects are independent, and averaging the correlation functions over the distribution of TLS parameters K, Δ and δ. The last is equivalent to a configurational average over all positions of TLS relative to the optical center.

In actual practice, it is extremely difficult to average the line shape as described, so the usual approximation is to take the *width* due to 1 TLS interacting with the optical center and averaging that over the distribution of K and Δ and summing over all TLS.

As is evident from Eq. (4.35), the width of the optical absorption line due to 1 TLS is determined by the real part of the eigenvalues of the matrix $i\boldsymbol{\omega} + \mathbf{T}$. Note that $\omega_{ca} = \omega_{eg} - \frac{1}{4}(\omega_{dc} - \omega_{ba})$ and $\omega_{db} = \omega_{eg} + \frac{1}{4}(\omega_{dc} - \omega_{ba})$, so that the ω_{eg} is just a constant term as is γ_{rad} in both transitions; both will be left out of the remaining formulas. That is, frequency will be measured with respect to ω_{eg} and widths given without the additive contribution of the radiative rate. There are two simple limits for which the linewidth can be easily found: (a) $(\omega_{dc} - \omega_{ba})$ large compared to the rates R_{ij} (b) small compared to these. Since $\delta \propto 1/r^s$, the nearby ions will give large δ and the ions far away from the chromophore will give small δ; there are, however, many more of the latter. Note that $\omega_{dc} - \omega_{ba} \approx 2\delta\Delta/(\Delta^2 + K^2)^{1/2} \equiv \delta'$.

Case (1): $\delta' > W_{ij}$. In this case the four level system will have two optical transitions, one at frequency $-\delta'/2$ with width $\frac{1}{2}$ ($W_{c \to d} + W_{a \to b}$) and one at frequency $\delta'/2$ with width $\frac{1}{2}$ ($W_{d \to c} + W_{b \to a}$). Using Eq. (4.32), the widths are

$$\frac{1}{2} G [\omega_{dc}^3 \{\bar{n}(\omega_{dc}) + 1\}] \frac{K^2}{\omega_{dc}^2} + \frac{1}{2} G [\omega_{ba}^3 \{\bar{n}(\omega_{ba}) + 1\}] \frac{K^2}{\omega_{ba}^2} \qquad (4.36a)$$

and

$$\frac{1}{2} G [\omega_{dc}^3 \bar{n}(\omega_{dc})] \frac{K^2}{\omega_{dc}^2} + \frac{1}{2} G [\omega_{ba}^3 \bar{n}(\omega_{ba})] \frac{K^2}{\omega_{ba}^2} \qquad (4.36b)$$

where G is a collection of constants.

Thus, as $T \to 0$, one of the lines becomes very narrow (i.e. width equal to γ_{rad}) while the other has a width governed by the *downward* flip rate of the TLS (i.e. phonon emission term), which is nonzero at $T = 0$. This formula for the widths agrees with the results of Small *et al.* [22], Lyo [37], Molenkamp and Wiersma [23], and Reineker *et al.* [43] before averaging over TLS parameters. This limit has been called the UPS (uncorrelated phonon-scattering) limit or the slow modulation limit. Note, that the widths are independent of δ in this limit.

The standard approximation employed to find the width from these expressions has been to average these widths over the initial populations of $|a\rangle$ and $|b\rangle$ to get the width Γ, and to assume $\omega_{ba} = \omega_{dc} = \varepsilon$ in all the subsequent expressions

$$\Gamma = G \frac{K^2}{\varepsilon^2} \cdot \varepsilon^3 \left\{ [\bar{n}(\varepsilon) + 1] \frac{e^{-\beta\varepsilon}}{1 + e^{-\beta\varepsilon}} + \bar{n}(\varepsilon) \frac{1}{1 + e^{-\beta\varepsilon}} \right\}$$

$$= \frac{G}{2} \frac{K^2}{\varepsilon^2} \left[\varepsilon^3 \coth \frac{\beta\varepsilon}{2} \right] \cdot \mathrm{sech}^2(\beta\varepsilon/2) \equiv \tau^{-1}(\varepsilon)\mathrm{sech}^2(\beta\varepsilon/2) \qquad (4.37)$$

where we have written the last form in terms of the "spin-lattice" relaxation rate of the TLS, $\tau^{-1} = \frac{1}{2} (R_\uparrow + R_\downarrow) = (GK^2/2) \varepsilon \coth(\beta\varepsilon/2)$.

Case (2): $\delta' \ll W_{ij}$. In this case, the TLS flipping is rapid compared to h/δ, so that fast modulation occurs. As we pointed out above (see the discussion after Eq. (4.33)), a usual assumption is to neglect δ in all of the relaxation rates, $T_{ij,kl}$ and only include δ in the frequency shift. Then the eigenvalues of $i\omega + T$ can be written down as

$$\frac{1}{2} (R_\uparrow + R_\downarrow) \pm \frac{1}{2} [(R_\uparrow + R_\downarrow)^2 - 2i\delta' (R_\uparrow + R_\downarrow) - \delta'^2]^{1/2} \qquad (4.38)$$

The relevant eigenvalue is the one with the smaller width, the other one leading to very little intensity. This eigenvalue is to order δ^2:

$$+ \frac{i\delta'}{2} \frac{(R_\uparrow - R_\downarrow)}{R_\uparrow - R_\downarrow} + \frac{\delta'^2 R_\uparrow R_\downarrow}{(R_\uparrow + R_\downarrow)^3} \qquad (4.39)$$

yielding a width $\delta^2 R_\uparrow R_\downarrow (R_\uparrow + R_\downarrow)^{-3} = \delta^2\tau/4 \, \mathrm{sech}^2 (\beta\varepsilon/2)$, in agreement with Lyo

[37], Molenkamp and Wiersma [23] and others in this limit. Note that these authors finally average this width over the distribution of TLS parameters to obtain the average width.

If one does *not* make the assumption that δ' can be neglected in the $T_{ij,\,k\ell}$, then additional terms (to order δ'^2) appear in the width. The relevant eigenvalue of the matrix is then given by

$$+ \frac{i\delta'}{2} \frac{(R_\uparrow - R_\downarrow)}{R_\uparrow + R_\downarrow} + \delta'^2 \frac{R_\uparrow R_\downarrow}{(R_\uparrow + R_\downarrow)^3} + \frac{\delta'^2}{\varepsilon^2} \frac{(R_\uparrow R_\downarrow)}{R_\uparrow + R_\downarrow} \left[\frac{\beta\varepsilon}{2} \coth \frac{\beta\varepsilon}{2} - 2 \right] \quad (4.40)$$

which differs from the last formula in the last term. The width of the optical transition then has two contributions. At low T, the first goes as csch $\beta\varepsilon$ while the second goes as $\beta\varepsilon$ csch $\beta\varepsilon$, so that which is dominant depends on the coupling constants, etc. At $T > \varepsilon/k_B$, the first term goes like $1/T$ (narrowing) while the second goes like T and can even be negative; thus, at high enough T, this formula will predict a nonsensical result. The breakdown of this approximation is due to the increasing importance of the other matrix elements σ_{da} and σ_{cb} in determining the dynamics. We are able to neglect them only if the mixing of the various transitions is small. When $T > \varepsilon/k_B$ the mixing can be large and we should examine the Redfield equations for the four coupled matrix elements, σ_{db}, σ_{ca}, σ_{da}, σ_{cb}. It turns out that these equations can be solved exactly, and Kassner and Reineker [43] did so. The resulting expressions are prohibitively complicated, but in the limit we are discussing turn out to yield (for the eigenvalue we are considering) the same imaginary part and a real part equal to

$$\delta'^2 \frac{R_\uparrow R_\downarrow}{(R_\uparrow + R_\downarrow)^3} + \frac{\delta'^2}{\varepsilon^2} \frac{R_\uparrow R_\downarrow}{R_\uparrow + R_\downarrow} \left(\frac{\beta\varepsilon}{2} \coth \frac{\beta\varepsilon}{2} - 1 \right). \quad (4.41)$$

This agrees with the earlier result at low T, and does not have the unphysical behavior discussed above at high T. In fact, both terms now yield a $1/T$ behavior at high T. Thus the high T narrowing is preserved. Happily, for the generally accepted parameter values of K, Δ, G, etc., the second term is much smaller than the first at all but the lowest temperatures.

This discussion shows that small differences in approximations can produce differences in the results (compare Eq. (4.39), (4.40) and (4.41)) which explain the differences in the various theoretical papers.

Once having determined the width due to a single TLS, we now sum over all the TLS for which $\delta' > W^{ij}$ which means r small (since $\delta \propto 1/r^s$, r the distance from TLS to chromophore, and $s = 3$ for dipole-dipole coupling, $s = 4$ for dipole-quadrupole coupling, etc.). In addition, we should average over the TLS parameters. Thus,

$$\langle \Gamma_{\text{total}} \rangle \propto \int_0^{\text{max}} d\varepsilon\, \rho(\varepsilon) \int_0^{R_c} dr\, r^2\, \Gamma \cdot n \quad (4.42)$$

where n is the spatial density of TLS, Γ is the form of Eq. (4.37), $\rho(\varepsilon)$ is the density of TLS splittings and R_c is a cutoff to ensure $\delta \geqslant \tau^{-1}$ so that the slow exchange limit holds. Thus $R_c \propto \tau^{1/s}$, and

$$\langle \Gamma_{\text{total}} \rangle \propto \int_0^{\varepsilon_{\max}} d\varepsilon \, \rho(\varepsilon) \, \tau^{-1 + 3/s} \operatorname{sech}^2 \frac{\beta \varepsilon}{2} . \tag{4.43}$$

If we now assume $\rho(\varepsilon) \propto \varepsilon^{\gamma}$ at low ε, then at low T

$$\langle \Gamma_{\text{total}} \rangle \propto T^{1 + \gamma + 3\left(1 - \frac{3}{s}\right)} \tag{4.44}$$

if we assume τ^{-1} varies as ε^3, as Lyo does. For dipole-dipole interaction, $\langle \Gamma_{\text{total}} \rangle \propto T^{1 + \gamma}$ at low T.

If we now take into account the far TLS for which $\delta' < W_{ij}$ (fast exchange), we must sum over the TLS from R_c to ∞. It is straightforward, using Eq. (4.39), to confirm that the same temperature dependence, $T^{1 + \gamma + 3(1 - 3/s)}$ is obtained.

Thus, it seems that, within the approximations mentioned, for dipole-dipole coupling, in order to obtain a $T^{1.3}$ behavior at low T, one must assert that $\gamma = 0.3$, i.e. the TLS density of states disagrees with the standard model. Other authors have derived temperature dependences consistent with $T^{1 + \gamma}$ in particular temperature regimes where the low and high temperature results are coming together.

There are however other calculations of the homogeneous linewidth due to TLS interactions that use the original density of states, and yield a temperature dependence close to $T^{1.3}$ over the relevant (i.e. very low T) range. For example, Kassner and Reineker [43] use the model we have set forth here; however, they derive a formula for the line shape of the full many body problem of one chromophone interacting with N TLS. Arguments are given that the low temperature line intensity is dominated by a few narrow lines, whose widths are calculated *numerically* as the sum over the exact linewidths of the chromophore 1 TLS problem. Care has to be taken to include all non-negligible contributions, so that a large interaction volume around the chromophore is considered. At low T, the contributions of quite weakly coupled TLS are important (since the TLS flip rates are so low), and only extremely weakly coupled TLS can be neglected. The numerical results [43] give a temperature dependence roughly consistent with $T^{1.3}$ over the relevant T range, in both the dipole-dipole and the dipole-quadrupole case, and using the original TLS distribution and the Debye phonon density of states. It is clear, from this numerical evaluation of the linewidth, that the average over the tunneling matrix elements (K or overlaps, $\exp(-\lambda)$) is subtle and can lead to a change in the temperature dependence, *especially* at very low T. It is this averaging which leads to a temperature dependence consistent with (but not exactly equal to) $T^{1.3}$ even with $\gamma = 0$. A drawback of this theory is that as yet no analytical formulas for the linewidth have been given, although approximate analytical expressions for the average over Δ and K are available. The complete average over δ, Δ and K has not been found, except numerically.

Lyo and Orbach [40] suggested a modification of Lyo's form based on the postulated existence of fracton models in disordered systems by Alexander and Orbach. The argument starts with the postulate that in a disordered system there exists a range of length scales (neither too long nor too short) for which the geometric structure is self-similar, i.e. the mass density scales with distance from any point like $r^{\tilde{d}-d}$ where $\tilde{d}$ is the fractal dimension and d the Euclidean dimension. Alexander and Orbach then examined the consequences of this for the vibrational modes of the system and found that in a certain frequency range (connected to the distance range over which the system is self-similar), the density of states of the localized modes (fractons) produced will scale as $\omega^{1/3}$ instead of the Debye form ω^{d-1}. Note that the existence of fractons does *not* mean that the long wavelength phonons are absent.

Now, Lyo and Orbach consider the TLS flipping rates to be caused by TLS-fracton interactions instead of TLS-phonon interactions, and suggest that these rates ($\tau^{-1}(\varepsilon)$) will then vary as $\varepsilon^{4/3}$ (i.e., $\varepsilon^{1/3}$ from the density of fracton states and ε from the fracton-TLS interaction). When this form is averaged in the same manner as before, these authors find $\langle\langle\Gamma\rangle\rangle \propto T^{1+\gamma+4/3(1-3/s)}$ so that for dipole-dipole interactions ($s = 3$) $\langle\langle\Gamma\rangle\rangle \propto T^{1+\gamma}$. These authors suggest that a possible explanation of the 1.3 exponent is that the relevant interaction is dipole-quadrupole ($s = 4$) and $\gamma \simeq 0$.

Osadko [42] has suggested that the TLS should be treated as a thermal bath in just the same manner that phonons are treated in the standard theories of line broadening. His theory is non-perturbative, but the essential flavor and results can be found using perturbation theory and the standard McCumber-Sturge [9] line shape formula.

Consider the chromophore to be a two state system ($|g\rangle$ and $|e\rangle$) as before and assume that the relevant interaction between the TLS and the chromophore is an operator which flips *two* TLS (note the resemblance to the low T linewidth due to Raman phonon processes), i.e.

$$V_{\text{TLS-chromophore}} = [|e\rangle\langle e| - |g\rangle\langle g|] \sum_{i \neq j} v_{ij}\sigma_i^+ \sigma_j^- \qquad (4.45)$$

where σ_i^+ is an operator which flips the ith TLS from the lower to the upper levels, while σ_i^- is the reverse and v_{ij} is the interaction matrix element. The physical picture is then that TLS dephase the optical transition by their constant flipping. The standard formula for the linewidth at low T then yields

$$\Gamma = \frac{1}{2}\int_{-\infty}^{+\infty} d\tau \langle [V_{ee}(\tau) - V_{gg}(\tau)][V_{ee}(0) - V_{gg}(0)]\rangle \qquad (4.46)$$

where the $V_{ee} - V_{gg}$ is the instantaneous change in the optical transition frequency due to the TLS (i.e. the matrix elements of $V_{\text{TLS-chromophore}}$) and the average is over the thermal distribution of TLS. Then

$$\Gamma = \frac{1}{2} \int_{-\infty}^{+\infty} d\tau \sum_{\substack{i,j \\ i \neq j}} |v_{ij}|^2 \langle \sigma_i^+(\tau) \sigma_i^-(0) \rangle \langle \sigma_j^-(\tau) \sigma_j^+(0) \rangle \tag{4.47a}$$

$$\Gamma \propto \int_0^{\varepsilon_{max}} d\varepsilon [\rho(\varepsilon)]^2 |v(\varepsilon)|^2 n_{TLS}^{\ell}(\varepsilon) n_{TLS}^{u}(\varepsilon) \tag{4.47b}$$

where we have assumed that (a) $|v_{ij}|^2$ is a function of $\varepsilon_i = \varepsilon_j$ only, (b) $\rho(\varepsilon)$ is the density of TLS energies and (c) $n_{TLS}^{\ell}(\varepsilon)$ is the thermal population of the lower TLS level while $n_{TLS}^{u}(\varepsilon)$ is the thermal population of the upper level. At equilibrium, $n_{TLS}^{\ell} = (1 + e^{-\beta\varepsilon})^{-1}$ and $n_{TLS}^{u} = e^{-\beta\varepsilon}(1 + e^{-\beta\varepsilon})^{-1}$, so that

$$\Gamma \propto \int_0^{\varepsilon_{max}} d\varepsilon \, \rho^2(\varepsilon) |v(\varepsilon)|^2 \, \text{sech}^2(\beta\varepsilon/2). \tag{4.48}$$

The temperature dependence now depends on the *assumed* form for $\rho^2(\varepsilon)|v(\varepsilon)|^2$. If this is assumed constant, then $\Gamma \propto T \tanh(\varepsilon_{max}/2k_BT) < T$; if it is assumed to vary as ε^x then $\Gamma \propto T^{1+x}$ at low temperature. Osadko can fit the experimental data with a particular (and perhaps reasonable) form for $|v(\varepsilon)|^2 \rho^2(\varepsilon)$.

The simplest procedure to obtain a $T^{1+\gamma}$ dependence of the linewidth is to assume that there are two mechanisms operating simultaneously. For example, Jackson and Silbey [39] assumed that the TLS contribution to the linewidth is linear in T (i.e. the dipole-dipole interaction, $\gamma = 0$ form) at low T while the local librational modes, known to dephase the optical transition in *crystals*, gave a typical optical phonon contribution. Thus, at low temperatures, where acoustic phonon direct contributions can be neglected:

$$\Gamma = AT + \sum_{\substack{\text{local} \\ \text{modes}}} \lambda_i \left(\frac{e^{-\beta\omega_i}}{1 - e^{-\beta\omega_i}} \right). \tag{4.49}$$

Here λ_i is the coupling constant to the ith local mode. This predicts that at low enough temperatures $(T < \omega_i/k_B)$ the linewidth is linear in T, but as the temperature is raised, Γ varies as $T^{1+\delta}$ where δ is small. Local mode frequencies of a few cm^{-1} and coupling constants taken from crystalline studies can fit most of the data in glasses.

The original Jackson-Silbey model assumed the local modes were *librations* of the chromophore; however, local modes of the glass give a contribution of exactly the same form. Thus, one may suggest that direct local mode contributions (of all sorts) when added to a linear T dependence can fit the data, at least for $T > 1$ K. This model, however, clearly suggests that for $T < \omega_i/k_B$ (i.e. temperatures well below the lowest local mode frequencies), the linewidth will be linear T (assuming $\gamma = 0$, as these authors do).

Since this paper was written, the *Journal of Luminescence* had an issue devoted to hole-burning. The interested reader should see the articles by Osadko [44], Huber [45], Small [46], Kassner and Silbey [47] as well as others, for remarks on the present state of the theory.

5. Concluding Remarks

In the last ten years, low temperature relaxation measurements in the microwave and optical domain have become possible so that we are learning about the dynamics of excitations in condensed phases at these temperatures. Various relaxation processes, including excitation transfer, phonon-excitation scattering, and even more exotic processes such as TLS and possibly fracton interactions with optical excitations are actively being studied. In this article, we have attempted to show how a single theoretical framework can be used to study these phenomena. A major assumption of the theory is that the relaxation of the system under study is slow compared to the relaxation processes in the thermal bath. As the experiments are pushed to lower temperatures and shorter times, we expect to run into situations in which this assumption fails because particular bath modes begin to relax very slowly. However, even in this case, it is possible, in principle, to remove these slow modes from the bath and treat them as part of the system. We look forward to these developments and the continued collaboration between experimenters and theorists this will engender.

References

1. Blum, K., *Density Matrix Theory and Applications* (Plenum Press, N.Y., 1981).
2. Redfield, A., in *Advances in Magnetic Resonance* **1** (1963), 1 . Albers, J. and Dutch, J. M., *J. Chem. Phys.* **55** (1971), 2613 .
3. Slichter, C. P., *Principles of Magnetic Resonance* (Harper and Row, N.Y., 1963).
4. Yoon, B., Deutch, J. M., and Freed, J., *J. Chem. Phys.* **62** (1975), 4687.
5. Zwanzig, R., in *Lectures in Theoretical Physics*, vol. III (Interscience, N.Y., 1961); van Kampen, N., *Stochastic Processes in Physics and Chemistry* (North-Holland, Amsterdam and N.Y., 1981); Oppenheim, I., Shuler, K., and Weiss, G., *Stochastic Processes in Chemical Physics* (M.I.T. Press, Cambridge, 1979).
6. Schiff, L., *Quantum Mechanics* (McGraw Hill, N.Y., 1968); Cohen-Tannoudji, C., Diu, B., and Laloe, F., *Quantum Mechanics* (J. Wiley, N.Y.).
7. Kubo, R., in *Advances in Chemical Physics*, vol. 15 (Interscience, N.Y., 1968); Kubo, R., in *Lectures in Theoretical Physics*, vol. 1 (Interscience, N.Y., 1958); Haken, H. and Strobl, P., *Z. für Phys.* **262** (1973) 185.
8. Wertheimer, R. and Silbey, R., *Chem. Phys. Lett.* **75** (1980), 243; Rips, I. B. and Capek, V., *Phys. Stat. Sol.* **B100** (1980), 451; Lindenberg, K. and West, B., *Phys. Rev. Lett.* **51** (1985), 1370; *Phys. Rev.* **A30** (1984), 568.
9. McCumber, D. E. and Sturge, M. D., *J. Appl. Phys.* **34** (1963) 1682.
10 Kittel, C., *Quantum Theory of Solids* (McGraw Hill, N.Y., 1963); Madelung, O., *Introduction to Solid State Theory* (Springer, N.Y., 1979).
11. Botter, B., Nonhof, C., Schmidt, J., and van der Waals, J. J., *Chem. Phys. Lett.* **43** (1976), 210; Botter, B., van Strien, A., and Schmidt, J., *Chem. Phys. Lett.* **49** (1977), 39; Verbeek, P. and Schmidt, J., *Chem. Phys. Lett.* **63** (1979), 384; Dietz, F., Konzelmann, U., Port, H., and

Schwoerer, M., *Chem. Phys. Lett.* **58** (1978), 565; Levinsky, H. and Brenner, H., *Chem. Phys.* **40** (1979), 111; Vollman, W., *Chem. Phys. Lett.* **57** (1981), 157.

12. Hutchison, C. A. and Mangum B., *J. Chem. Phys.* **34** (1961), 907. Hutchison, C. A. and King, J., *J. Chem. Phys.* **58** (1973), 392.

13. McConnell, H., *J. Chem. Phys.* **28** (1958), 430.

14. Van 't Hof, C. and J. Schmidt, *Chem. Phys. Lett.* **36** (1975), 460; *ibid.*, **42**, (1976), 73; Harris, C. B., *J. Chem. Phys.* **67** (1977), 5607.

15. (a) R. Silbey, in *Organic Molecular Aggregates*, edited by P. Reineker, H. Haken, and H. C. Wolf (Springer, N.Y., 1984). (b) P. Reineker, J. Kohler, U. Schmid, and R. Silbey, *J. Chem. Phys.* **83** (1985), 623.

16. Harris, C. B. and Francis, A., *Chem. Phys. Lett.* **9** (1972), 181, 188; A. van Strien, J. Schmidt, and R. Silbey, *Mol. Phys.* **49** (1982), 151; J. F. C. van Kooten, van Strien, A., and Schmidt, J., *Chem. Phys. Lett.* **90** (1982), 337.

17. Schmidt, J., in *Organic Molecular Aggregates*, edited by P. Reineker, H. Haken, and H. C. Wolf (Springer, N.Y., 1983); Schmidt, J., this volume.

18. Davydov, A. S., *Theory of Molecular Excitons* (McGraw Hill, N.Y., 1962); Knox, R. S., *Theory of Excitons* (John Wiley, N.Y., 1966).

19. Benk, H. and Silbey, R., *J. Chem. Phys.* **79** (1983), 3487.

20. van Kooten, J. F. C., Munowitz, M. G., and Schmidt, J., *Mol. Phys.* **52** (1984), 1397.

21. Jackson, B. and Silbey, R., *J. Chem. Phys.* **77** (1982), 2763.

22. (a) A. A. Gorokhovskii, R. K. Kaarli, and L. A. Rebane, *JETP Lett.* **20** (1974), 216. (b) J. M. Hayes and G. J. Small, *J. Chem. Phys.* **27** (1978), 151; Hayes, J. M., Stout, R. P., and Small, G. J., *J. Chem. Phys.* **73** (1982), 4229; *ibid.*, **74** (1983), 4266; Small, G., in *Spectroscopy and Excitation Dynamics of Condensed Molecular Systems*, edited by V. Agranovitch and R. Hochstrasser (North-Holland, Amsterdam, 1983); Carter, T. R., Fearey, B. L., Hayes, J. M., and Small, G. J., *Chem. Phys. Lett.* **102** (1983), 272.

23. (a) H. de Vries and D. Wiersma, *Phys. Rev. Lett.* **36** (1976) 91. (b) K. Duppen, L. W. Molenkamp, J. B. W. Morsink, D. A. Wiersma, and H. P. Trommsdorff, *Chem Phys. Lett* **84** (1981), 421; L. A. Molenkamp and D. A. Wiersma, *J. Chem. Phys.* **80** (1984), 3054. (c) L. W. Molenkamp and D. A. Wiersma, *J. Chem. Phys.* **83** (1985), 1.

24. A. Szabo, *Phys. Rev. Lett.* **25** (1970), 924; P. J. Selzer, D. L. Huber, D. S. Hamilton, W. M. Yen, and M. J. Weber, *Phys. Res. Lett.* **36** (1976); J. Hegarty and W. M. Yen, *Phys. Rev. Lett.* **43** (1979), 1126.

25. S. Völker. R. M. Macfarlane, R., and J. H. van der Waals, *Chem. Phys. Lett.* **53** (1973), 8; S. Völker and J. H. van der Waals, *Mol. Phys.* **32** (1976), 1703; S. Völker, R. Macfarlane, A. Genack, H. P. Trommsdorff, and J. H. van der Waals, *J. Chem. Phys.* **67** (1977), 1759; S. Völker and R. Macfarlane, *J. Chem. Phys.* **73**, (1980), 4476; A. I. M. Dicker, J. Dobkowski, and S. Völker, *Chem. Phys. Lett.* **84** (1981), 415.

26. Friedrich, J., Swalen, J., and Haarer, D., *J. Chem. Phys.* **73** (1980), 705; Friedrich, J., Wolfrum, H., and Waarer, D., *J. Chem. Phys.* **77** (1982), 2309; Breinl, W., Friedrich, J., and Haarer, D., *J. Chem. Phys.* **80** (1984), 349; *ibid.*, **81** (1984). 3915.

27. H. P. H. Thijssen, A. I. M. Dicker, and S. Völker, *Chem. Phys. Lett.* **92** (1982), 7; H. P. H. Thijssen, S. Völker, M. Schmidt, and H. Port, *ibid.*, **94** (1983), 537; H. P. H. Thijssen, R. van den Berg, and S. Völker, *ibid.*, **97** (1983), 295; *ibid.*, **103** (1983), 23; *ibid.*, **120** (1985), 496; *ibid.*, **120** (1985), 503.

28. *Persistent Spectral Hole-Burning: Science and Applications*, edited by W. E. Moerner (Springer Verlag, N.Y., 1988).

29. Hegarty, J., Broer, M. M., Golding, B., Simpson, J. R., and MacChensy, J. B., *Phys. Rev. Lett.* **51** (1983), 2033.

30. Anderson, P. W., Halperin, B. I., and Varma, C., *M. Phil. Mag.* **25** (1972), 1; Phillips, W. A., *J. Low Temp. Phys.* **7** (1972), 351.

31. Black, J. L. and Halperin, B. I., *Phys. Rev.* **B16** (1977), 2879.

32. Lasjaunias, J. G., Ravex, A., Vandorpe, M., and Hunklinger, S., *Solid State Comm.* **17** (1975) 1045.

33. S. Hunklinger, in *Phonon Scattering in Condensed Matter*, eds. W. Eisenmenger, K. Lassmann, and S. Doettinger (Springer Verlkag, N.Y., 1984).

34. C. M. Varma, R. Dynes, and J. Banaver, *J. Phys. C.* **15** (1982), L1221.

35. Lyo, S. K. and Orbach, R., *Phys. Rev.* **B22** (1980), 4223.

36. Reineker, P. and Morawitz, H., *Chem. Phys. Lett.* **86** (1982), 359; Morawitz, H. and Reineker, P., *Sol. State Comm.* **42** (1982), 609.

37. Lyo, S. K., *Phys. Rev. Lett.* **48** (1982), 688; Lyo, S. K., *Organic Molecular Aggregates*, edited by P. Reineker, H. Haken, and H. C. Wolf (Springer, N.Y., 1983).

38. Hayes, J. M., Stout, J. P., and Small, G. J., *J. Chem. Phys.* **74** (1981), 4266; see also ref. 22.

39. Jackson, B. and Silbey, R., *Chem. Phys. Lett.* **99** (1983), 381.

40. Lyo, S. K. and Orbach, R., *Phys. Rev.* **B29** (1984), 2300.

41. Maynard, R., *Journal de Phys.* **C7** (1985), 325.

42. Osadko, I. S., *JETP Letters* **39** (1984), 426; *Chem. Phys. Lett.* **115** (1985), 411.

43. P. Reineker, H. Morawitz, and K. Kassner, *Phys. Rev.* **B29** (1984), 4546; K. Kassner and P. Reineker, *Chem. Phys.* (in press, 1986); also in *Optical Properties of Glasses*, ed. by I. Zschokke-Gränacher (Reidel, 1986).

44. I. S. Osadko and A. Shtygashev, *J. Lum.* **36** (1987), 373.

45. D. L. Huber, *J. Lum.* **36** (1987), 307.

46. R. Jankowiak, L. Shu, M. J. Kenney, and G. J. Small, *J. Lum.* **36** (1987), 293.

47. K. Kassner and R. Silbey, *J. Lum.* **36** (1987), 283.

INDEX

277